AF725574

Polymers for Light-Emitting Devices and Displays

Scrivener Publishing
100 Cummings Center, Suite 541J
Beverly, MA 01915-6106

Publishers at Scrivener
Martin Scrivener (martin@scrivenerpublishing.com)
Phillip Carmical (pcarmical@scrivenerpublishing.com)

Polymers for Light-Emitting Devices and Displays

Edited by

Inamuddin, Rajender Boddula, Mohd Imran Ahamed and Abdullah M. Asiri

Scrivener Publishing

WILEY

This edition first published 2020 by John Wiley & Sons, Inc., 111 River Street, Hoboken, NJ 07030, USA and Scrivener Publishing LLC, 100 Cummings Center, Suite 541J, Beverly, MA 01915, USA
© 2020 Scrivener Publishing LLC
For more information about Scrivener publications please visit www.scrivenerpublishing.com.

All rights reserved. No part of this publication may be reproduced, stored in a retrieval system, or transmitted, in any form or by any means, electronic, mechanical, photocopying, recording, or otherwise, except as permitted by law. Advice on how to obtain permission to reuse material from this title is available at http://www.wiley.com/go/permissions.

Wiley Global Headquarters
111 River Street, Hoboken, NJ 07030, USA

For details of our global editorial offices, customer services, and more information about Wiley products visit us at www.wiley.com.

Limit of Liability/Disclaimer of Warranty
While the publisher and authors have used their best efforts in preparing this work, they make no representations or warranties with respect to the accuracy or completeness of the contents of this work and specifically disclaim all warranties, including without limitation any implied warranties of merchantability or fitness for a particular purpose. No warranty may be created or extended by sales representatives, written sales materials, or promotional statements for this work. The fact that an organization, website, or product is referred to in this work as a citation and/or potential source of further information does not mean that the publisher and authors endorse the information or services the organization, website, or product may provide or recommendations it may make. This work is sold with the understanding that the publisher is not engaged in rendering professional services. The advice and strategies contained herein may not be suitable for your situation. You should consult with a specialist where appropriate. Neither the publisher nor authors shall be liable for any loss of profit or any other commercial damages, including but not limited to special, incidental, consequential, or other damages. Further, readers should be aware that websites listed in this work may have changed or disappeared between when this work was written and when it is read.

Library of Congress Cataloging-in-Publication Data

ISBN 978-1-119-65460-5

Cover image: Pixabay.com
Cover design by Russell Richardson

Set in size of 11pt and Minion Pro by Manila Typesetting Company, Makati, Philippines

10 9 8 7 6 5 4 3 2 1

Contents

Preface

Polymer light-emitting diodes (PLEDs) or organic light-emitting diodes (OLEDs) are organic semiconductor light sources that emit light in response to an electric current. These PLEDs are promising devices with the aforementioned features to convert electrical energy to light energy, which is the necessary component of any display technology. OLEDs have received increasing attention since they were first developed in 1989. The development of OLEDs has attracted considerable interest in innovations for our daily life and future. They have promising applications in flat panel displays, electronic products, automotive, flexible displays, industrial products, and future wearables due to unique electrical and optical properties including low-cost, easy processing, low-operating voltage, energy-saving, eco-friendliness, thinner and smaller in size, lightweight, flexibility, and cost-effective fabrication process.

Polymers for Light-Emitting Devices and Displays provides an in-depth overview of fabrication methods and unique properties of polymeric semiconductors, and its potential applications for LEDs, organic electronics, displays, optoelectronics, and so on. Engineers, chemists, material science, and research scholars, students, and faculty members working in the area of organic electronics will benefit by understanding the materials used in optoelectronics. Based on thematic topics, the book edition contains the following eight chapters:

Chapter 1 is a detailed summary of the working principles of PLEDs. Different polymers used in PLEDs and limitations of PLEDs in the illumination system where the intensity of light is very high compared with displays like televisions and laptops are discussed.

Chapter 2 presents an overview of the newest polymeric materials and processes beyond the classical structure of PLED, leading to the low-cost and all-solution processed devices with enhanced parameters, which are closer to commercial production line requirements and custom needs.

Chapter 3 seeks to obtain a better understanding of the fluorescence quenching behavior and intramolecular charge transfer (ICT) character of two kinds of cyclopentadithiophene (CPDT) derivatives. A comparative study based on the optoelectronic properties of CPDT dimers for their highly efficient blue emitters in OLEDs is developed using the density functional theory (DFT) approach.

Chapter 4 deals with conjugated polymers and their application in the light-emitting diodes (OLEDs and PLEDs) as optoelectronic devices. It provides basic information on the classification of polymers and their modification via functionalization, copolymerization, doping, etc., for device fabrication, function, and use of conjugated polymers in blue, red, green, and multicolored light-emitting diodes along with challenges and their future perspectives. Additionally, the chapter focuses on the advantages/disadvantages and application of OLED technology in various fields.

Chapter 5 discusses the novel and noteworthy work carried out on electrospun nanofibers used for LEDs. It mainly focuses on the fabrication technology for producing electrospun nanofibers and how metal oxide semiconducting, perovskite, rare earth ion-doped, and coordination polymeric electrospun nanofibers are useful in designing smart clothes and LEDs.

Chapter 6 summarizes the roles of diversified architectures, layers, components, and their structural modifications in determining efficiencies and parameters of PLEDs as high-performance devices. Additionally, some recently developed materials and concepts, including white PLEDs, quantum dots, thermally activated delayed fluorescence, and transparent PLED, are discussed in detail.

Chapter 7 gives a general idea of polymer liquid crystal devices (PLCs), their synthesis, and applications in various liquid crystal devices (LCs) and displays.

Chapter 8 reviews the state-of-art of materials and technologies to manufacture hybrid white light-emitting diodes based on inorganic light sources and organic wavelength converters. It takes stock of the benefits—but also the weak spots—of the hybrid technology to envisage its future impact among the well-established inorganic lighting technologies.

Editors
Inamuddin
Rajender Boddula
Mohd Imran Ahamed
Abdullah M. Asiri

1

Applications of Polymer Light-Emitting Devices and Displays

D. Prakash Babu[1], S. Naresh Kumar[1], N. Suresh Kumar[2], K. Chandra Babu Naidu[3*] and D. Baba Basha[4]

[1]*School of Applied Sciences, REVA University, Bangalore, India*
[2]*Department of Physics, JNTUA, Anantapuramu, India*
[3]*Department of Physics, GITAM Deemed to be University, Bangalore, India*
[4]*Department of Physics, College of Computer and Information Sciences, Majmaah University, Al'Majmaah, Saudi Arabia*

Abstract
This chapter gives information of polymer light-emitting diodes (PLEDs) and their applications. Besides, background, types, and the development of PLEDs also discussed. Further, the behavior of different PLEDs has been discussed with respect to various parameters, brightness, color purity, light conversion efficiency, and color stability are discussed.

Keywords: Polymer, light-emitting diodes, efficiency, color purity

1.1 Introduction

In the past one decade, the display technology has undergone several technological advancements and industries and household are looking for low cost, flexible, power efficient, and durable displays. Polymer light-emitting diodes (PLEDs), which convert electric energy into light, are promising devices with aforementioned features to convert electrical energy to light energy, which is the necessary component of any display technology. High temperature resistance, short response time, smooth brightness, and a large viewing angle are the additional advantages with PLEDs [1].

**Corresponding author*: chandrababu954@gmail.com

Inamuddin, Rajender Boddula, Mohd Imran Ahamed and Abdullah M. Asiri (eds.) Polymers for Light-Emitting Devices and Displays, (1–14) © 2020 Scrivener Publishing LLC

These special characteristics of PLEDs give the scope to use them in the applications where a large array of displays is required [2]. At present, inorganic light-emitting diodes are widely used. The advancement of technology demands advancement of display also, some times, the display device needs to be flexible, this flexibility can be easily provided by PLEDs. In this chapter, the basic structure of PLED, the mechanism of light emission, and different applications are discussed.

1.2 Background

In 1990, an article was first published in Nature on "Light-emitting polymers" by J. H. Burroughes, Richard Friend, and others [3].

Basic structure of PLED

The basic structure of a PLED is illustrated in Figure 1.1. It consists of thin layers of light-emitting polymers film sandwiched between a transparent electrode which is anode and a non-transparent electrode which is cathode. Indium tin oxide (ITO) layer coated on glass substrate is most commonly used as the transparent anode. The glass provides the mechanical support for the PLED. ITO being transparent to light allows the light photon created inside the diode to escape from the device. There are two polymer layers in a typical PLED structure; among them is the hole transporting layer and the other is the light-emitting layer. Generally, the metal cathode is deposited over of the polymers by means of thermal evaporation.

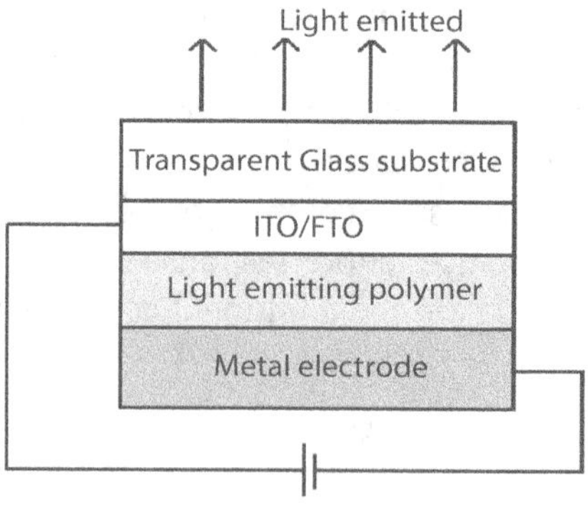

Figure 1.1 The basic structure of a PLED.

1.3 The Mechanism of Light Emission

Electron-hole recombination causes emission of a photon in visible region. Electrons are injected from the cathode to the LUMO (lowest unoccupied molecular orbit) and the holes are injected from the anode to the HOMO (highest occupied molecular orbit) of a conducting polymer. The reliability and the efficiency of the diode are strongly influenced by the materials which form the cathode, anode, and the emissive layers. A typical PLED may either be a single- layer device or a multilayer device. One example of PLED is the one fabricated from conjugated polymers including polyacetylene, polythiophene polypyrrole (PPy), poly (para-phenylene vinylene), and polyaniline (PANI) [4]. Another example of the active element used in PLED is the poly (p-phenylene vinylene) (PPV).

PLEDs like polymer light-emitting diodes and polymer light-emitting electrochemical cells gain huge interests owing to their high capabilities to serve as next generation illuminants and displays. Contrasted to inorganic light-emitting materials, conducting polymers possess very good film-forming behavior enabling the deposition uniformly by solution-based techniques, for example, screen printing and spin-coating that are competent of upscaling to industrial scale manufacturing. Polyfluorene, poly(p-phenylene vinylene), polycarbazole, and poly(p-phenylene) are widely researched; their solubility, morphology, stability, doping, etc., are proved to increase the device performance.

For example, poly(p-phenylene) films are widely synthesized by precursor methods because of its insolubility in commonly used organic solvents. Its solubility can be increased by the preparation of conducting polymers (ladder type) which leads to improved co-planarity [5]. Furthermore, a complete color display may be achieved through adjusting the structure of the molecule to regulate the energy gap of the HOMO-LUMO. Also, small amount of molecule doping proved to give desired luminous properties. Other than light emitting, color changes (electrochromic devices) and strain (electromechanical actuator) can also be stimulated by applying electric energy. Further, the electromechanical actuators directly convert the electrical energy to mechanical energy. These materials find applications in fabricating robotics, artificial muscles, etc. Electrochromic devices produce revocable color variation in reaction to the applied electric field, which makes it suitable for electronic skins and smart windows.

Owing to the tunable redox states under electricity, the conducting polymers are the fascinating materials for high performance electrochromic devices and electromechanical actuators. Further enhancement of the electrochromic or actuating ability of the polymer and the response speeds is improved by incorporating the graphene and other nanocarbon materials. For example, the multiwalled carbon nanotubes incorporated with polyaniline through an electrochemical deposition technique to aid as composite electrodes which can exhibit large conductivity ranges from 100 to 1,000 S/cm^{-1} that enables reversible and rapid electrochromic developments within short time [6]. PLEDs are not loaded with only merits. They do have a main disadvantage of weathering of the polymers with time. The disadvantage of the PLED technology is the sensitivity of the organic light-emitting materials to the atmospheric oxygen and water vapor. Hence, to protect PLEDs, a weather proof transparent polymer which is chemically and physically stable must be used for encapsulation.

1.4 Widely Used Polymers in PLED Applications

Among wide choices of polymers, some particular polymers gained special attention owing to their processability and functionality advantages that are discussed below. The advancement of important polymer material groups has been discussed here. This report gives the group of materials which can exhibit utmost potential till date to be espoused as the emissive materials in PLED applications, for example, the poly(fluorene)s, the poly(phenylenevinylene)s, etc. Polyfluorene homo- and co-polymers are purposefully emphasized because they are not well re-viewed, much progress has been made only recently, and this group of polymers are rapidly developed as a most promising viable LED polymeric material of widespread commercial interest.

1.4.1 Polyfluorene-Based Luminescent Polymers

Fukuda *et al.* reported the first fluorene-based polymers, by ferric chloride oxidative polymerization of 9-alkyl-fluorene and 9,9-dialkylfluorene [7, 8]. Their molecular weight was relatively low, with some-extent of separating and non-conjugated connections across locations except 2 and 7 [7, 8]. By using the transition-metal-catalyzed reactions of monomeric 2,7-dihalogenatedfluorenes, researchers introduced the homo-polymers for minimization of branching, improving regiospecificity. Further, Suzuki and co-workers discovered the palladium-catalyzed synthesis of mixed biphenyls from aryl bromide and phenylboronic acid [9, 10].

1.4.2 Polyfluorene Homo-Polymers

In general, polyfluorenes with substituents C6 or C9 are solvable in traditional organic diluters like aromatic hydro-carbons, chlorinated-hydrocarbons, etc. [11]. The polymers with large molecular weight does not consist separate glass transition. The polymers with straight alkyl substituents exhibit liquid crystallinity and tend to be semicrystalline. For example, F8, exhibits constant liquid crystallinity up to the temperature 270°C [12]; however, the polymer-materials having diverged alkyl-substituents exhibit non-crystallinity. Further, entire polymers while excited with UV emits a strong blue light, either in solution or in their solid state. They have a wide and drab absorption spectrum, whereas the photoluminescence spectrum shows distinct vibrionic-structures [13]. In general, the Stoke's shift lower than 50 MeV indicates a prolonged conformation.

1.4.3 Polyfluorene Alternating Copolymers

Tertiary aromatic amines are very good hole-transport materials, viable for photoconductors and LEDs. Preparation of large molecular weight, varying co-polymers containing different aromatic amines and 9,9-dialkylfluorene is possible through the Pd-catalyzed polymerization process. These alternating polymers are all soluble in conventional organic solvents, excellent film formers, and are good blue emitters. These polymer films exhibit discrete and adjustable oxidation capacities through cyclic-voltammetry that could be cycled exclusive of any significant alteration. The mobilities of positive charge carriers of the above said polymers are relatively large (3×10^{-4} to 1×10^{-3} cm^2/Vs) [14–16]. Due to these large mobilities of holes, these polymeric materials recommended for the applications in photoconductors also in LEDs for transportation of holes. Attempting to create polymers with distinctive properties, the alternating copolymer approach has been extended to other conjugated monomers, such as triarylamine, thiophene, etc. [11]. The co-polymers consisting large molecular weight exhibit high photoluminescence and emission spectra of the co-polymers may be associated with degree of co-monomers delocalization, e.g., the copolymers of bithiophene produce the spectra in yellow region, cyano-stilbene produces the spectra in green region, and thiophene in bluish green region [11].

1.4.4 Derivatives of PPV

The emission of yellow-green light by PPV under electrical stimulation was discovered in past decade, since then, several researchers focused on optimization of PPV and to make this as a potential material [17, 18, 19]. Most importantly, some of the advancements have taken place in preparation, regulating the balance in charge carriers, improving the efficiency in power, and enhancing the life-time, also in adjusting the emission of wavelength. Owing to the vinylene linkages, photo-oxidatively, the PPV chemical structure is unstable, also there is some restriction in the improving the saturated blue rich and red rich emitters. Even though these problems continuing to encounter the initiation of PPV into the display devices commercially, noteworthy development has been made towards the controlling and optimization of the PPV materials to make these are potential aspirants for the applications in PLED devices [11].

1.4.5 Soluble Precursors of PPV

For poly (arylene vinylene) series, the parent structure is PPV owing to absence of functional groups to improve solubility, rigid structure, and propensity to develop crystalline morphology, these materials are stubborn and directly not processable form the solution. Meanwhile for polymeric emission systems solvent process ability is a necessary characteristic; further, the soluble precursors to the PPV which can be molded as films then transformed to PPV through heating have been established.

1.4.6 Derivatives of PPV for Solution-Processing

PPVs are generally difficult to process but possess properties that are capable of good PLED candidates. To overcome the processing difficulty, plenty of research is devoted towards the advancement of soluble PPVs. Making of thin films is easy with soluble PPVs, exclusive of successive thermal-conversion. In 1991, Heeger *et al.* reported the applications of PPVs particularly 2,5-dialkoxy functional PPVs; according to them, the alkoxy groups having at least one bulky or long polymer groups are soluble in diverse organic solvents which include xylene, chloroform, etc. The functionality of the bulkier materials has been described to interrupt the propensity of PPV in order to increase the efficiency of EL.

1.4.7 Polyphenylenes

In the area of polymer light-emitting devices, there exist another class of conjugated polymer group called PPP (poly(1,4-phenylene)); these PPP materials consist large bandgap and permit blue light emission. Subsequently, the design of blue emitters which are having high efficiency and long life time endures a major task in advancement of the polymers. Therefore, activities of research in focused on PPP to emphasize the methods towards PPP thin-films through solvable precursor polymer materials which are thermally converted, in addition to the improvement of soluble PPPs. These class polymers have high molecular weights that are sufficient to mold films including excellent integrity that have been accomplished besides diodes with blue-emission that have been built and reported with considerable efficiencies [20].

1.5 Parameters to be Considered for Display Applications

The following parameters are considered for making different display technologies:

i. Color purity and brightness,
ii. Light conversion efficiency, and
iii. Durability.

1.5.1 Color Purity and Brightness

Entire spectrum of colors (and infrared) is possible with different PLEDs. For orange and green emitting PLEDs, life cycles of over 10,000 hours have been reported. However, till now the data is not available, blue devices with high cyclic life. This causes the loss of blue component of light emitted with time and the PLED will eventually lose the entire blue light, and hence, the visible light emitted cannot maintain color purity with time. So, a blue emitting polymer with lifetime equal to orange and green ones is barely needed in realizing a display device to maintain the emission color through the lifetime of the device.

High luminance values may be achieved at small voltages. For orange PLEDs, the observed starting value of emission is of around 1.79 V which is above the bandgap. A brightness of 100 cd/m^2 is reported around 2.5 V for the same PLED. It might be associated with a distinct brightness of 60 cd/m^2 for computer and laptop displays. Even a 50-nm thin film polymers are reported to emit light. If the layer thickness increases, for instance, 100 nm, the voltage rises approximately by 1V. The low-voltage process enables device operation possible in ordinary less-expensive integrated chips. Up to 10,000 cdm^{-2} brightness can be achieved at as low as 6 V. In pulsed open, even 100,000 cdm^{-2} brightness is possible to achieve. Even some groups proved laser action is possible in polymer devices with intensity more than 1,000,000 cdm^{-2}.

1.5.2 Light Conversion Efficiency

The efficiency of a PLED depends on the external efficiency which is calculated in forward direction. However, the value of external efficiency observed in forward direction is large compared to the values observed using integrated spheres. In addition, sometimes, the samples behave like an optical fiber, in which total internal reflection takes place when the angle of incidence is greater than the critical angle, due to this in PLEDs considerable quantity of light which is produced in the emissive-layer escapes through the sideways of the sample. Ching Tang *et al.* [21] proposed a simple solution to determine the whole value in terms of candela per ampere. Green PLEDs exhibit highest efficiency of 75 cd/A. Further enhancements can be achieved through rising in the efficiency of photoluminescence and also by enhanced electron-injection. In general, PLEDs' efficiencies are far superior than normal bulbs and LEDs. In current display technologies, PLED is far superior and simple to fabricate.

1.5.3 Color Stability

Stability of polymers and their properties are main concerns regarding this polymer technology. In the present scenario, PLEDs' displays and backlights for LCDs can meet customer specification only in orange light-emitting materials. Quick evolution is happening to grow blue and green emitting polymers to similar extent of constancy. Some researchers proved that it's possible to fabricate device on elastic substrates as an alternative of rigid glass-substrates. But these devices are reported to have short lifetime around 1 day. These devices are ruined by diffusion of water through the plastic film because of the sensitivity of the device for O_2 and H_2O, it may lead to rapid corrosion of the positive electrode of the device. Recently, advancement in the production of elastic films consisting good water barrier properties, flexible light-emitting films, and displays was made possible.

1.6 Applications in Large and Small Area Devices

1.6.1 Displays

PLED technology can be employed to fabricate small and simple unicolor segmented displays to complicated and large full-colored displays. The application ranges are typically classified based on the size of the display.

1.6.1.1 Matrix and Small Segmented Displays, ≤25 cm²

The appliances are where the information displayed is limited, for example, in car dash boards, professional-equipment, etc. Utmost of the display area is typically monochrome. Comparing PLEDs with other technologies, reflective LCD is advantageous as a function of power consumption, but it has a poor contrast, it has a dull visual aspect, it has low viewing angle, and it is not readable in the dark. PLED exhibits a discrete advantages in slimness and an improved power consumption factor in the range of 10 to 100, combining with a backlight and LCD. Some of the advantages of PLED devices are thin, response time is fast, graphics resolution is very high, display brightness is high, and contrast is also high.

1.6.2 Thin and Flat Light Sources

In present scenario, application of PLEDs as sources of light, apart from some special cases, is questionable. When compared with power efficiency of 19.99 Lm/W for florescent sources and 60 Lm/W for incandescent tubes and the PLED power efficiency is as low as of 4 to 10 Lm/W. Also, the lifetime of polymer-based LED devices is limited by the high intensity of light needed for illumination. Nevertheless, PLED devices might be utilized in all types of applications in signaling such as brake lights for cars, decorative light sources, potentially rear lights, etc. LCD backlighting is one explicit application where the source of the light is really lightweight, thin, and flat. The usage PLEDs in comparatively less in large-area applications, less than 100 cm^2, can be distinguished based on the purpose for which they are used. Also, the backlight not only used to increase the contrast of the display in daylight but also used to illuminate the display in dark for example car stereo, radio sets, etc.

1.6.3 Cloth-Type PLEDs

The formation top emission OLEDs on a substrate of fabric is an easy way in advancement of OLEDs. The substrates of fabrics are categorized thru spatial voids and through these voids water can easily pass. Besides the assembly of constituent fibers forms the uneven surface; unfortunately, this is not compatible with the OLEDs. Hence, to prepare OLED-compatible fabric substrates, it is necessary to introduce the methods to eliminate the spatial voids then reformation of surface to be even [22–24]. There exist two stages to form a glass-like surface which can be used as fabric substrate to embed OLEDs [24]. Foremost, for partial planarization, i.e., fill the valleys, low viscous ductile polyurethane (PU) is spin coated on self-assembled fabric substrate. Next, to reduce the roughness of the surface, a high viscous polyurethane is deposited on the fabric before clean guide substrate is transported to the subsequent fabric substrate *via* lamination at room temperature. Subsequently, the distinguished methods of top-emission OLED production and multibarrier encapsulation, extremely robust wearable OLEDs were attained on the modified fabric substrate [24]. Even after this promising achievement, remaining practical problems should be overcome to use in realistic wearable displays. The predicted planarization procedure for fabric substrate well-suited with OLEDs weakens the nature of the fabric like softness, breathability, etc.

Furthermore, accomplishing consistent wash ability of the device might be another critical obstacle in the perspective of wearable electronic clothes. In view of the ability to concede actual light-emitting fabrics, the fiber-shaped OLEDs are much closer to cloth-type display concept, due to tiny and curved substrate, it is difficult to prepare OLEDs on the fiber substrate. B. O'Conner *et al.* reported a technique that the deposition of layers of OLEDs on the fiber is retain the rotation of the fiber in vacuum deposition [25]. Another reported technique is dip-coating method which is cost-effective and simple and can be employed for solution-based PLED-fibers [26]. Whereas both techniques surely proved the possibility of OLEDs on fibers, advanced studies on electrical addressing and reliable encapsulation schemes for the accumulated fibers are required for working as a display device [27].

1.6.4 PLEDs in Wearable Electronics

S. Choi *et al.* [28] have fabricated a light-emitting fabric, which is efficient and flexible, making it suitable for displays which can be worn like clothes. The PEN fibers are used for weaving the fabric, and the thermal lamination is used to form planarization layer onto this fabric. The surface roughness of $R_q = 2.073$ nm, these fabrics make them appear very smooth. Organic light-emitting diodes were placed thru thermal evaporation and the additive protective layers by transparent-flexible encapsulation effectively block the saturation of water and oxygen. Besides, the prepared device exhibits the luminance of around 35,844 Cd/m^2 and maximum current efficiency of about 70.43 Cd/A. In addition, the device on the material was found to operate stably after harsh bending, even at a bending radius of 2 mm for 3,000 cycles and a bending radius of 1 cm after 30,000 cycles. However, more bending of these fabrics results in leakage current within the device and cracks on the fabric. These fabrics can find various electronic textile industrial applications like curtain manufacturing, serves as functional and table clothes in and healthcare, fashion, as well as in the automobile industries.

1.7 Conclusion

The construction and working principles of PLEDs is discussed in detail. Different polymers used in PLEDs are discussed. The important parameters that are the key factors to be considered while adopting the polymer electroluminescent materials for PLEDs like brightness, color purity, light conversion efficiency, and color stability are discussed. Finally, the application of the PLEDs to small, midsize, and large size displays are discussed. Also, their application in flexible displays and cloth type wearable displays are discussed. These PLEDs are undergoing a rapid development and may soon be available in all forms of displays. The present limitations of PLEDs in illumination system where the intensity of light is very high compared with displays like television and laptops will soon be rectified with improved polymers.

References

1. Liang, J., Lu, L., Xiaofan, N., Zhibin, Y., Qibing, P., Elastomeric polymer light-emitting devices and displays. *Nat. Photonics*, 7, 817–824, 2013.

2. Hameed, S., Predeep, P., Baiju, M., Polymer light emitting diodes - A review on Materials and techniques. *Rev. Adv. Mater. Sci.*, 26, 30–42, 2010.
3. Burroughes, J.H., Bradley, D.D.C., Brown, A.R., Marks, R.N., Mackay, K., Friend, R.H., Burns, P.L., Holmes, A.B., Light-emitting diodes based on conjugated polymers. *Nature*, 347, 539–541, 1990.
4. Belgacem, M.N. and Gandini, A., *Monomers, Polymers and Composites from Renewable Resources*, University of Aveiro, CICECO, Chemistry Department, Portugal, Amsterdam, 2008.
5. Tasch, S., Niko, A., Leising, G., Scherf, U., Efficient white light-emitting diodes realized with new processable blends of conjugated polymers. *Appl. Phys. Lett.*, 68, 1090, 1996.
6. Chen, X., Lin, H., Chen, P., Guan, G., Deng, J., Peng, H., Smart, stretchable supercapacitors. *Adv. Mater.*, 26, 4444–4449, 2014.
7. Fukuda, M., Sawaka, K., Yoshino, K., Electronic Characterization of New Bright-Blue-Light-Emitting Poly(9,9-dioctylfluorenyl-2,7-diyl)-End Capped With Polyhedral Oligomeric Silsesquioxanes. *Jpn. J. Appl. Phys.*, 28, 1433, 1989.
8. Fukuda, M., Sawaka, K., Yoshino, K., Synthesis of fusible and soluble conducting polyfluorene derivatives and their characteristics. *J. Polym. Sci., Polym. Chem. Ed.*, 31, 2465, 1993.
9. Miyaura, N., Yanagi, T., Suzuki, A., The Palladium-Catalyzed Cross-Coupling Reaction of Phenylboronic Acid with Haloarenes in the Presence of Bases. *J. Synth. Commun.*, 11, 513–519, 1981.
10. Friend, R., Gymer, R., Holmes, A., Burroughes, J., Marks, R., Taliani, C., Bradley, D., Dos Santos, D., Bredas, J., Logdlund, M., Salaneck, W., Electroluminescence in conjugated polymers. *Nature*, 397, 121–128, 1999.
11. Bernius, T., Inbasekaran, M., O'Brien, J., Wu, W., Progress with Light-Emitting Polymers. *Adv. Mater.*, 12, 1737–1750, 2000.
12. Grell, M., Bradley, D., Inbasekaran, M., Woo, E., A glass-forming conjugated main-chain liquid crystal polymer for polarized electroluminescence applications. *Adv. Mater.*, 9, 798–802, 1997.
13. Bernius, M., Inbasekaran, M., Woo, E., Wu, W., Wujkowski, L., Polyfluorenes. *J. Mater. Sci.: Mater. Electron.*, 11, 111, 2000.
14. Redecker, M., Bradley, D., Inbasekaran, M., Wu, W., Woo, E., High Mobility Hole Transport Fluorene-Triarylamine Copolymers. *Adv. Mater.*, 11, 241–246, 1999.
15. Redecker, M., Bradley, D., Inbasekaran, M., Woo, E., Mobility enhancement through homogeneous nematic alignment of a liquid-crystalline polyfluorene. *Appl. Phys. Lett.*, 74, 1400, 1999.
16. Redecker, M., Bradley, D., Baldwin, K., Smith, D., Inbasekaran, M., Wu, W., Woo, E., An investigation of the emission solvatochromism of a fluorene-triarylamine copolymer studied by time resolved spectroscopy. *J. Mater. Chem.*, 9, 2151–2154, 1999.
17. R. Wessling and R. Zimmerman, US Patent 3 401 152, 1968.

18. R. Wessling and R. Zimmerman, US Patent 3 706 677, 1972.
19. Braun, D., Heeger, A., Kroemer, H., Improved efficiency in semiconducting polymer light-emitting diodes. *J. Electron. Mater.*, 20, 945–948, 1991.
20. Yang, Y., Pei, Q., Heeger, A., Efficient blue polymer light-emitting diodes from a series of soluble poly(paraphenylene)s. *J. Appl. Phys.*, 79, 934, 1996.
21. Tang, C.W., Direct interaction with large-scale display systems using infrared laser tracking devices. *Sot. Inf. Display Conf. Proc.*, p. 181, 1996.
22. Kim, W., Kwon, S., Lee, S.M., Kim, J.Y., Han, Y., Kim, E., Choi, K.C., Park, S., Park, B.C., facilitated embedding of silver nanowires into conformally-coated iCVD polymer films deposited on cloth for robust wearable electronics. *Org. Electron.*, 14, 3007–3013, 2013.
23. Kim, H., Kwon, S., Choi, S., Choi, K.C., Solution-processed bottom-emitting polymer light-emitting diodes on a textile substrate towards a wearable display. *J. Inf. Disp.*, 16, 179–184, 2015.
24. Kim, W., Kwon, S., Han, Y.C., Kim, E., Choi, K.C., Kang, S.H., Park, B.C., Wearable Electronics: Reliable Actual Fabric-Based Organic Light-Emitting Diodes: Toward a Wearable Display. *Adv. Electron. Mater.*, 2, 1600220, 2016.
25. O'Connor, B., An, K.H., Zhao, Y., Pipe, K.P., Shtein, M., Fiber shaped light emitting device. *Adv. Mater.*, 19, 3897–3900, 2007.
26. Kwon, S., Kim, W., Kim, H., Choi, S., Park, B.C., Kang, S.H., Choi, K.C., High Luminance Fiber-Based Polymer Light-Emitting Devices by a Dip-Coating Method. *Adv. Electron. Mater.*, 1, 1500103, 2015.
27. Lee, S.M., Kwon, J.H., Kwon, S., Choi, K.C., A review of flexible OLEDs towards highly durable unusual displays. *IEEE Trans. Electron Devices*, 64, 5, 2017.
28. Choi, S., Kwon, S., Kim, H., Kim, W., Kwon, J.H., Lim, M.S., Lee, H.S., Choi, K.C., Highly Flexible and Efficient Fabric-Based Organic Light-Emitting Devices for Clothing-Shaped Wearable Displays. *Sci. Rep.*, 7, 6424, 2017.

2

Polymer Light-Emitting Devices by Solution Processing

Mariya Aleksandrova

Technical University of Sofia, Department of Microelectronics, Sofia, Bulgaria

Abstract

Solution processing of organic-based light-emitting devices (OLEDs) has been widely studied as a technology for low-cost, large area fabrication of optoelectronic devices. This chapter focuses on the usage of polymer films deposited by solution processing for the manufacturing of different single-colored or fill-colored electroluminescent displays. The survey focuses on the polymer molecules engineering and devices architecture for creating highly efficient polymer light-emitting devices (PLEDs) with parameters compatible to the commercially available OLED using small molecules. The relevance of the materials selection and deposition process modes are highlighted, in terms of energy bands alignment, smoothening of the films surfaces, and lack of intermixing at the layer interfaces. All these factors are crucial for achieving stable and efficient PLED. Some of the main achievements and challenges reported in the last few years are summarized and discussed in relation with the turn-on voltage, current density, maximum brightness, luminous efficiency, and possibility for commercialization of the devices. New light-emitting, hole and electron transporting materials, as well as electrodes and substrates materials are considered.

Keywords: All-solution processing, polymer light-emitting device, hole transporting layer, color tuning, polymer emissive layer, PEDOT:PSS, flexible substrates, energy level alignment

Email: m_aleksandrova@tu-sofia.bg

Inamuddin, Rajender Boddula, Mohd Imran Ahamed and Abdullah M. Asiri (eds.) Polymers for Light-Emitting Devices and Displays, (15–50) © 2020 Scrivener Publishing LLC

2.1 Introduction

The organic electronics is one of the most advanced branches of the engineering science. At present, it is possible to synthesize a great species of organic molecules that could be conductive, insulating, or semiconducting. Based on this classification, there are a huge number of organic materials, according to the band gap width and therefore to the electrical properties, which can be easily modified by replacing any of the major functional groups in the organic compound or by doping. By tailoring the molecular structure of the materials, it is possible to achieve great variety of properties and applications with easy tunable characteristics. Such flexibility in the capabilities of organic materials is very useful in the field of optoelectronics [1, 2].

Organic optoelectronics is a multidisciplinary field that includes physics of solid-state matter, synthetic chemistry, thin film deposition technologies, methods for structural characterization, and last, but not least, electronic engineering for processing the control signals to the device, or the useful signal generated from the device [3].

Organic semiconductors are divided into two major classes: low molecular weight compounds (organic crystals) and high molecular weight compounds (polymers). Recently, the research interest has been focused mainly on polymers. One of the major advantages of conductive polymers over the small molecules is their ability to be easily processed and applied as thin films from solution. This makes their manufacturing simple, fast, and cost-effective, because the solution processes are vacuum-free, not requiring complex and expensive equipment [4, 5]. Usually, printing and other solution-based processes are used for the polymeric films deposition. Since the most polymeric devices include several thin films, organic, as well as inorganic, the interface formation between them is also very important for the devices operation [6].

Nowadays, the consumer requirements to the display devices impose the need for development of increasingly attractive displays with qualities such as high brightness, a broad range of colors, color saturation, a wide viewing angle, small sizes, and low power consumption. Organic-based displays meet these requirements. Currently, the commercial organic light-emitting devices (OLEDs) could be found in the multi-color flat panel displays in TVs, cell phones, and digital cameras [7, 8]. Actually, all OLED displays on the market today use small molecules, and they are produced by expensive vacuum evaporation process [9]. Polymer light-emitting devices (PLEDs) technology also steps in the competition to make displays of commercial devices, like some of the latest MP3 players and audio systems, which are however low-volume series. Thus, the polymer-based products currently have only an advertising function to demonstrate trends in the development of the above-mentioned products of the modern electronics [10].

Because PLED displays have their own illumination and no backlighting is required, they tend to be even thinner and lighter than liquid-crystal displays (LCD). At the PLED matrix, it is possible control of the individual pixels; therefore, the consumption is significantly lower than at the LCD, where the entire panel should be continuously illuminated. PLEDs also provide higher contrast, more "true" colors and a wider viewing angle. These advantages are the same like for OLED, but additionally, PLED films can be deposited over a large area and at room temperature by spin-coating, inkjet printing, or spray deposition. The use of such techniques allows the application of the polymer layers to various types of substrates, including flexible foils, textile, and even paper.

2.1.1 Materials, Design, Main Parameters, and Characteristics of PLEDs

PLED is an electroluminescent (EL) device, at which the functional polymer layers are deposited between two electrodes. When applying a DC voltage, electrons and holes are injected from the electrodes into the light-emissive polymeric film. When the opposite charge carriers meet at this film, they recombine, resulting in light emission with wavelength depending on the band gap of the used polymer, which is semiconductor. At least one of the electrodes in the display structure must be transparent, in order to fluently emit the light generated inside the structure without optical losses, and at the same time, it must be conductive in order to have good injection properties (Figure 2.1a). Additionally, the electrodes' work functions must be

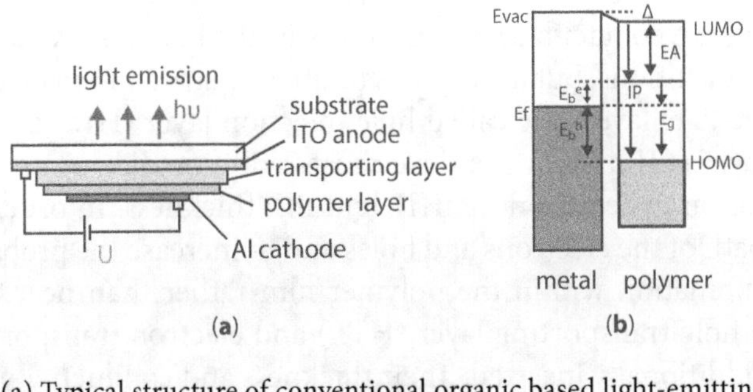

Figure 2.1 (a) Typical structure of conventional organic based light-emitting device; (b) Band diagram, showing the energy levels alignment at the interface electrode/polymer layer.

close in energy to one of the energy levels of the polymer—to the energy of the Highest Occupied Molecular Orbital (HOMO) for the anode and to the energy of the Lowest Unoccupied Molecular Orbital (LUMO) for the cathode (Figure 2.1b) [11]. In this term, the most suitable and commonly used material for anode is Indium Tin Oxide (ITO) and for cathode is aluminum. This represents the classical structure of a PLED. It involves vacuum deposition processes (sputtering and thermal evaporation) increasing the total price of the device and making more or less senseless the using of solution, low cost processes for the polymer films deposition only. That's why, recently, the efforts have been focused on the finding of suitable materials and processes as alternative to the typical electrodes. An overview of these materials and technologies is made in this chapter.

2.1.2 Main Problems at PLEDs and How the Solution Processes Can Affect Them

The nature of the most EL polymers is such that their energy levels are not close to the work functions of the suitable electrode materials. This leads to formation of high energy barriers between the layers, which impede the injection of charge carriers into the bulk of the emissive material. As a result, the amount of charge carriers that recombine radiatively is strongly reduced and the quantum efficiency also decreases. Therefore, higher voltages should be applied to overcome the height of the injection barriers and to achieve feasible luminance of radiation, but this increases the power consumption and again reduces the efficiency of the structure. The approaches used to improve the current efficiency (CE) (in cd/A) and luminous efficiency (in lm/W) are as follows:

1) To reduce the injection barriers at the interfaces between the layers by "ladder effect" [12]. It means insertion of intermedium films with energy levels taking the middle energy position between the energies determining the contact barriers. In this way, the barriers are divided into few smaller partitions, but it supposes multilayer structure to be produced. The intermediate layers are called hole injection layer (HIL) and electron injection layer (EIL). Their presence causes additional delay in the charge carriers motion depends on the HIL and EIL thickness. In order to make equal the path of the electrons and holes and to increase the probability for their recombination within the polymer film rather than near the interface zones, hole transporting layer (HTL) and electron transporting layer (ETL) are additionally inserted. Their thickness and conductivity are compliant with the corresponding charge carriers' mobility (Figure 2.2) [13]. It is preferable if all these layers are polymeric; otherwise, the device may

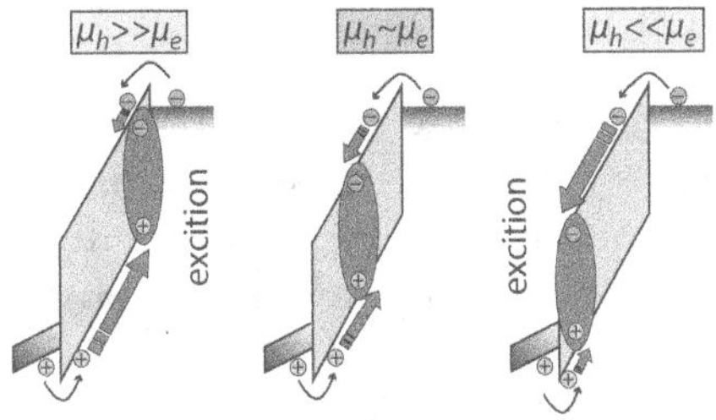

Figure 2.2 Physical processes revealed in the organic-based LED, related to the molecules chemical nature and properties (charge carriers mobility is considered).

be overheated and thermal damaged from the multiple high-temperature vacuum deposition processes that is has to be subjected on. As a consequence, a need arises for development of new classes of polymeric materials, which are not light-emitting, but supplementary in the PLED structure and they must be soluble for the needs of solution processing technologies. An overview of the newest materials and technologies for producing the above-mentioned layers is made and presented in this chapter.

2) To increase the uniformity of the polymeric layers involved in the PLED structure and to decrease the defects concentration [14]. Since the deposition is not carried out in a vacuum environment, the layers are exposed to contamination from the ambient, which act as traps that affect adversely the radiative recombination by quenching it. There are also problems with the uniformity of the solution deposited films at nanoscale (Figure 2.3) and mutual penetration of separate films that must have well defined interfaces. Depending on the solution deposition process (spray coating, spin-coating, ink-jet printing, etc.), the traps concentration and energy distribution strongly vary and affect at different extend the electro-optical characteristics. In this chapter, the most recent reports on the solution deposition processes are summarized with analysis and discussion of their effect on the films quality, PLED efficiency, and the justified balance between performance and cost efficiency.

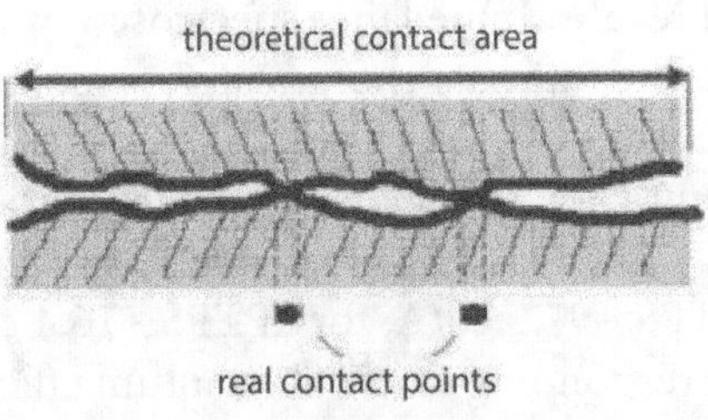

Figure 2.3 Contact points between functional layers in PLED.

2.1.3 Aim of This Chapter

The aim of this chapter is to make an overview on the newest polymeric materials and processes beyond the classical structure of PLED, leading to low-cost solution and all-solution processed devices with enhanced parameters, which are closer to the commercial production line requirements and the custom needs.

2.2 Materials for Fabrication of PLEDs and Their Performance at Solution Processing

2.2.1 New Polymers for Light-Emissive Layers and for Supplementary HTL and ETL

A recent approach to fulfill the industrial requirements for large-area color displays at a low- cost fabrication is to make a hybrid polymer-inorganic material. New tetrahedral silicon-based hosts have been synthesized by Suzuki cross-coupling reaction, as a composition of a Si atom and a phenyl or m-pyridyl phenyl moiety [15]. Silicon atoms are responsible for the broad band gap, covering blue emission spectrum and quenching of the triplet, non-radiative states. Polymer constituent is characterized with good morphological stability and electron transporting properties leading to the desired balance between the charge carriers. The new host materials diphenylbis(3-(pyridine-3-yl)phenyl)silane (3PTPS), diphenylbis(3-(pyridine-2-yl)phenyl)silane (2PTPS), and diphenylbis(3-(pyridine-4-yl)phenyl)silane (4PTPS) having pyridine moieties with different orientations are soluble. The host material, belonging to the class of small molecules {2,2,2-(1,3,5-benzenetriyl)tris-[1-phenyl-1H-benzimidazole] (TPBI)} is also soluble and can be combined with the typical orange-red, blue, green, and white emissive polymers for solution processed device fabrication. Spin-casting has been used for the tested devices. For some of the cases, the CE exceeds 100 cd/A, which is beyond previously reported results. Impedance spectroscopy measurements have been applied in order to estimate charge carriers mobility and it has been found balanced mobility of $\sim 10^{-4}$ $cm^2/(V{\cdot}s)$ for the both charge types, which also contributes for the superior electro-optical characteristics, shown in Figure 2.4. The current efficiencies are as follows for 2PTPS ~95.7 cd/A, for 3PTPS ~97.5 cd/A, for 4PTPS ~76.4 cd/A, and for TPBI ~60.6 cd/A. The corresponding external quantum efficiencies (EQEs) are 35.4%, 35.5%, 28.0%, and 21.2%.

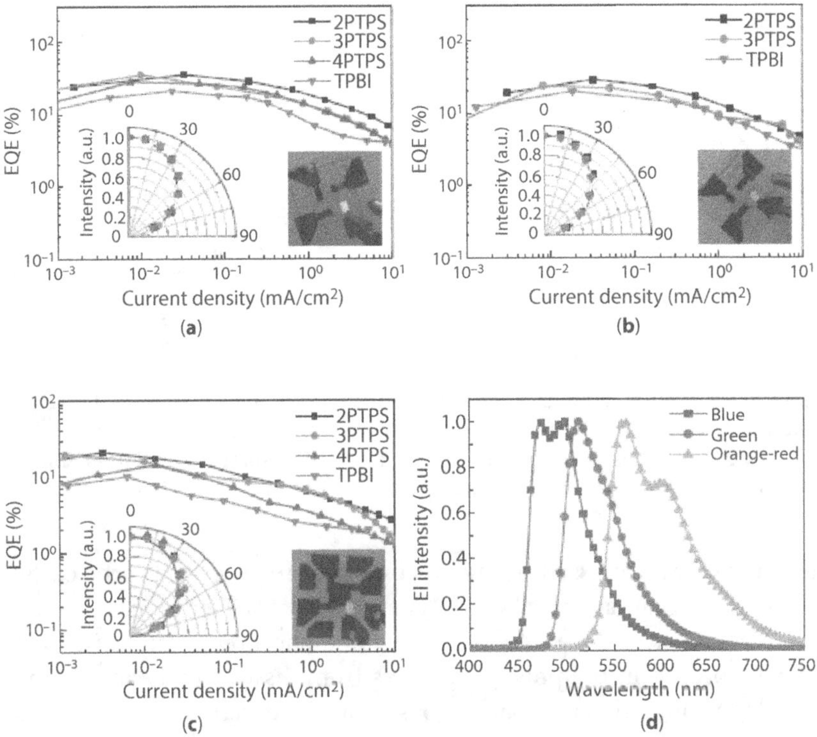

Figure 2.4 (a–c) PLEDs with various electron-transporting host materials. Insets: left—EL distribution, according to viewing angle; right—photographs of devices and their color emission; (d) EL spectra of the solution-processed PLEDs [15]. (Reprinted from Science Advances, 2: e1601428, Han *et al.* Copyright (2016) with permission of AAAS).

To overcome the problem with the solution processed PLED to dissolute the first polymer layer during deposition of a next one, recently, electron confining cross-linkable polymer 9,9′-bis (4-vinylphenylmethylen)[3,3′] bi-carbazole (VPMC) has been synthesized [16]. It serves as a HTL and supports the function of the HIL from poly (3,4-ethylenedioxythiophene)-poly (styrenesulfonate) (PEDOT:PSS), preventing it from solvent damage (Figure 2.5). As a result of the high quality interface, remarkable brightness value of 15,200 cd/m^2 and more than 20% increase in the power efficiency have been demonstrated, as compared with the HIL only device. The emissive layer is N,N′- Dicarbazolyl-4,4′-biphenyl (CBP) and has been spin-coated the same as the rest two polymer materials. However, for better results, before light-emitting layer (EML) deposition, a solution has been prepared with the host and guest molecules mixed in specific ratio. In this way, no energy barrier exists between the HTL and EML and triplet-triplet

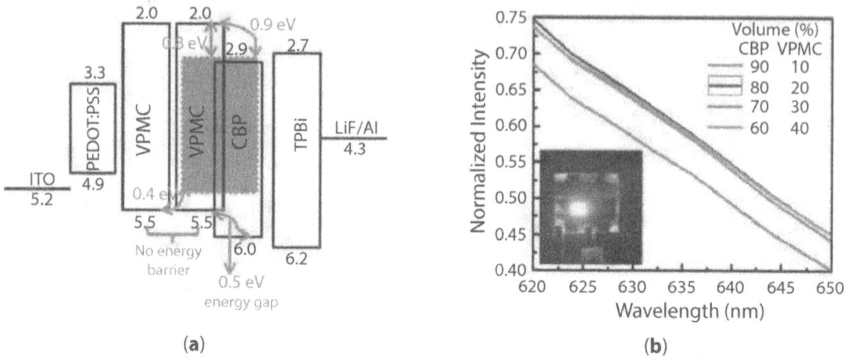

Figure 2.5 (a) Energy bands diagram of the solution produced light-emitting device with co-host structure; (b) The EL emission slightly moved towards higher wavelength as the concentration of co-host was increased from 10 to 20 wt% [16]. (Reprinted from Thin Solid Films, 660, 956–960, Dubey *et al.*, Copyright (2018), with permission from Elsevier).

transitions are suppressed. Therefore, excitons are generated at the co-host system provoking radiative host-to-guest energy transfer process.

Reports are available in the literature for application of semi-orthogonal solvents to prevent the polymeric layers from dissolving during the next layer deposition. This approach is based on the different solubility of the polymers in solvents with different polarity [17]. Because naturally, not all polymers subjected to processing with semi-orthogonal solvents are suitable for combined work in a PLED due to unfavorable energy level alignment, novel polymers have to be developed. Poly(9,9-dioctylfluorene) end-capped with a triphenylamine derivative (PFO-TF) and the emissive polymer PPF-SO15, or PPF-SO15-BT1, have been used and spin-coating over PEDOT:PSS electrodes was used for solution processed device. The side groups of the constituents are responsible for the limited mixture of the films after contact with the next layer solvent during the wet deposition process. Favorable energy diagram can be seen with negligible hole injection barriers (Figure 2.6a). That's why the blue PLED is characterized with luminescence efficiency greater than 5 cd/A, which is better than the exhibited from a device without hole transporting layers (~3.5 cd/A) or device with conventional poly (9-vinylcarbazole) as a transporting layer (~4.7 cd/A). The turn-on voltage for the optimal structure is reduced to ~3V (Figures 2.6b and 2.6c).

One of the modern approaches to build efficient all-solution processed PLED is application of quantum dot (QD) layer, protected by solvent erosion resistant HTL [18]. The advantages of QDPLED are related to emission wavelength tunable by the size of QD, narrow emission spectrum,

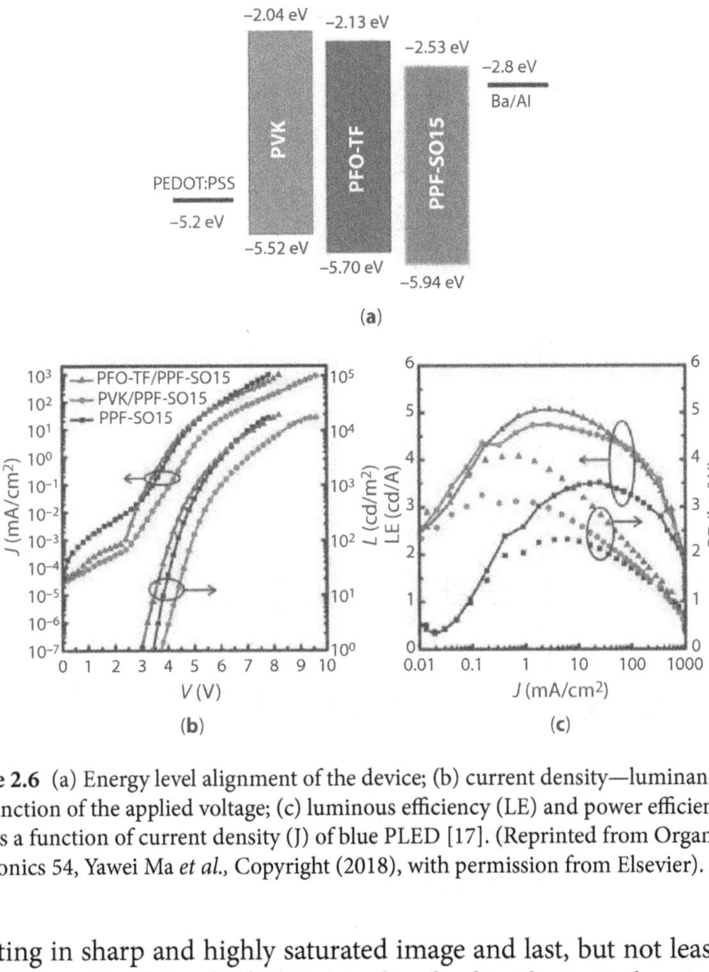

Figure 2.6 (a) Energy level alignment of the device; (b) current density—luminance plots as a function of the applied voltage; (c) luminous efficiency (LE) and power efficiency (PE) as a function of current density (J) of blue PLED [17]. (Reprinted from Organic Electronics 54, Yawei Ma *et al.*, Copyright (2018), with permission from Elsevier).

resulting in sharp and highly saturated image and last, but not least, they are solution dissolving [19]. QDs are dissolved in the typical organic solvents, like chloroform (CF) for example, in the same way as most of the HTL materials. Therefore, it could be easy dissolved by the solvent of the next HTL coating, if the device design is inverted [it is related to thin film transistor (TFT) pixel control circuit]. Cross-linkable agents and polymeric surface modifiers have been used before to prevent the dissolution of the QD layer, but still the quantum efficiency is not any greater than few percent. However, it has been found that the polar orthogonal solvent 1,4-dioxane used for poly (9-vinlycarbazole) (PVK) as a HTL, successfully blocks the QD layer damage. Due to lack of mutual penetration, each film is characterized with excellent uniformity and the high quality physical contact also contributes for significant improvement of the device behavior in terms of switch on voltage reduction, as can be seen from Table 2.1.

Table 2.1 Main parameters of the all-solution processed inverted PLED with QD [18].

	Devices	Voltage (V)		Luminous efficiency (cd/A)		EQE (%)	Luminance (cd/m^2)
		V_T	At 150 mA/cm^2	Max	At 150 mA/cm^2	Max	
Red	Single-layer HTL	3.3	15.1	19.6	13.2	11.2	2.27×10^4
	Double-layer HTL	2.7	6.9	22.1	21.9	12.7	9.80×10^4
Green	Single-layer HTL	5.1	15.4	15.5	11.6	3.83	1.99×10^4
	Double-layer HTL	2.7	6.1	21.4	20.6	5.29	8.00×10^4
Blue	Single-layer HTL	5.3	16.8	1.55	0.90	4.69	1.28×10^3
	Double-layer HTL	4.1	8.6	1.99	1.62	5.99	4.14×10^3

(Reprinted with permission from ACS Nano, vol. 12, no. 2, 1564-1570. Copyright (2018) American Chemical Society).

Device with PVK film only exhibited turn-on voltages of 3.3, 5.1, and 5.3 V with peak brightness of 22,700 cd/m^2, 19,900 cd/m^2, and 1,280 cd/m^2 for RGB QLEDs, respectively (Figure 2.7a). Additionally, the RGB QLEDs are characterized with current efficiencies of 19.6 cd/A, 15.5 cd/A, and 1.55 cd/A. In this way, quantum efficiency of 11.2%, 3.83%, and 4.69% have been obtained for inverted all-solution processed PLEDs with QDs. For optimization of the device performance, second HTL could be inserted, leading to additional breaking down the height of the interfacial energy barrier and quantum efficiency increase to 12.7%, 5.29%, and 5.99%. All films in the structure have been spin-coated.

Although some optimization of the PLED performance has been obtained with utilization of QDs for certain colored emission, integration of the basic colors (RGB) in single device by QDs incorporation is still challenging. It means that optimized performance of white PLED is difficult achievable. In general, the problem that appeared is the unwanted energy transfers between the QDs and light-emitting polymer, which is crucial for the high colors purity and radiative recombination. Recently, it has been reported application of various QDs as emitters. All-solution processed white QDs EL devices have been demonstrated [20]. They consist of trichromatic mixture of RGB emitting colloids and a homo-polymer of (poly [9,9-dioctylfluorenyl-2,7-diyl] ended with N,N-Bis (4-methylphenyl)-aniline, producing blue emission (ADS329BE). The ADS329BE matrix polymer is chosen due to absorbance overlapping with that of the QDs. This device exhibited maximum luminescence of ~16 × 10^3 cd/m^2,

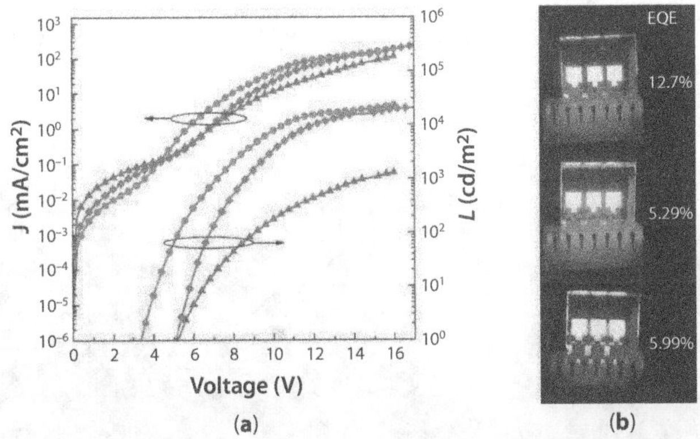

Figure 2.7 (a) J-V-L characteristics of PLED with quantum dots with single HTL (right pointed arrow) and double HTL (left pointed arrow); (b) Photos of the PLED emitting different colors [18]. (Reprinted with permission from ACS Nano, vol. 12, no. 2, 1564–1570. Copyright (2018) American Chemical Society).

maximum CE of 1.43 cd/A, and turn-on voltage of 2 V, according to the ratio between the blend components (Figure 2.8a). In this way, tailoring of the HOMO and LUMO energies can be controlled, respectively, the transport properties at the layer boundary are improved (Figure 2.8b). All four layers, excluding ITO anode and Al cathode, have been deposited by spin-coating technique, and for this reason, the device is ascribed to the solution processed.

One of the most reliable and effective ways to precisely tune the properties of the light-emitting polymers and thus to control PLED characteristics is tailoring of the molecular structure during the dyes synthesis, rather than to make the device architecture complex. This could be realized by central dyes and side chain attached fluorine oligomers, preventing luminescent quenching between the emitting cores [21]. Depending on the central dye 1–4, as is shown in Figure 2.9a, it is possible to control the

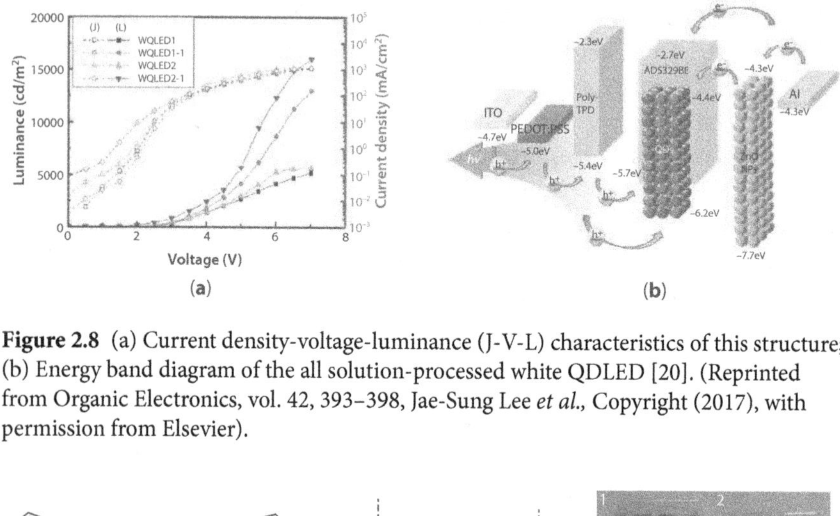

Figure 2.8 (a) Current density-voltage-luminance (J-V-L) characteristics of this structure; (b) Energy band diagram of the all solution-processed white QDLED [20]. (Reprinted from Organic Electronics, vol. 42, 393–398, Jae-Sung Lee *et al.*, Copyright (2017), with permission from Elsevier).

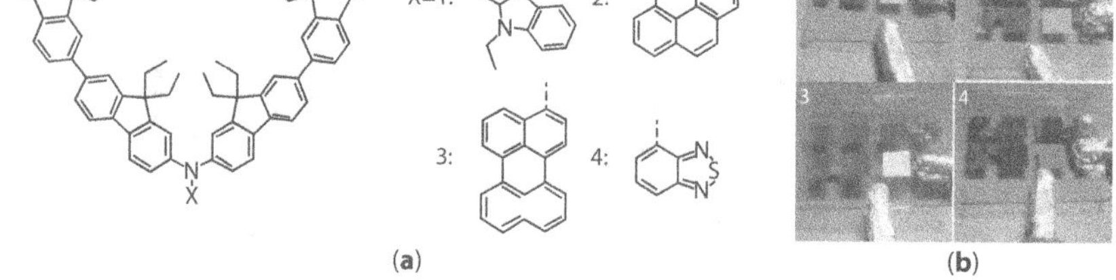

Figure 2.9 (a) Solution-processable fluorescent compounds; (b) Corresponding EL devices with color, depending on the dyes involved as emissive films [21]. (Reprinted from Journal of Photopolymer Science and Technology 26(3):403–410, Yong-Jin Pu *et al.*, Copyright (2013), with permission from The Society of Photopolymer Science and Technology).

luminescent spectrum of the compounds from blue to red color (Figure 2.9b). Introduction of suitable host in the main substances is important issue, because it provides the necessary charge transport properties and the required high triplet energy level, in order to reduce the quenching and recombination of charges away from the interface zones at the emissive and supplementary films. M-carbazolylbenzene (mCB) is such host material, which dendrons could attach to each phenyl ring. Materials engineering helps for the solubility of the complexes and suppression of the unwanted interactions between the core moieties. In this way, significantly higher efficiency can be achieved in comparison with the approaches relying of device architecture engineering. For example, the above-mentioned combination results in power efficiency of 19 lm/W, luminous efficiency of 32 cd/A and EQE of 12% at 100 cd/m^2. If the host material is cross-linkable (for example, by thermal activation), then the layer would not be affected by the solvent of the next layer during the wet deposition. Investigation of the UV-vis absorption spectra of such films produced by spin-coating shows that the spectra are not changed, which is clear indication for solvent resistance.

Not only tailoring of the emitter materials molecules, but solvent engineering employing synergetic solvents, is also applied to control the films quality of the solution-processed PLEDs [22]. The aim is to promote the charge carriers injection due to the improved film morphology during the wet deposition process, which results in a smaller number of traps and reduced contact resistance. The synergetic solvents contain CF and chlorobenzene (CB), which ensure good solubility, as well as smooth film surfaces (roughness about 1 nm), as can be seen from Figure 2.10. Therefore, contribution

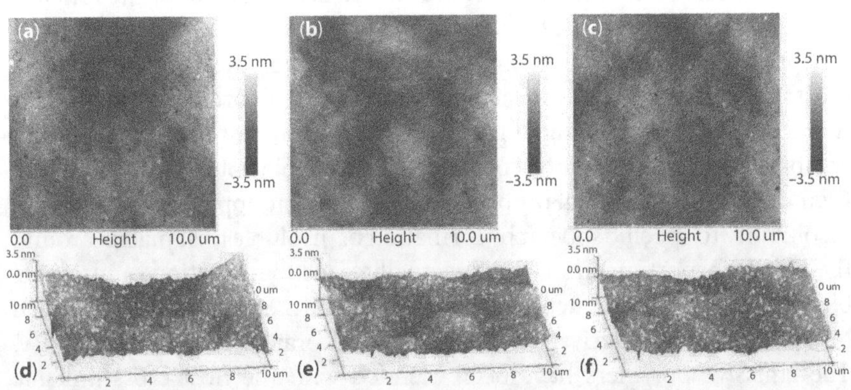

Figure 2.10 AFM 2D and 3D topography of the emission layer prepared by synergetic solvents including CF/CB mixed with ratio of (a, d) 1:0, (b, e) 0.8:0.2, and (c, f) 0.5:0.5 [22]. (Reprinted from Organic Electronics 71, 1–6, Jian Yang *et al.*, Copyright (2019), with permission from Elsevier).

for the solution-processed emissive device performance is reflected as efficiency of 15.76 cd/A, reported for blue PLED. The materials involved in the structure are bis [4-(9,9-dimethyl-9,10-dihydroacridine)phenyl] solfone (DMACDPS), bis [2-(diphenyl-phosphino)phenyl] ether oxide (DPEPO), 1,3,5-tris(1-phenyl1-Hbenzimidazol-2-yl)benzene (TPBi), poly (3,4-ethylenedioxythiophene)-poly (styrene sulfonate), and 9-(4-tert-butylphenyl)-3,6-bis(triphenylsilyl)-9H-carbazole (CzSi). The polymers have been spin-casted, and their solutions are based on mixture of CF and CB in different percent ratios.

The best results in terms of solubility and smoothness are obtained at equal presence of both solvents, as the roughness decrease with almost 40% in comparison with ratio 0.8:0.2, or single solvent usage. Further, the reference (or standard) device used for comparison has been prepared with pure CF and the modified uses CF:CB = 0.5:05. Electron-only and hole-only devices have been studied separately in order to prove that the electron or hole current flows are higher for modified device in comparison to the standard one (Figure 2.11 a,b). This could be related to the better junction between the surrounding coatings in the stack and this EML. Transient EL measurement (not shown) proves that solvent engineering reduces the traps near the interfaces, thus preventing charges accumulation which blocks the charges injection from the electrodes. The driving voltage of 6.5 V has been determined from the current-voltage-brightness characteristics (Figure 2.11c) for the optimal solvent mixture, which is not significantly improved as compared to the non-optimized device. However, due to the improved interface contact between the smooth films, the conditions for charge carriers injection are improved, which is the reason for the remarkable CE of almost 16 cd/A (Figure 2.11d).

Appropriate blends between typical emissive polymer and commercial cross-linkable matrices to tune the solubility is reported as an effective way to prevent intermixing of layers in solution-processed PLEDs without chemical modification of the commercially available polymers and altering their electrical and optical properties [23]. The basic approach is to achieve insolubility to specific solvent by formation of insoluble host matrix around the hole transport material, which actually leads to formation of insoluble blend layer. The core structure is HTL of PEDOT:PSS, a blue emitter based on spirobifluorene derivative (SPB-02T) and orange emitter MEH-PPV. The emissive polymers have been blended with the host cross-linkable matrix. Afterward, the polymer poly (9,9-di-n-octylfluorenyl-2,7-diyl) (PFO) was applied as a hole-blocking layer. All layers have been deposited by spin-coating and sandwiched between standard electrodes. As an indication of possible intermixing, EL spectrum has been considered for

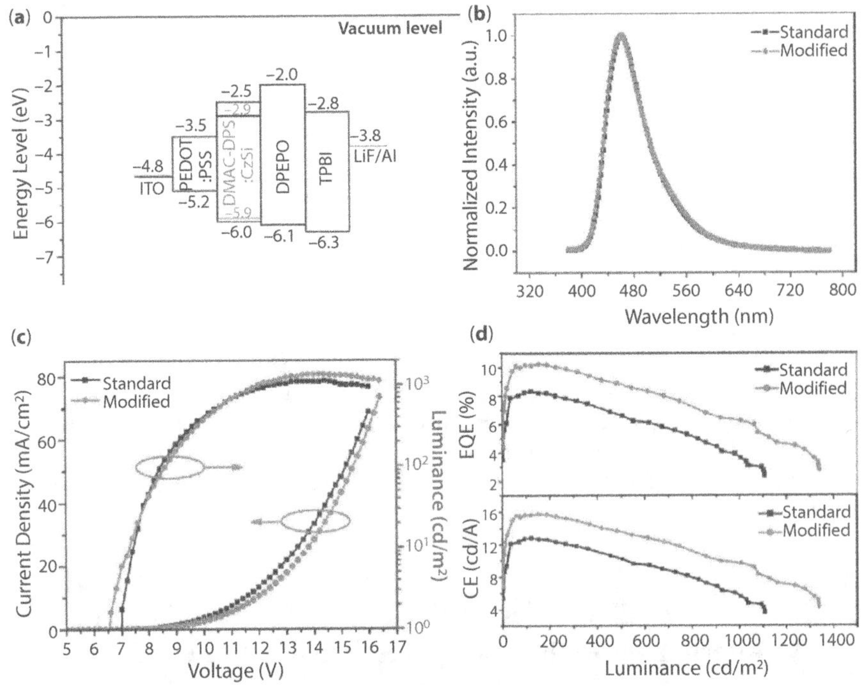

Figure 2.11 (a) Band diagram, (b) normalized EL spectra, (c) the current-voltage-brightness curves, and (d) current and external quantum efficiencies versus luminance of devices prepared from pristine CF solvent (standard) or CF/CB synergetic solvents (modified) [22]. (Reprinted from Organic Electronics 71, 1–6, Jian Yang *et al.*, Copyright (2019), with permission from Elsevier).

monitoring and the result has been compared with simulation revealing ideal structure without mutual penetrating of materials in the recombination zone. As well, if there is a mixture, the dominant emission will be shifted toward the adjacent host and the emission would not be enhanced with the voltage increase, due to the narrow recombination zone within the emitter. It has been found that, if the blend ratio is 80:20, the pristine and blended layers exhibit very similar transport properties.

The MEH-PPV emission arises between 550 and 650 nm. In the range 400–550 nm, there are additional peaks, corresponding to the SPB-02T blue color. The peaks have increased with the voltage rise from 4–7V, which is indication for emission shift from orange to blue-orange (looking white) (Figure 2.12a). This process stops at 7V. If there has been mixing between the layers, only one peak (the orange one of MEH-PPV) would be measured, as well as the measured emission would coincide with the calculated one, according to the simulations, assuming sharp interface (Figure 2.12b).

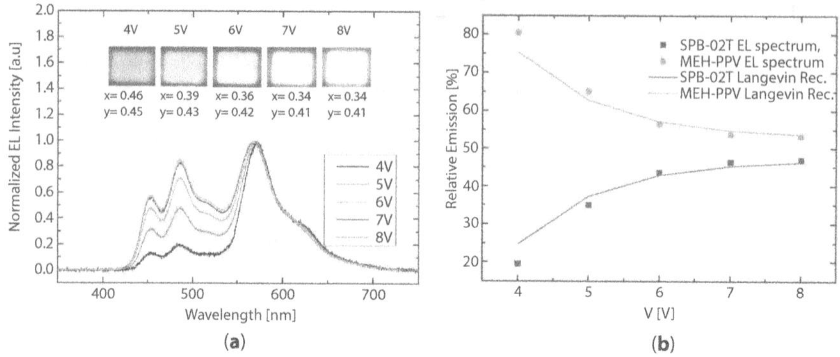

Figure 2.12 (a) Normalized EL spectrum of the SPB-02T/MEH-PPV/PFO structure for different applied voltages; (b) Measured relative emission obtained from the EL spectrum and the calculated relative emission obtained by simulation [23]. (Reprinted from C. Kasparek, P. W. M. Blom, *Appl. Phys. Let.* 110, 023302 (2017), with the permission of AIP Publishing).

2.2.2 ITO-Free Electrodes—Solution Processed and Polymer Alternatives to the Transparent Conductive Oxides

With the appearing of the flexible electronics and particularly flexible displays, reliability problems have begun to arise. One of the major issues is the poor stability of ITO and other TCO (transparent conductive oxide) anodes against multiple bending. It has been found that ITO film deposited on plastic substrates exhibits longitudinal cracks under relatively small tensile strain (up to 10%) and small number of repeating bending cycles (<100) [24]. First, the sheet resistance increases with few orders of magnitude, the function of the electrode is blocked and then physical degradation of the coating happens.

First step to replace the brittle ITO electrodes and the expensive vacuum deposition process is to find suitable alternative material, with sufficient electrical conductivity and transparency for the visible light. According to the current standards in the field of optoelectronics, the sheet resistance of the conductive films should be near to 40 Ohms per square and the transparency—at least 95%, but not lower than 90% [25]. The conductive polymer PEDOT:PSS has been extensively studied in the recent years as possible alternative to ITO, spin-coated onto a UV treated flexible polyethylene terephthalate (PET) substrate [26]. The system UV treated PET/PEDOT:PSS exhibits superior flexibility estimated by the negligible sheet resistance change of PEDOT:PSS after 20,000 cycles of bending, which is 3.8%, versus 23.5% for the conventional ITO electrode (Figure 2.13a).

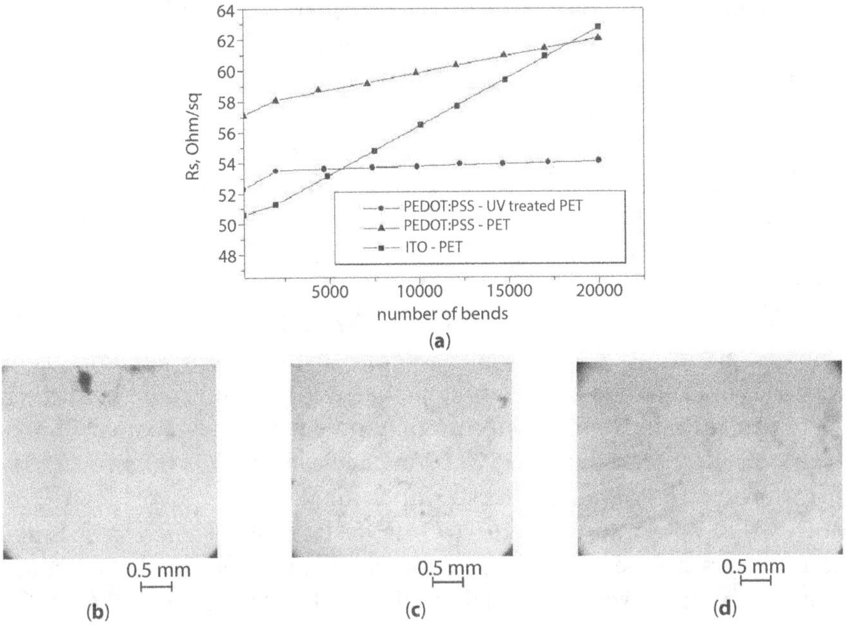

Figure 2.13 (a) Sheet resistance variation at different bending cycles for PEDOT:PSS deposited on UV treated and untreated PET. ITO on PET has been used as a reference; Microscopic images of films' surfaces after 5,000 bends; (b) PEDOT:PSS on UV treated PET; (c) PEDOT:PSS on untreated PET; (d) ITO on PET [26]. (Reprinted from Microelectronic Engineering 145, 112–116, M. Aleksandrova *et al.*, Copyright (2015), with permission from Elsevier).

PLED is produced by using poly[2-methoxy-5-(2-ethylhexyloxy)-1,4-phenylenevinylene] (MEH-PPV) solution, which is spin-coated onto the PEDOT:PSS without intermixing of the layers, because of their solubility in different solvents. Optical microscopic images show that the mechanical load of PEDOT:PSS film on untreated PET resulted in a dark spots on the surface. It is well known that the multiple bending (~5,000) provoke cracking of the ITO films, which is longitudinally distributed to the stress applied direction. PEDOT:PSS film grown on the PET substrate treated by UV exposure remains unaffected (Figures 2.13b–d). The transparency of the fabricated films is relatively poor (81%) and the turn-on voltage is not yet satisfactory (8 V), because of the energy barrier height at the anode interface; however, due to the lower cost and higher bending durability, this combination has been considered as an alternative transparent electrode that is going to replace ITO in the flexible optoelectronic devices in the near future.

Elegant approach to overcome the disadvantages of PEDOT:PSS electrode is to mix metal ions and a solid electrolyte with light-emitting polymers like in an electrochemical cell for simultaneously doping and PN-junction formation [27]. As a result, the need of work function tailoring is eliminated and PEDOT:PSS can be used as anode without loss of efficiency. In addition, fully solution process fabrication of PLED could be realized, including ink-jet printing, which is favorable for direct patterning of the coatings, thus avoiding photolithography. The lack of chemical or plasma etching makes possible fabrication of polymeri- based display on specific substrates like plastic and textile. The devices have exhibited stable emission at voltage of 7 V (Figure 2.14a), which is lower in comparison with the structure without *in situ* doped active layer and could be further decreased with the doping concentration increment. The maximal CE has been 1.32 cd/A at brightness ~108 cd/m^2, achieved at 15 V (Figure 2.14b).

Unfortunately, although successful results have been achieved for PEDOT:PSS as an alternative to TCO anode, its relatively high work function embarrasses the electrons injection into the polymer semiconductors.

It has been proven that single-walled carbon nanotubes (SWCNTs) can also serve as electrodes in a metal-free PLED device. Electrode films, made in this way, have shown network formation of the nanotubes, establishing work function unsuitable for cathode, because of the high-aspect-ratio structures. However, with a proper construction of the PLED device, called polymer light-emitting electrochemical cell (PLECs), containing p-i-n junction, it is possible to use the nanotubes for cathode and for anode without energy losses at both interfaces [28]. In the same time,

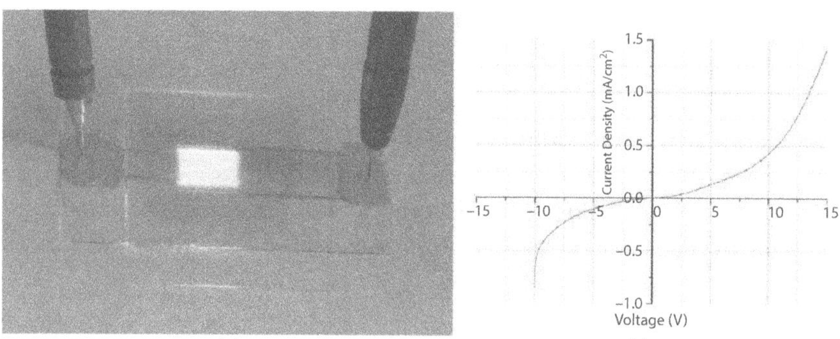

(a) (b)

Figure 2.14 (a) Photo of all-solution processed yellow PLED of electrochemical cell type; (b) Current-voltage characteristics of the PLED in forward and reverse direction [27]. (Reprinted from Materials Today: Proceedings, vol. 4, Z. Shu *et al.* Copyright (2017), with permission from Elsevier).

the brittle metallic films have been replaced from the cathode and the flexibility of the device has been strongly enhanced—it could stand multiple bending to a 2.5-mm radius without worsening the device performance in the same extend like for PLED with ITO or a conducting polymer as electrodes (Figure 2.15). For all components, solution processes like printing of SWCNT dispersion and spin-casting of polyfluorene-based copolymers emitting different colors have been used.

The application of graphene as an electrode material makes also possible the exclusion of vacuum deposition processes from the PLED manufacturing. Not long ago, the graphene performed only anode function due to the easy p-doping procedure. It has not been possible to serve as a cathode, due to the lack of reliable process for n-doping. For several years, there is a way to obtain all-solution processed light-emitting device with graphene cathode [29]. Although the pristine graphene is CVD grown, its transfer on the substrate is realized in aqueous medium. The n-doping is also done in the solution with insertion of liquid n-type dopant of CsF. 1,4,5,8,9,11-hexaazatriphenylene hexacarbonitrile (HAT-CN) has been used as HTL, 2,7-disubstituted fluorene-based triaryldiamine (VB-FNPD) as a cross-linked HTL, poly (N-vinylcarbazole) (PVK) derivative has been the emissive layer and poly [(9,9-bis(30-(N,N-dimethylamino) propyl)-2,7-fluorene)alt-2,7-(9,9-dioctylfluorene)] (PFN) has been applied as an EIL (Figure 2.16a). Spin-coating has been used for all polymer layers. To increase the electrical conductivity of the cathode, the graphene sheets are multiple. For keeping the work function low and closer to the LUMO of EIL, every graphene sheet has been subjected to n-doping at each transferring to the substrate. The maximum brightness of 1,034 cd/m^2 at 13 V

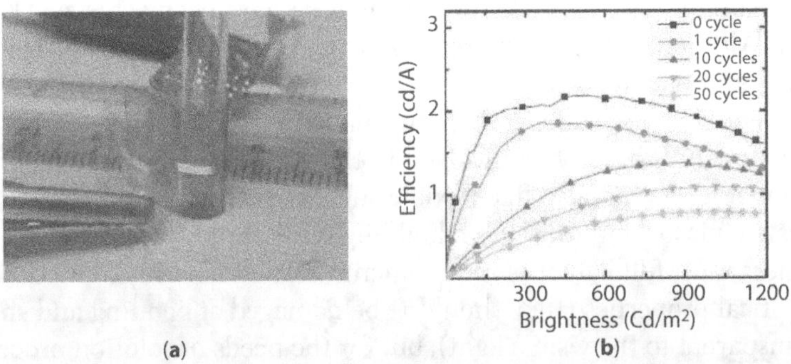

Figure 2.15 (a) Blue PLEC bent to 2.5-mm-radius of curvature; (b) Efficiency-brightness curves of a blue PLEC in a flat shape and after multiple cycles of bending to 2.5-mm radius [28]. (Reprinted from J. Photonics for Energy, Vol. 1 (1), 011003, Yu *et al.* Copyright (2011), with permission from SPIE).

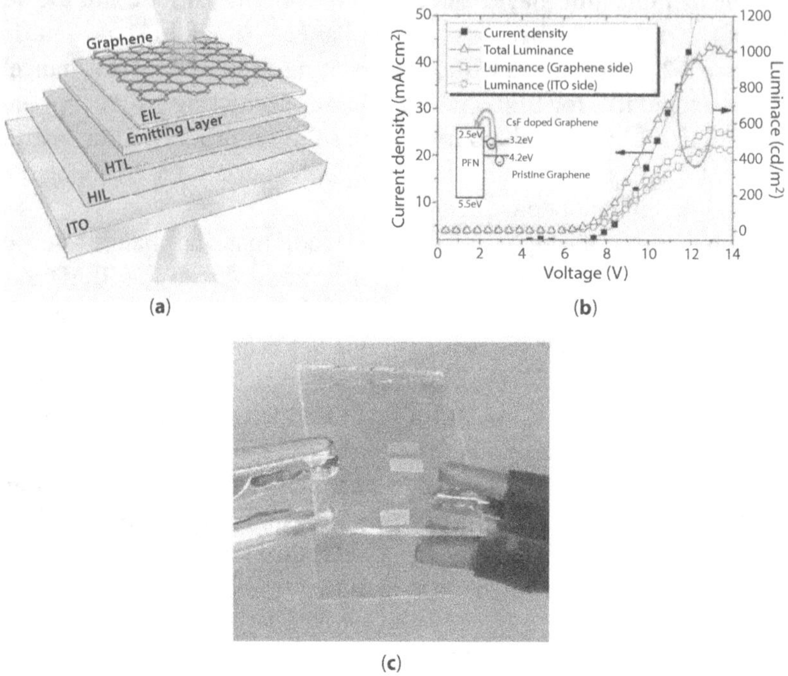

Figure 2.16 (a) Device structure: ITO/HAT-CN/VB-FNPD/PVK5Firpic/PFN5CsF/multilayer graphene; (b) Current density-voltage-luminance (J-V-L) curve of the device. The inset shows the electron injection in case of pristine and CsF-doped graphene; (c) Photo of the fully transparent PLED with graphene cathode [29]. (Reprinted from Scientific Reports, vol. 5: 9693, Copyright 2015, with permission due to Creative Commons Attribution 4.0 International License).

and maximum CE of 3.1 cd/A at 8.3 mA/cm^2 have been achieved (Figure 2.16b). Although the results are poorer as compared to the device with thermal evaporated Al cathode, the proposed design has been considered as cost-effective and sufficiently performance effective (Figure 2.16c).

The polymers have been involved in the PLED structures not only like functional layers but also like flexible substrates. The typical flexible substrates are plastic and their basic building units are polymer chains. These polymers must fulfill a lot of requirements, related not only to mechanical and optical properties (they shouldn't be damaged at bending and should be transparent to the visible light), but for the needs of solution processed PLED, they must exhibit also good wettability and good chemical stability toward the polymeric solutions responsible for the functional layers growth [30]. The wetting ability of the polymer substrates can be controlled by different surface modifications, including UV-ozone treatment [31],

Corona treatment [32], and laser treatment [33]. Although flexible substrate fulfilling all requirements has not been synthesized yet, the polymers polyethylene terephthalate (PET) and polyethylene naphthalate (PEN) are currently used in the wearable electronic devices, because of their mechanical and optical properties, satisfying thermal stability, very good solvent, and oxidation resistance, as well almost no penetration of moisture. Table 2.2 summarizes the most important properties of the commercial flexible substrates for PLEDs, including also polyimide (PI) and polycarbonate (PC) [34].

In addition to the conventional polymers, new materials suitable for plastic substrates of solution processed PLEDs are developed, called hybrimers. The plastic reinforcing composited (or hybrimers) are novel compounds, developed for application in semi-flexible microelectronic devices. They are synthesized by advanced technology for hybridization between silicon dioxide and hydro-carbon-oxide (a kind of polymer chain). Unique structure of hybrimer/PEDOT:PSS/Au/PEDOT:SS/emissive layer/metal has been reported in [35]. It has showed negligible degradation of the layer interfaces due to the improved mechanical strength of the hybrimer, in comparison with the typically used PET, and due to the enhanced durability of the polymer-based anode. Spray coating technology has been used for deposition of the conductive polymer from the system PEDOT:PSS/Au/PEDOT:SS and for the yellow-orange light-emitting polymer BCHA-PPV. The turn-on voltage has been about 13.5 V, the current density about 8.7 μA/cm^2, breakdown voltage about 21.3 V, and the maximum luminescence has been about 272cd/m^2 (Figure 2.17a). Although the performance of the PLED is relatively poor before bending, as compared to the current state-of-the-art, it is characterized with significant mechanical durability

Table 2.2 Main properties of the commercially available 100 μm polymer substrates for PLEDs application.

Properties	**PET**	**PEN**	**PI**	**PC**
Flexibility (Tensile strength, MPa)	230	200	150	75
Elasticity (Young modulus, GPa)	2.8–3	3–5	2–3	2.4
Transparency in the visible region, %	>90	88	65–70	90
Water absorption, %	0.6	0.4	2.9	0.3
Solvent resistance	Good	Good	Good	Poor
Wetting ability to organic solutions	Poor	Poor	Poor	Poor

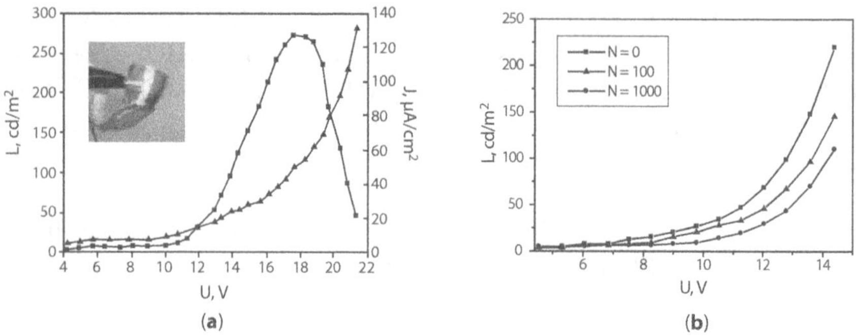

Figure 2.17 (a) Current-voltage and brightness-voltage characteristics of the prepared flexible PLED with hybrimer substrate before bending; (b) Stability of the luminescence at different bending cycles [35]. © [2017] IEEE. (Reprinted, with permission from M. Aleksandrova *et al.*, 2017 40th International Spring Seminar on Electronics Technology (ISSE 2017, Sofia, Bulgaria), 1–6).

after multiple bending (folding/releasing cycles) (Figure 2.17b). The reduction of the luminous intensity is with 15.1% after 100 bending cycles and with 31.1% after 1,000 bending cycles. Compared to flexible PLEDs with other plastic substrates and ITO anodes, approximately six times improvement of the electro-optical behavior has been achieved. These results can be explained with the better elasticity of the polymeric-based electrode, as has been proved earlier [36]. In this way, the plastic deformations cannot change dramatically the continuity of the films, in contrast to the case of metal-oxide coatings, where cracking occurs at smaller than 100 numbers of bends due to their brittleness. Possible reason could be also the strong adhesive coupling of the PEDOT:PSS layer with the hybrimer substrate, which prevent inner stress arising, that usually accelerates the layers degradation, strongly affects the dipoles orientation and distribution near the interfaces, and influences the processes of charge carriers motion and traps occurrence.

The spin-coating is simple technique for distribution of liquid substances to form thin films, but it gives relatively high non-uniformity of the coating when large and square in shape substrates should be used, which is the case at the display technology. Therefore, possible reasonable alternative comparable in cost efficiency could be ink-jet printing. The limitations for the substrate area and shape have been eliminated. Moreover, direct patterning of the solution processed material can be realized, omitting in this way the photolithography [37]. Furthermore, this process offers precise solution presence control, which is very important for the patterns quality. The deposition is conducted at near the room temperature, enabling application onto plastic or textile substrate. As well, small widths of the cathode lines with intermediate fine isolating lines could be obtained which is the greatest advantage of ink-jet printing over the spray coating, screen printing, and other vacuum-free wet processes (Figure 2.18a). Not only the EML polymers could be ink-jet printed, but also conducting nanoparticles serving as cathodes [38]. In addition, patterned poly [(9,9-bi(3'-(N,N-dimethylamino)propyl)-2,7-fluorene)-alt-2,7-(9,9-dioctylfluorene)] ($PFNR_2$), serving as an EIL with suitable electron affinity has been obtained and mixed with a curable epoxy adhesive. After thermal treatment, the epoxy cross-links and form a stable copolymer, which is insoluble in the typical organic solvents using for PLED. As can be seen from Figure 2.18b, the cathode lines are well defined, defect-free, and uniform. Figure 2.18c demonstrates the improvement of the quality of the EIL with increasing the content of the polymer ($PFNR_2$) over the epoxy—the films have become denser and more uniform. All this has led to improvement of

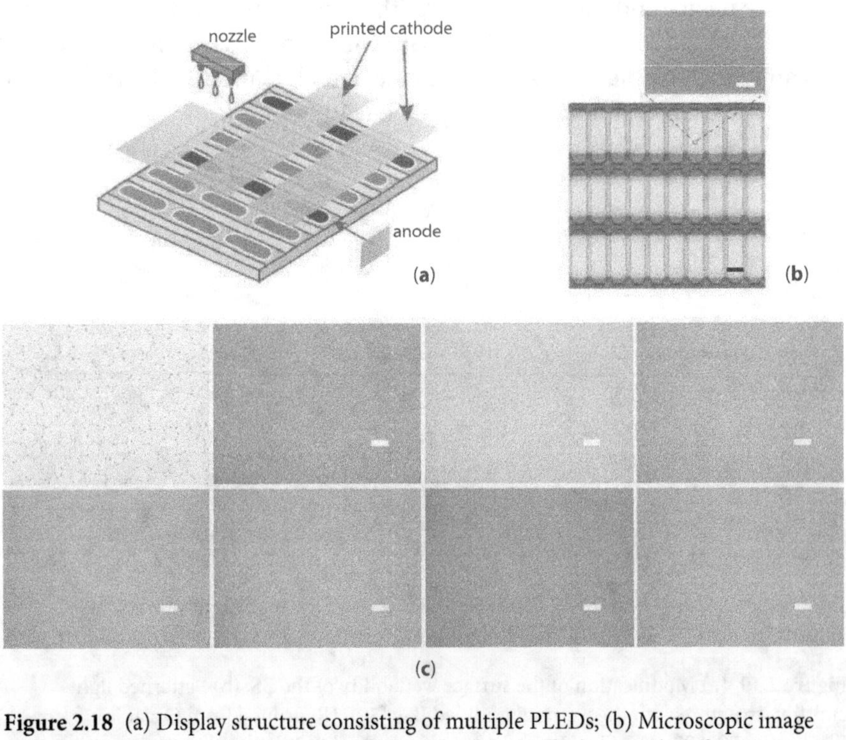

Figure 2.18 (a) Display structure consisting of multiple PLEDs; (b) Microscopic image of the display pixels; and (c) Microscopic images of the buffer layer with different components ratios. Scale bar 10 μm [38]. (Reprinted by permission from Nature: [Hua Zheng *et al.*, Nature Communications, vol. 4, pp. 1971–1978], Copyright (2013)).

the current-brightness-voltage characteristics, with turn-on voltage lower than 5 V, maximum brightness exceeding 9,000 cd/m^2, and luminous efficiency over 9 cd/A.

An approach competitive to ink-jet printing in terms of high resolution direct patterning is scalable wetting modification of bank structures (BSs) or pixel separators [39]. After UV ozone modification of the electrode and BSs' surfaces, their hydrophility has been increased and polymer solution can fluently fill the gaps between two adjacent BSs, thus improving the uniformity of the coating (Figure 2.19a). A mold of a polydimethylsiloxane (PDMS) stamp has been used as a base and the functional polymer has been inserted by injection in the channels between the PDMS mold and the flexible polyethylene naphthalate (PEN) substrate due to the capillary forces (Figure 2.19b). The hole injection PEDOT:PSS film and the emissive coating of phenyl-substituted poly (para-phenylene vinylene) co-polymer have been spin-coated on the modified surface.

The current-voltage-brightness curves of two different emissive structures are compared (Figure 2.20 a,b). One of them is without the BS, called reference device and the other one is with BS of 5 μm. The results are comparable, which is the proof that the patterning process doesn't reduce significantly the soluble PLED performance, which is usually the case when other patterning technologies are applied (Figure 2.20 c,d).

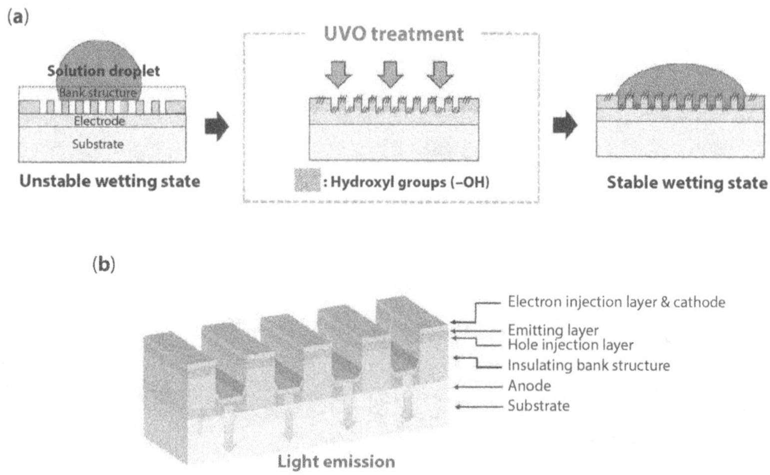

Figure 2.19 (a) Modification of the surface wettability of the BS; (b) Patterned light-emitting structure with the wetting modified BSs [39]. (Reprinted from Organic Electronics 73, 332–336, Sin-Hyung Lee *et al.*, Copyright (2019), with permission from Elsevier).

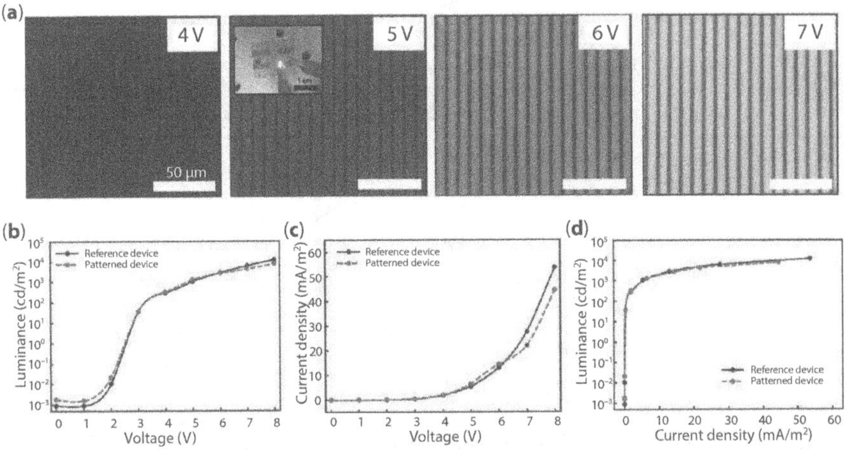

Figure 2.20 (a) Microscopic images of the patterned emissive structures 5 μm in width, at different driving voltages, (b) Brightness-voltage, (c) current-voltage, and (d) brightness-current characteristics comparison of the reference and patterned devices [39]. (Reprinted from Organic Electronics 73, 332–336, Sin-Hyung Lee *et al.*, Copyright (2019), with permission from Elsevier).

2.3 Specific Phenomena at PLED—Energy Transfers, Traps, Excitons Formation, and Color Tuning

The colored emission is one of the most important features of the display devices. Red, green, and blue are the main colors for representation of the images because they serve as a basis for all other colors. Polymers, emitting red and green colors, have been already developed and could be directly implemented into a real display device [40, 41]. However, achieving of blue light-emitting polymers with competitive efficiency is problematic because of lack of proper polymeric hosts to get the high triplet state of the blue phosphorescent dopants. Thus, the transfer of triplet excitons from the guest to the host materials has been prevented. In order to overcome this challenge, novel polymer host has been recently developed for blue PLED [42]. It is based on phenyl-substituted carbazole (PCz) and a heteroatom-containing bulky core [triphenylphosphine (TPP)/triphenylphosphine oxide (TPPO)] for obtaining polymeric host material with high energy of the triplet state, high thermal stability, and good solubility (Figure 2.21). Both constituents have been responsible for the charge carriers' balance and the bipolar character of the structure, which is a condition for the device stability. The chemical synthesis has followed the procedure of the Suzuki

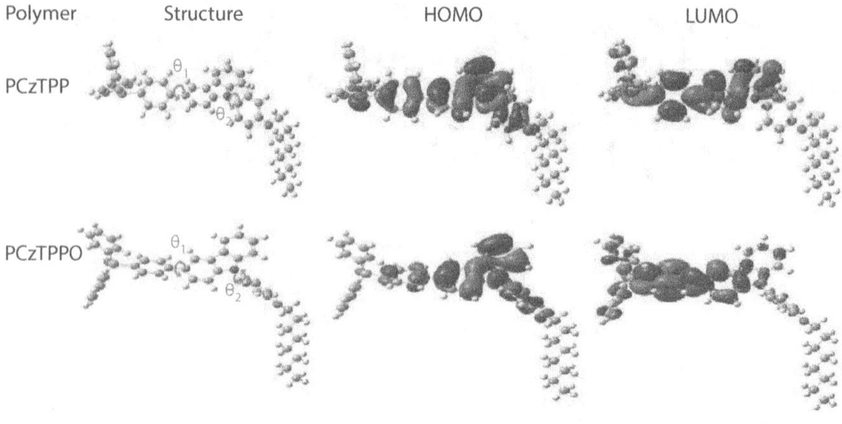

Figure 2.21 Calculated spatial distribution of HOMO and LUMO of PCzTPP and PCzTPPO [42]. (Reprinted from Organic Electronics vol. 43, 1–8, Wang *et al.* Copyright (2017), with permission from Elsevier).

polymerization. Although typical electrodes have been used for the emissive device configuration (ITO and Ca/Al both using vacuum deposition process), the emissive polymer layers, as well as the additional hole and electron transport polymer layers have been produced by solution process—spin-coating. It has been reported maximum CE of more than 10 cd/A, and a maximum EQE of 6.6%, which are significantly higher than these of PVK only devices. The results could be ascribed to the reduced interface energy barrier and balanced injection of recombining pairs (Figure 2.22).

Emerging technology is solution processed independently tunable PLED, designed with two vertical polymer emitters as multilayer stack [43]. The operational principle requires incorporation of silver nanowires as an

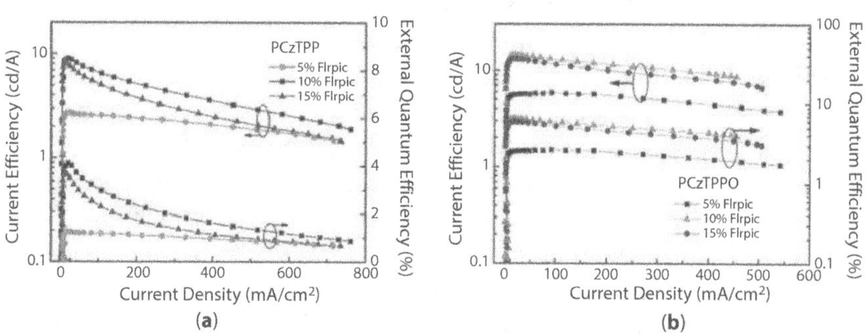

Figure 2.22 (a) Efficiency-current density curves of improved devices based on PCzTPP; (b) PCzTPPO as hosts [42]. (Reprinted from Organic Electronics vol. 43, 1–8, Wang *et al.* Copyright (2017), with permission from Elsevier).

intermediate accessible contact, enhancing charge injection and forming three-terminal PLED. In this way, the control over the emission spectrum and intensity of the stacked devices has been facilitated. This kind of engineering is possible due to the semitransparent nature of the polymeric thin films, which can be deposited to form several sub-diodes or sub-pixels one over another. The nanowires cannot absorb significant part of the light emitted and can be solution-deposited, like the emitters. The materials that have been involved are red-emitter (Livilux, SPR-001), green-emitter (Livilux, SPG-01T), blue-emitter PFO, PEDOT:PSS solution, and a AgNW dispersion (Figure 2.23a). The high length-to-diameter aspect ratio of the nanowires has resulted in a low sheet resistance of 20 Ω/sq and high transparency of 86%—parameters rather close to those reported for ITO electrodes. All of the solution-processed layers have been deposited using a doctor blade in an ambient atmosphere. SEM image of the cross-section has shown lack of intermixing between the individual layers (Figure 2.23b). The authors have reported about similar turn-on voltages of the new devices to their reference simple, single layer cell, which is an indication for negligible effect of the subsequent solution deposited layers on the optoelectronic performance and the cross-talk lack between the neighboring pixels of this complex structure (Figure 2.24).

In order to achieve full control over the basic display parameters, such as color, turn-on voltage, brightness, and quantum efficiency, it is necessary to deeply understand the fundament of the processes revealed inside the polymer films at molecular level. For this purpose, study of the mechanisms of excitons generation upon excitation, charge distribution, excitons quenching, and polaron-exciton interactions is needed. Solution processed

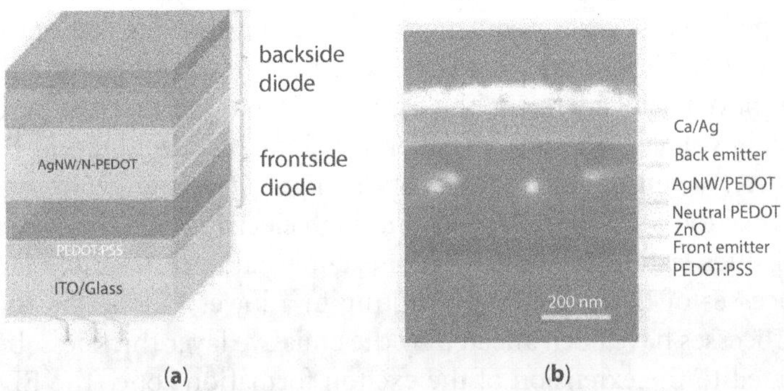

Figure 2.23 (a) Device architecture of the three color-tunable PLED; (b) Cross-sectional SEM image of the stacked device [43]. (Reprinted from Light: Science & Applications 6, e17094, Fei Guo *et al.* Copyright (2017), with permission from Nature).

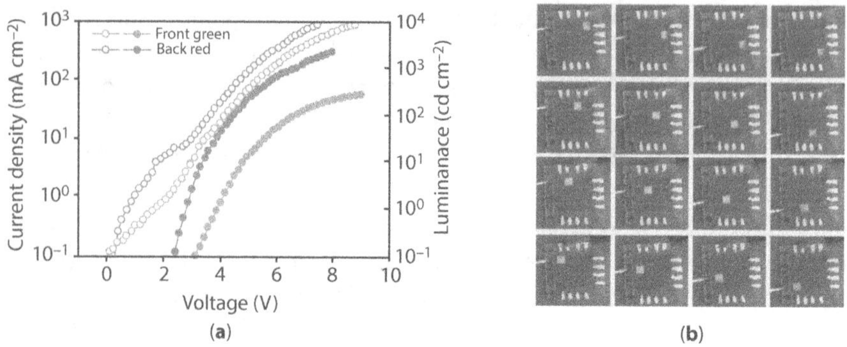

Figure 2.24 (a) Current-brightness-voltage characteristics of one of the three PLED cell combinations (green and red); (b) Demonstration of the operation of the individual front green pixels [43]. (Reprinted from Light: Science & Applications 6, e17094, Fei Guo *et al.* Copyright (2017), with permission from Nature).

PLEDs with Super Yellow (SY) emitting polymer with high EQE of 4.1% have been fabricated and subjected to specialized delayed EL measurement in order to record the electric responses of fluorescent states in the emitting material, which would provide information for the internal processes dynamics [44]. To activate specific processes in the PLED, it has been supplied with periodic forward and reversed pulses and the transient current value has been estimated. The forward pulses are first constant in value (6 V), providing stable emission away of the breakdown zone. The reverse pulses are supplied with different values and in different moments—immediately after light emission quenching and after some short time (Figures 2.25a,b). It has been found that the reverse bias removing doesn't lead to restoring of the EL, which is an indication that the triplet-triplet annihilation (TTA) is not the main process in the structure and probably the trapped charge carriers during the forward pulses are responsible for the delayed EL. Therefore, this is evidence that more singlet than triplet excitons have been generated. The transient EL has increased as quickly as the driving voltage is higher and after switching off, the EL decay lifetime gets shorter when the current density is higher (Figure 2.25c). It means that the exciton quenching is weaker at high electric field. The higher current densities lead to accumulation of excitons and polarons and facilitate the processes of EL quenching, resulting in a lower PLED efficiency. All these processes have been affected by the emissive layer thickness, because it is related to the extension of the exciton formation zone. The EL decay time in this case has increased at greater SY thickness, which means that the singlet excitons are dominant in the process of light emission.

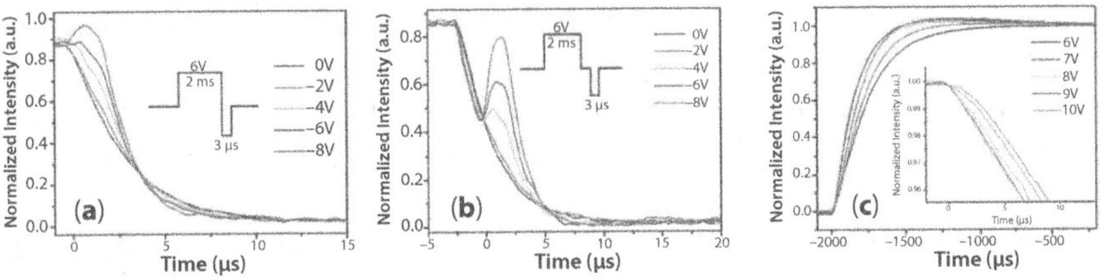

Figure 2.25 Time-resolved EL responses of the PLED: the reverse voltages (from 0 V to −8 V) are applied in the off-state (a) immediately; (b) after 3 μs; (c) time-resolved turn-on EL responses of the devices with various driving voltages. Inset: time-resolved turn-off electroluminescence responses [44]. (Reprinted from Organic Electronics 68, 45–49, Peng Wang *et al.*, Copyright (2019), with permission from Elsevier).

The general picture gets more complex when white solution processed PLED is in question. To investigate energy transfer processes among different chromophores used in these PLEDs, time-resolved photoluminescence (PL) has been measured [45]. Star-like molecular structure has been designed for suppression of the EL quenching, as compared to the linear polymer system involved in the multilayer and blend systems (Figure 2.26). More concretely, the energy transfer between the host and

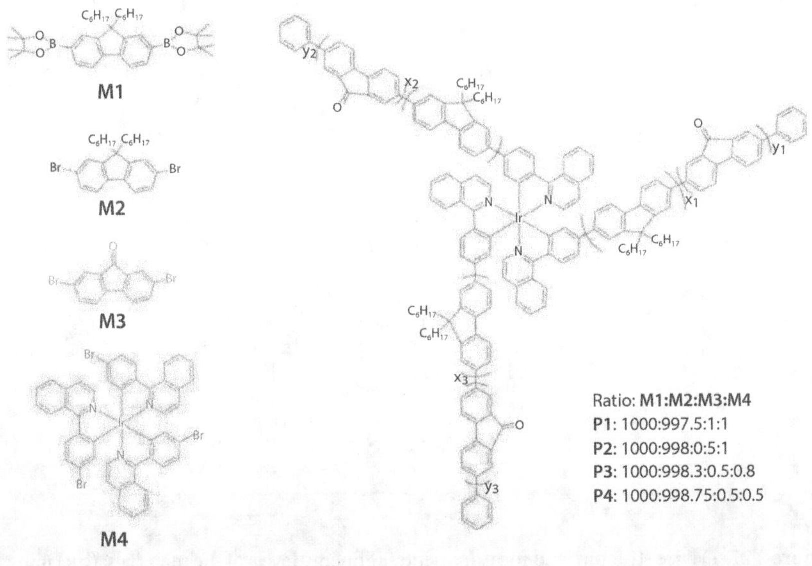

Figure 2.26 Chemical structures of the monomers and star-shaped white-emission copolymers. M1–M4 are four kinds of monomers synthesized into the star-shaped copolymers P1–P4 [45]. (Reprinted from Materials, 11, 1719, Hui He *et al.*, Copyright (2018), with permission from MDPI).

guest chromophores in three-color star-shaped white-light copolymers has been investigated. Star-shaped copolymers have been used to produce solution processed white PLEDs (Figure 2.27a). They consists of core unit of Ir(piq)$_3$ and poly(9,9-dioctylfluorene) (PFO) arms connected at the ends to fluorenone unit (FO).

Devices, using the polymers P1–P4 as emitters have been fabricated by spin-coating of the corresponding polymer solutions. The maximum luminance of the devices has varied between 3,627 and 10,960 cd/m^2. The highest value corresponds to the highest content of the red and green units, which fulfill the requirements of the band diagram for similarity of the energy levels at the interfaces and facilitate charge carriers recombination away from the interface region with increased traps concentration.

According to the time resolved PL measurements with activation by laser pulse, the fluorescence decay has been fitted with an exponential decay

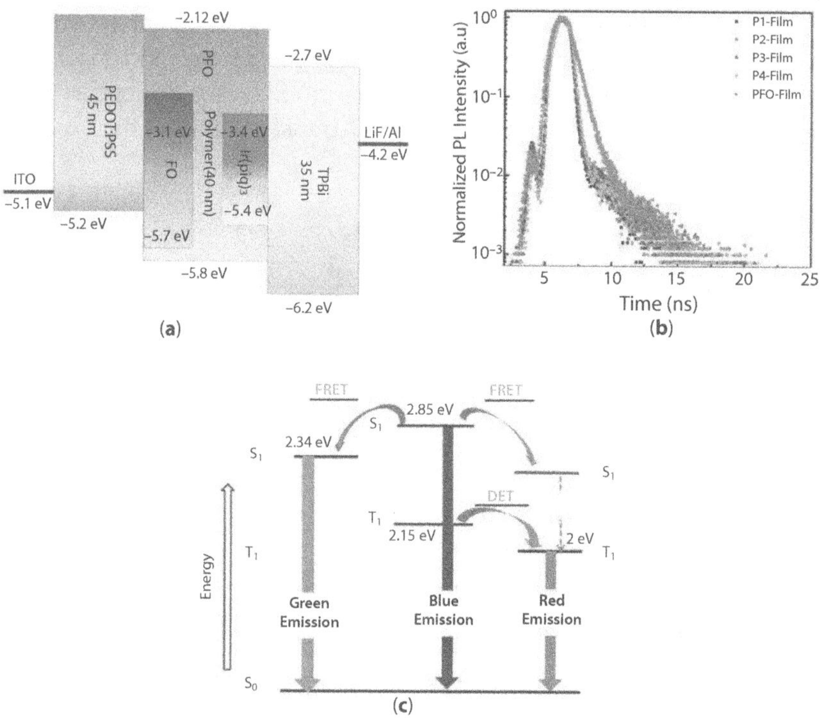

Figure 2.27 Device structure and measurement: (a) Energy levels of the materials; (b) Time-resolved photoluminescence (PL) transients of P1–P4 and poly(9,9-dioctylfluorene) (PFO); (c) Exciton energy diagram of PFO, FO and Ir(piq)3 and energy transfer processes [45]. (Reprinted from Materials, 11, 1719, Hui He *et al.*, Copyright (2018), with permission from MDPI).

curve. The PL life time has varied between 220 and 260 ps for the P1–P4 devices with meaning of fast relaxation and a short lifetime (Figure 2.27b). This can be ascribed to the successful energy transfer between the PFO chains, Ir(piq)$_3$, and FO chromophores. As the distance between the molecules is smaller than the Förster resonant energy transfer (FRET) radius, the energy transfer between the PFO, Ir(piq)$_3$, and FO molecules are negligible. In this co-polymer, singlet excitons formed in the FO units have migrated to the red- and green-emission chromophores through intramolecular FRET processes (Figure 2.27c). The existing of intra- and intermolecular transfer results in decrease of the fluorescence lifetime.

2.4 Conclusions

Polymer light-emitting devices have become a very attractive due to the low-cost deposition of the functional layers and possibility for manufacturing of large area devices. Their ability to precisely tune the polymer molecular structure and thus to control the color and the other electrical and optical parameters typical for the displays will make them preferable in the near future as competitive device to OLED with small molecules. The wet processes used for PLED manufacturing are environmental friendly, can be can be easy re-scaled, allow treatment of different substrates (plastic, textile, cellulose, etc.), and some of them give directly patterned coating. The great interest of this technology has provoked the researchers to study some fundamental physical and chemical processes, responsible for the excitons generation, polaron-exciton interactions, energy transfer, and interface conditions. Understanding the relation between them and the device performance will offer much better control over the PLED's efficiency. As the applications become closer to the market, the requirements for PLED are more demanding in terms of the used materials and technologies. In this chapter, some of the modern methods to enhance the PLED performance have been reviewed. As a general conclusion, it could be noted that the efforts of the research groups are more focused on the materials engineering, rather than on the device design, new deposition technologies, or optimization of the previously known technologies. The main approaches for achieving high quantum efficiency (high brightness at low current density and low voltage) are energy level alignment at the films interfaces and decreasing the roughness of the grown coatings. The former approach is realized by polymer emissive or transporting layers synthesis, solvent engineering, new combination of materials in typical PLED device structure or novel device structures with specifically patterned films.

The latter is realized by optimization of the deposition modes and treatment of the surfaces for their better wettability and smooth films formation. However, for now, the efficiency cannot be concurrent to the OLEDs, so more efforts need to be made for commercialization of PLEDs, mainly in the field of materials science.

References

1. von Hauff, E., The Role of Molecular Structure and Conformation in Polymer Electronics, in: *Semiconductors and Semimetals*, pp. 231–260, Elsevier, San Diego: Academic Press, USA, 2011.
2. Irfan, A., Chaudhary, A.R., Muhammad, S., Al-Sehemi, A.G., Bo, H., Mumtaz, M.W., Qayyum, M.A., Tuning the optoelectronic and charge transport properties of 2,5-di(pyrimidin-5-yl)thieno3,2-b]thiophene by oligocene end cores substitution. *Results Phys.*, 11, 599, 2018.
3. Sun, S.-S. and Dalton, L.R., *Introduction to Organic Electronic and Optoelectronic Materials and Devices*, CRC Press, Inc., Boca Raton, FL, USA, 2008.
4. Chang, M., Lim, G.T., Parkand, B., Reichmanis, E., Control of Molecular Ordering, Alignment, and Charge Transport in Solution-Processed Conjugated Polymer Thin Films. *Polymers*, 9, 212, 2017.
5. Surve, N.S., Fabrication and characterization of Polymer Light Emitting Diodes (PLEDs) for display application, Thesis, Rochester Institute of Technology, Rochester, New York, 2011.
6. Jogi, V.K., High Performance Polymer Light-Emitting Devices. *World J. Nano Sci. Eng.*, 4, 16, 2014.
7. Koden, M., *OLED Displays and Lighting*, John Wiley & Sons, Chichester, UK, 2016.
8. LG Company, 2014, http://www.lgdisplay.com/eng/product/commercial, accessed 12.08.2019.
9. Spindler, J.P., Hamer, J.W., Kondakova, M.E., OLED Manufacturing Equipment and Methods, in: *Handbook of Advanced Lighting Technology*, pp. 417–441, Springer, Switzerland, 2017.
10. Polymer OLEDs (PLED): Introduction and market status, OLED-info. 2019, https://www.oled-info.com/pled, accessed 12.08.2019.
11. Atxabal, A., Braun, S. *et al.*, Energy Level Alignment at Metal/Solution-Processed Organic Semiconductor Interfaces. *Adv. Mater.*, 29, 1606901, 2017.
12. Mu, H., Li, W., Jones, R., Steckl, A., Klotzkin, D., A comparative study of electrode effects on the electrical and luminescent characteristics of Alq3/TPD OLED: Improvements due to conductive polymer (PEDOT) anode. *J. Lumin.*, 126, 225, 2007.

13. Aleksandrova, M., Interfaces optimization at molecular optoelectronic devices, Thesis, Technical University of Sofia, Bulgaria, 2010.
14. Gassmann, A., Yampolskii, S.V., Klein, A., Albe, K., Vilbrandt, N., Pekkola, O., Genenko, Y.A., Rehahn, M., von Seggern, H., Study of electrical fatigue by defect engineering in organic light-emitting diodes. *Mater. Sci. Eng. B*, 192, 26, 2015.
15. Han, T.-H., Choi, M.-R., Jeon, C.-W., Kim, Y.-H., Kwon, S.-K., Lee, T.-W., Ultrahigh-efficiency solution-processed simplified small-molecule organic light-emitting diodes using universal host materials. *Sci. Adv.*, 2, e1601428, 2016.
16. Dubey, D.K., Yadav, R.A.K., Tavgeniene, D., Grigalevicius, S., Jou, J.-H., Crosslinkable hole-transporting small molecule as a mixed host for efficient solution-processed red organic light emitting diodes. *Thin Solid Films*, 660, 956, 2018.
17. Ma, Y., Peng, F., Guo, T., Jiang, C., Zhong, Z., Ying, L., Wang, J., Yang, W., Peng, J., Cao, Y., Semi-orthogonal solution-processed polyfluorene derivative for multilayer blue polymer light-emitting diodes. *Org. Electron.*, 54, 133, 2018.
18. Liu, Y., Jiang, C. *et al.*, Highly Efficient All-Solution Processed Inverted Quantum Dots Based Light Emitting Diodes. *ACS Nano*, 12, 1564, 2018.
19. Lee, K.H., Han, C.Y., Kang, H.D., Ko, H.J., Lee, C.H., Lee, J.H., Myoung, N.S., Yim, S.Y., Yang, H.S., Highly Efficient, Color-Reproducible Full-Color Electroluminescent Devices Based on Red/Green/Blue Quantum Dot-Mixed Multilayer. *ACS Nano*, 9, 10941, 2015.
20. Lee, J.-S., Kang, B.-H. *et al.*, All-solution-processed high-brightness hybrid white quantum-dot light-emitting devices utilizing polymer modified quantum dots. *Org. Electron.*, 42, 393, 2017.
21. Pu, Y.-J., Chiba, T., Aizawa, N., Sasabe, H., Kido, J., Multilayered Organic Light-Emitting Devices by Solution-Process. *J. Photopolym. Sci. Tec.*, 26, 403, 2013.
22. Yang, J., Song, D., Zhao, S., Qiao, B., Xu, Z., Wang, P., Wei, P., Highly efficient and bright blue organic light-emitting devices based on solvent engineered, solution-processed thermally activated delayed fluorescent emission layer. *Org. Electron.*, 71, 1, 2019.
23. Kasparek, C. and Blom, P.W.M., Solution-processed multilayer polymer light-emitting diode without intermixing. *Appl. Phys. Lett.*, 110, 023302, 2017.
24. Lewis, J., Material challenge for flexible organic devices. *Mater. Today*, 9, 38, 2006.
25. Chauhan, R.N., Singh, C., Anand, R.S., Kumar, J., Effect of Sheet Resistance and Morphology of ITO Thin Films on Polymer Solar Cell Characteristics. *Int. J. Photoenergy*, 2012, 879261, 2012.
26. Aleksandrova, M., Kurtev, N., Videkov, V., Tzanova, S., Schintke, S., Material alternative to ITO for transparent conductive electrode in flexible display and photovoltaic devices. *Microelectron. Eng.*, 145, 112, 2015.

27. Shu, Z., Kemper, F., Beckert, E., Eberhardt, R., Tünnermann, A., ITO-Free, Fully Solution Processed Transparent Organic Light-Emitting Electrochemical Cells on Thin Glass. *Mater. Today – Proc.*, 4, 5039–5044, 2017.
28. Yu, Z., Liu, Z., Wang, M., Sun, M., Lei, G., Pei, Q., Highly flexible polymer light-emitting devices using carbon nanotubes as both anodes and cathodes. *J. Photon. Energy*, 1, 011003, 2011.
29. Chang, J.-H., Lin, W.-H., Wang, P.-C., Taur, J.-I., Ku, T.-A., Chen, W.-T., Yan, S.-J., Wu, C.-I., Solution-processed transparent blue organic light-emitting diodes with graphene as the top cathode. *Sci. Rep.*, 5, 9693, 2015.
30. Si, Y., Yu, C., Dong, Z., Jiang, L., Wetting and spreading: Fundamental theories to cutting-edge applications. *Curr. Opin. Colloid Interface Sci.*, 36, 10, 2018.
31. Aleksandrova, M., Cholakova, I., Bodurov, G., Kolev, G., Dobrikov, G., Influence of UV Treatment on the Electro-optical Properties of Indium Tin Oxide Films Used in Flexible Displays. *Int. J. Mater. Metall. Eng.*, 6, 1068, 2012.
32. Louzi, V.C. and de C. Arvalho Campos, J.S., Corona treatment applied to synthetic polymeric monofilaments (PP, PET, and PA-6). *Surf. Interfaces*, 14, 98, 2019.
33. Sorgato, M., Masato, D., Lucchetta, G., Orazi, L., Effect of different laser-induced periodic surface structures on polymer slip in PET injection moulding. *CIRP Ann.*, 67, 575, 2018.
34. Aleksandrova, M., Specifics and challenges to flexible organic light-emitting devices. *Adv. Mater. Sci. Eng.*, 2016, 4081697, 2016.
35. Aleksandrova, M., Dobrikov, G., Singh, A.K., Videkov, V., Kolev, G., Flexible optoelectronic device with polymer based electrode on hybrimer substrate - Impact of the bending on the interfacial processes. *40th International Spring Seminar on Electronics Technology (ISSE 2017), Sofia, Bulgaria*, p. 1, 2017.
36. Aleksandrova, M., Videkov, V., Ivanova, R., Singh, A.K., Thool, G.S., Highly flexible, conductive and transparent PEDOT:PSS/Au/PEDOT:PSS multilayer electrode for optoelectronic devices. *Mater. Lett.*, 174, 204, 2016.
37. Mashayekhi, M., Winchester, L., Evans, L., Pease, T., Laurila, M.-M., Mäntys, M., Evaluation of Aerosol, Superfine Inkjet, and Photolithography Printing Techniques for Metallization of Application Specific Printed Electronic Circuits. *IEEE Trans. Electron Devices*, 63, 1246, 2016.
38. Zheng, H., Zheng, Y. *et al.*, All-solution processed polymer light-emitting diode displays. *Nat. Commun.*, 4, 1971, 2013.
39. Lee, S.-H., Kim, H., Park, H.-L., Kang, S., Lee, S.-D., Solution-processed organic light-emitting diode in high-resolution line patterns by scalable wetting modification. *Org. Electron.*, 73, 332, 2019.
40. Liu, Z., Zhang, L., Gao, X., Zhang, Q., Shi, S., Synthesis and characterization of novel red-emitting conjugated polymers based on triphenylaminesilole-carbazole-fluorene. *Mater. Chem. Phys.*, 212, 208, 2018.

41. Jongseong, K., Shanmugasundaram, K., Chitumalla, R.K., Jang, J., Choe, Y., Phenothiazine derivatives as an easily accessible emitter for green light-emitting electrochemical cells. *J. Lumin.*, 197, 383, 2018.
42. Wang, J., Xu, X., Peng, J., Yao, C., High triplet, bipolar polymeric hosts for highly efficient solution-processed blue phosphorescent polymer light-emitting diodes. *Org. Electron.*, 43, 1, 2017.
43. Guo, F., Karl, A., Xue, Q.-F., Tam, K.C., Forberich, K., Brabec, C.J., The fabrication of color-tunable organic light-emitting diode displays via solution processing. *Light – Sci. Appl.*, 6, e17094, 2017.
44. Wang, P., Huang, Q., Zhao, S., Qin, Z., Xu, Z., Song, D., Qiao, B., Investigating the evolution of excitons in polymer light-emitting diodes by transient measurement. *Org. Electron.*, 68, 45, 2019.
45. He, H., Liao, X., Cheng, J., Li, Y., Yu, J., Li, L., Investigation of Energy Transfer in Star-Shaped White Polymer Light-Emitting Devices via the Time-Resolved Photoluminescence. *Materials*, 11, 1719, 2018.

3

DFT Computational Modeling and Design of New Cyclopentadithiophene (CPDT) Derivatives for Highly Efficient Blue Emitters in OLEDs

Rania Zaier[1], Said Hajaji[1], Masatoshi Kozaki[2] and Sahbi Ayachi[1]*

[1]Laboratory of Physico-Chemistry of Materials (LR01ES19), Faculty of Sciences of Monastir (FSM), University of Monastir, Monastir, Tunisia
[2]Laboratory of Physical Organic Chemistry, Graduate School of Science, Osaka City University, Sugimoto, Sumiyoshi-ku, Osaka, Japan

Abstract

Two types of linear p-conjugated cyclopentadithiophene (CPDT) derivatives, having the same backbone but bridging positions with >C=O or >C=C(CN)2 groups were designed. The first type is the homo-dimer (D1) of 2-(4H-cyclopenta[2,1-b:3,4-b']dithiophen-4-ylidene)malononitrile and the second type is the co-dimer (D2) incorporating the 4H-cyclopenta[2, 1-b:3,4-b']dithiophen-4-one as subsequent subunit. CPDTs Molecular properties were studied in both ground and excited states using, respectively, the Density Functional Theory (DFT) and its Time-Dependent extension (TD-DFT) at B3LYP functional with 6-311g(d,p) basis set in acetonitrile. The observed PL quenching effect for D2 is attributed to intra-molecular charge transfer (ICT). Based on ionization potential (IP), electronic affinity (EA) and reorganization energies, we show the enhanced charge transport of the studied dimer molecules.

In typical OLED device structure, the layers include glass ITO as anode, NPD (hole-injection layer (HIL), 40 nm), emissive layer (D1 or D2, 30 nm), Alq3 (electron injection layer (EIL), 40 nm) and Al (100 nm) as cathode. The adjusted

**Corresponding author*: ayachi_sahbi@yahoo.fr

Inamuddin, Rajender Boddula, Mohd Imran Ahamed and Abdullah M. Asiri (eds.) Polymers for Light-Emitting Devices and Displays, (51–76) © 2020 Scrivener Publishing LLC

thickness values are able to improve the carrier's injection efficiency. Therefore, the simulated I(V) characteristics are analyzed for both studied compounds.

Judging from the computed results, we believe that the CPDT derivatives have interesting optical properties, which are useful for constructing novel blue-light emitting materials for OLEDs.

Keywords: CPDT derivatives, DFT, electronic properties, ICT, OLED

3.1 Introduction

Organic light-emitting diodes (OLEDs) have received an increasing attention since they were first developed [1–3]. This is due to their unique electrical and optical properties including low cost, easy processing, low-operating voltage, and flexibility [4–6]. Recently, OLEDs have made an extensive progress because of their potential applications in solid state lighting technologies and in full color flat-panel display applications [7–10]. The OLEDs are a relatively new display technology and its progress is still very fast [11]. For this reason, OLEDs based on organic electroluminescent (EL) materials have been considerably investigated [12, 13]. In fact, the purpose of developing full-color displays and white lighting is to find materials that can emit pure red (R), green (G), and blue (B) light called RGB emitters [14–17]. It remains a great challenge to produce high performance blue molecular emitter because of the intrinsically wide band gap, which is still required for further application of OLEDs [18].

Blue emission has been investigated in a number of organic materials including anthracene derivatives [19, 20], carbazole derivatives [21, 22], fluorene derivatives [23, 24], and pyrene derivatives [25, 26]. However, efficient and stable blue emitters with good color purity are still rare due to their large optical band gap [27].

To achieve novel high blue light-emitting materials, judicious molecular design is required to attain claimed emission properties, wide energy band gap, and charge mobility, and so forth [28, 29]. Among the materials on which OLEDs are based, thiophene derivatives have shown to be the most prospective functional ones thanks to their intrinsic electron-rich nature, sterling thermal stability, excellent optical and electronic properties, and the handy adjustment of their properties through the insertion of functional substituents at different positions of their backbones [30–32]. It is worth noting that the charge injection efficiency determines charge balance and efficiency in practical OLED optimization. To improve the charge balance, novel device structures with the different electron transporting materials having high electron mobility and hole blocking ability are recommended.

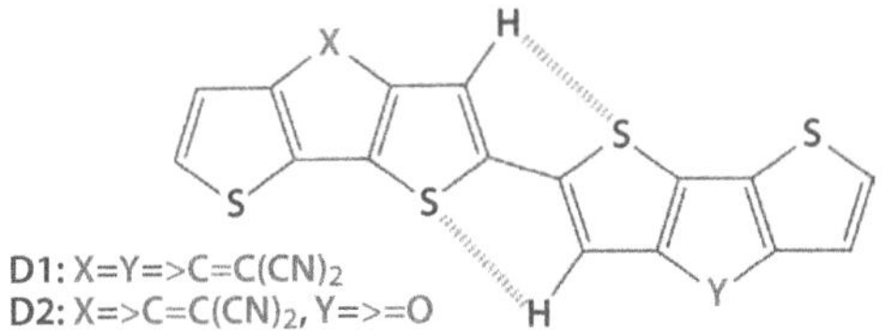

Scheme 3.1 Molecular structures of the studied CPDT dimers.

In order to probe the properties that are difficult to measure and to provide new links necessary for the understanding of the structure-property relationships of materials, the theoretical approach was used as a helpful tool to understand specific phenomena. It also helps scientists make predictions before running the actual experiments. Indeed, DFT-based calculations constitute not only an ideal compromise tool but also an essential analysis method in the same way as other spectroscopic methods of analysis. In a previous work [33], our group has developed, for the first time, a theoretical investigation on homo-dimer from CPDT-bridged carbonyl groups. The present work aims at understanding the fluorescence quenching behavior and intramolecular charge transfer (ICT) character for two kinds of CPDT derivatives, denoted D1 and D2 whose corresponding chemical structures are presented in Scheme 3.1. Both dimer molecules having the same backbone but incorporating >C=O or $>C=C(CN)_2$ as bridged groups were designed and theoretically tested that improve the power efficiency of blue OLEDs. To get deeper insights into the photo-physical and fluorescence properties of D1 and D2, their structural and molecular geometries, molecular orbital transitions, as well as simulated optical properties (UV-VIS and emission) and radiative life time were studied using density functional theory (DFT) and time-dependent density functional theory (TD-DFT).

3.2 Computational Methods

First of all, the optimized geometries of the two examples of CPDT derivatives using the Gaussian 09 package [34], in their ground states, have been developed employing the density functional theory (DFT) at the B3LYP level of theory. Particularly, the used 6-311 g(d,p) basis set was chosen as a compromise between high-computational cost and theoretical calculation quality. The electronic properties including the ionization potential (IP), the electron affinity (EA), and the reorganization energies have been conducted using the same methods of calculation based on optimized neutral and charged states of the dimer molecules in acetonitrile medium by means of the conductor-like polarizable continuum model (CPCM). The total energy, highest occupied molecular orbital (HOMO), lowest unoccupied molecular orbital (LUMO) energies, energy gap, and dipolar moment were also calculated at the same level of theory.

The S1 excited-state geometry was calculated using *ab initio* CIS/6-311g(d,p), from the optimized ground-state geometry. Then, the transition energy, oscillator strength, as well as the electronic transition were calculated at both ground and excited states using the time-dependent extension of DFT (TD-DFT). This allows the simulation of optical and emission spectra and helps in studying charge transfer/recombination process in organic electronic devices.

The crucial electronic parameters for achieving high electroluminescence efficiency are IP, EA, efficiency of charge carrier injection, electron and hole mobility, as well as light absorption and emission. To give a more quantitative description that can be related to structure of organic electronic materials, it is so important to appreciate how changes in molecular structure affect the electronic properties and device performance.

International Commission on Illumination (CIE) was used to obtain emission colors coordinates. Finally, the current-voltage (I-V) characteristics of OLED devices based dimer (D1 or D2), as emitting layer, have been simulated using SILVACO software [35].

3.3 Molecular Geometry

First of all, we can find that for both compounds, highly planar structures due to the presence of intra-molecular non-covalent interactions can apparently increase the rigidity of the molecular backbone (see Scheme 3.2). The simulated parameters for the ground state optimized structures are given in Table 3.1.

As depicted in Table 3.1, the dipole moment of D1 is nearly zero (0.002 Debye) due to the high symmetry of the molecular structure. The dipole

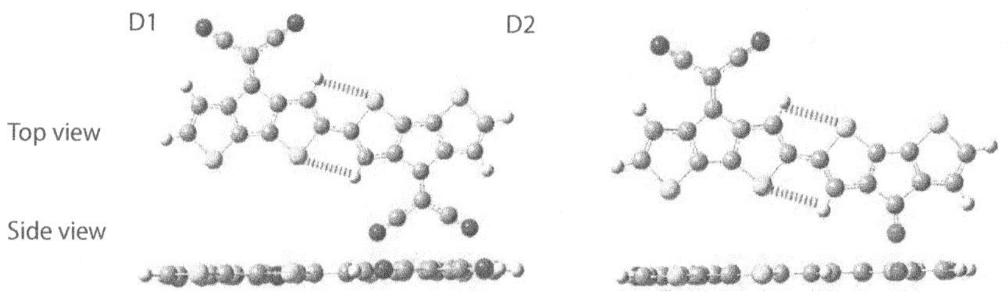

Scheme 3.2 Optimized molecular structures of CPDT dimers.

Table 3.1 Calculated total energies and dipole moments at DFT//B3LYP/6-311g(d,p) level.

Compounds	Total energy (Hartree)	Dipole moment (Debye)
D1	−2730.1535	0.002
D2	−2581.5587	3.917

moment of D2 was found at about 3.9172 Debye which arose from the difference between bridged groups within the molecule. The total energies of D1 and D2 at optimized geometries were found to be −2,730.1535 and −2,581.5587 Hartree, respectively, indicating that D1 is more stable than D2 due to the high structure symmetry of D1.

Accordingly, Mulliken atomic charge distribution is a good approach to get deeper insights into the molecular structure properties. Regarding the geometries under investigations, it is clear that D1 is a symmetric molecule for containing identical unit blocks. Whereas, D2 is an asymmetric molecule, which contains different bridged groups. These observations are well confirmed by means of Mulliken atomic charge distribution of D1 and D2 as illustrated in Figure 3.1 at ground and excited states.

At the ground states of dimers, CPDT becomes partially positively charged, while the bridging groups (>C=O and >C=C(CN)$_2$) are partially negatively charged. As for the excited states of molecules, a redistribution of electronic charges was detected and strictly related to the photo-induced electron transfer. More accurately the involved n→π* electronic transition originated from the carbonyl group and the arrangements can result in strong π-π interaction within the CPDT fragments.

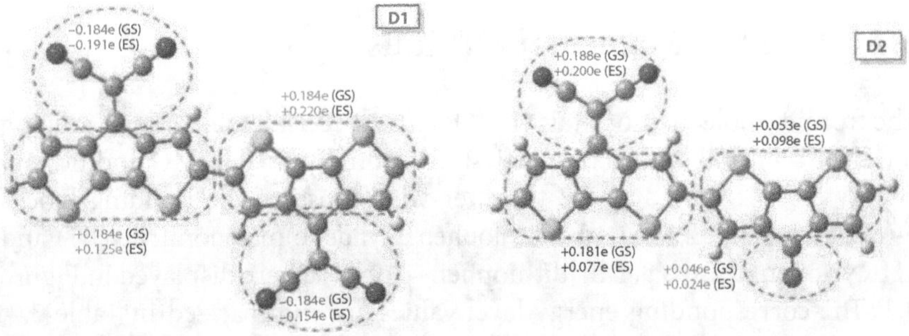

Figure 3.1 Mulliken atomic charge distribution at ground and excited states of CPDT dimers.

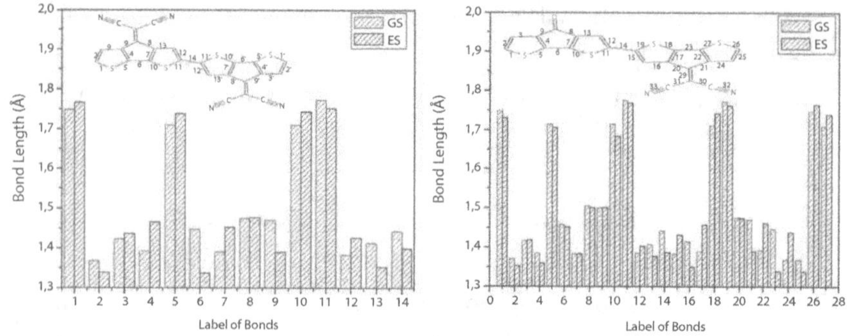

Figure 3.2 Bond lengths at optimized ground (GS) and excited (ES) states of D1 (at left) and D2 (at right).

As shown in Scheme 3.2, highly planar structures of dimer molecules originated from the S---H intra-molecular non-covalent interactions were detected, that enhance the rigidity of the molecular backbone [36]. The dihedral angles between CPDT units are nearly zero (planar structures) for both CPDT dimers. As already described [37, 38], non-covalent bond lengths of S---H at ground state were found to be around 2.90 Å for both CPDT dimers. Further, infrared vibrational modes for the S---H interactions were calculated at approximately 35 cm^{-1} in both cases.

The different bond lengths of the studied compounds in optimized ground and excited states geometries are illustrated in Figure 3.2. As can be seen, there is a slight discrepancy in terms of bond lengths between ground and excited states as a result of the modification in electron delocalization within the molecular structures. Particularly, the bond length numbered 14 (linkage between the two CPDT subunits) is dramatically reduced to form a double bond favoring the intra-molecular charge transfer between subunits.

3.4 Frontier Molecular Orbitals

The frontier molecular orbitals (FMOs) play a considerable role in judging the electronic and optical properties. Accordingly, the HOMO and LUMO energy levels of D1 and D2 together with those of the building blocks 2-(4H-cyclopenta[2,1-b:3,4-b′]dithiophen-4-ylidene)malononitrile (B1) and 4H-cyclopenta[2,1-b:3,4-b′]dithiophen-4-one (B2) are displayed in Figure 3.3. The corresponding energy level values are summarized in Table 3.2. From the schematic electronic diagrams, we note that the building block B2 contributes mainly to the HOMO energy level, while B1 contributes chiefly to the LUMO energy level, which indicates the importance of each part in the electronic properties of the title materials.

For the investigated compounds D1 and D2, the distributions of the HOMO levels are the same. Where, we found that they are mainly located on the whole main framework excepting the carbonyl (>C=O) and >C=C-$(CN)_2$ groups. Otherwise, there is a discrepancy between the HOMO and LUMO distributions of the title molecules. For instance, the LUMO of D1 is located over all the molecular structure, revealing an intra-molecular charge transfer (ICT) that takes place within molecule from π-conjugated backbone to the dicyano groups. While, in the case of D2, the LUMO is delocalized on the >C=C-$(CN)_2$ part, which is confirmed by an ICT occurring from the carbonyl group. Consequently, the shortened carbonyl bond, upon excitation, is related to the n→π*

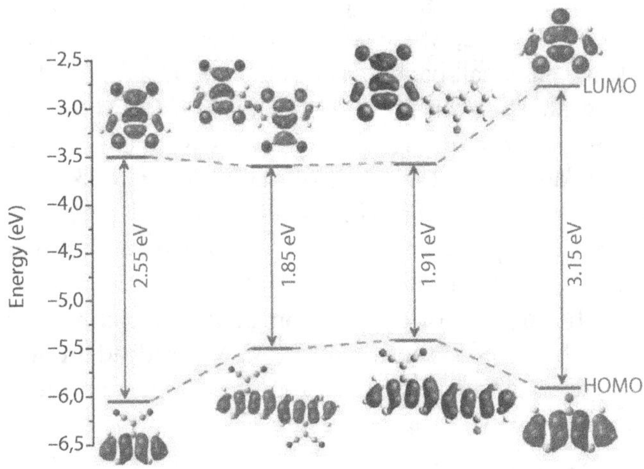

Figure 3.3 Frontier molecular orbitals (FMOs) and energy levels at ground state using DFT//B3LYP/6-311g(d,p) method of CPDT dimers.

Table 3.2 Calculated energetic parameters at DFT//B3LYP/6-311g (d,p) level.

Compounds	ε_{HOMO} (eV)	ε_{LUMO} (eV)	ΔE_{gap} (eV)	ε_{HOMO-1} (eV)	ε_{LUMO+1} (eV)
B1	−6.05	−3.50	2.55	−7.20	−1.07
B2	−5.93	−2.78	3.15	−7.42	−0.67
D1	−5.51	−3.60	1.91	−6.65	−3.58
D2	−5.43	−3.58	1.85	−6.58	−2.93

electronic transition [39, 40]. These results reveal the importance of the bridged group in influencing the electronic properties. The band energy gap separating the HOMO and LUMO energy levels of D1 and D2 are also calculated as illustrated in Table 3.2. The band gap energies of D1 and D2 are almost the same with the values of 1.91 eV and 1.85 eV, respectively. In fact, the slight difference between energy band gaps (0.06 eV) could be explained by the high similarity of the molecular structures of the compounds under investigation.

Further, the density of states (DOS) corresponding to D1 and D2 were simulated. The simulated results are shown in Figure 3.4. It is relevant to note that no sharp peaks were observed in DOS for both dimers indicating the absence of impurity.

It is well known that the DOS represents the energy level distribution of electron. The characteristic features of the electronic structure for both compounds reveal that a full splitting of the energy levels of HOMO−n and LUMO+n with n ≥ 2, which significantly contributed to charge transport [41]. Thus, the small band gaps of D1 and D2 along with the strong electronic delocalization within the conjugated chain led to an important optical absorption properties resulting from HOMOLUMO transitions.

The ICT is related to the formation of charge density difference between ground and excited states that can be visualized through the electronic density difference (EDD) plots. The EDD plots of the studied compounds were simulated. The simulated results are illustrated in Figure 3.5. As displayed in this figure, the regions of the electron density depletion presented in blue are mostly localized at the carbonyl (>C=O) bridged group, while the regions of the electron density increment presented in purple are largely aligned with the >C=C(CN)$_2$ groups [42, 43].

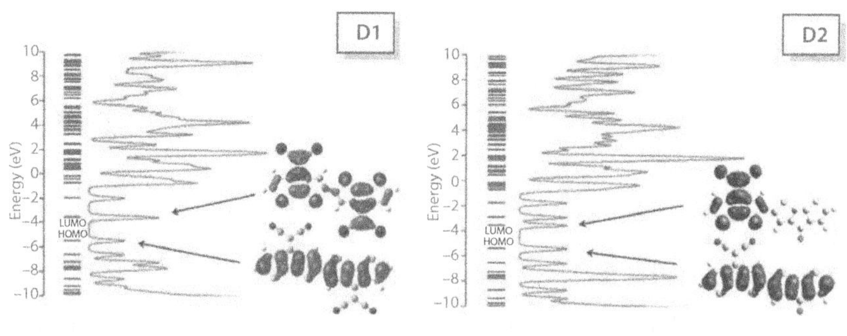

Figure 3.4 Density of state (DOS) plots of D1 and D2.

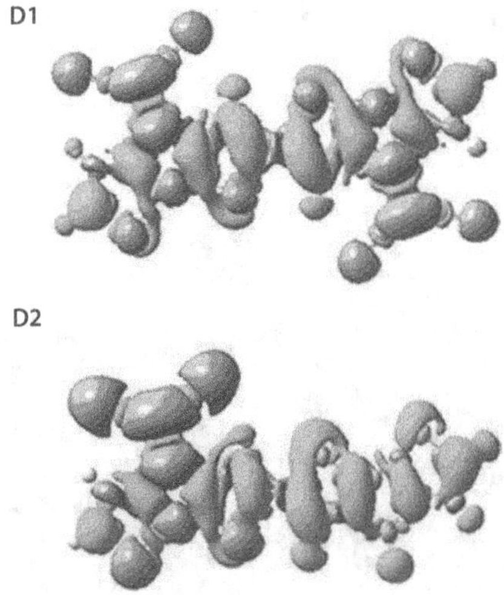

Figure 3.5 Electronic different density (EDD) between ground and excited states of the studied D1 and D2 compounds.

3.5 Molecular Electrostatic Potential Maps

The Mulliken charges and surface electrostatic potentials usually shed light on the electron distribution and electrostatic interaction within molecules. Therefore, molecular electrostatic potential (MEP) is a practical descriptor for chemical reactivity and site selectivity of electrophilic and nucleophilic attacks of the studied molecules as long as it is directly related to the electron density. Here, the results are depicted in Figure 3.6.

As already described in [33], in ground states of CPDT dimers, it revealed that negative charges cover the $>C=C(CN)_2$ groups for D1. As for D2, the negative charges cover both the $>C=C(CN)_2$ and the $>C=O$ groups. The positive charges are placed over the entire main conjugated framework. From the combination results of the FMOs analyses and the MEP plots, it is possible to predict the site of interaction through the electronic charges distribution over the molecules [44].

3.6 Optical Absorption and Emission Properties

3.6.1 UV-Vis-NIR Optical Absorption Properties

In order to elucidate the optical properties of the studied compounds, UV-Vis-NIR optical absorption spectra were simulated in acetonitrile at

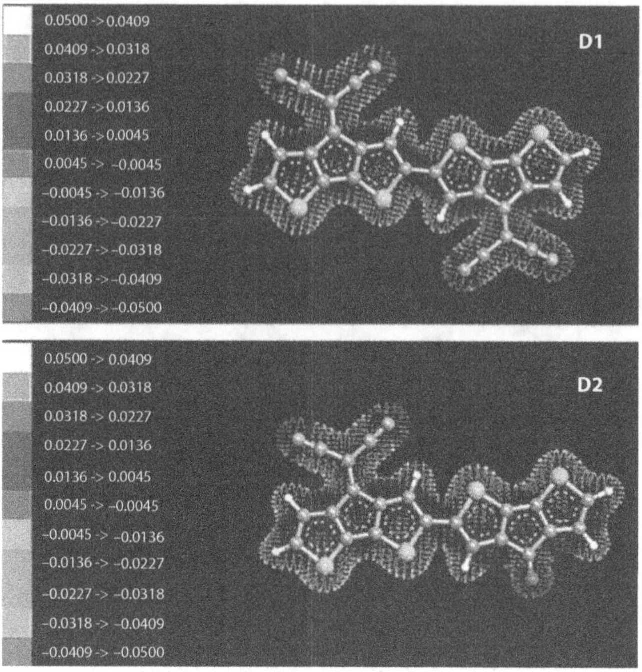

Figure 3.6 Molecular electrostatic potential surface (MEPs) of D1 and D2.

Table 3.3 The vertical transition energies (nm) and their oscillator strengths of optical absorption from the ground to the first excited states ($S_0 \rightarrow S_1$) of CPDT dimers calculated by TD-DFT//B3LYP/6-311g(d,p) level of theory.

Compounds	λ_{max}^{Abs} (nm)	E (eV)	f (a.u)	Electronic transition
D1	357	3.47	0.7434	H-2→L+1 (28%)
	402	3.08	1.0941	H→L+2 (93%)
	493	2.51	0.0247	H-1->L+1 (95%)
	947	1.31	0.0780	H→L (96%)
D2	387	3.20	0.8221	H→L+2 (89 %)
	649	1.91	0.1671	H→L+1 (96 %)
	984	1.26	0.0386	H→L (96 %)

B3LYP/6-311g(d,p) level of theory. The matching photo-physical data are summarized in Table 3.3.

Absorption spectra with oscillator strengths are illustrated in Figure 3.7, to get better insides into the main electronic transitions. As illustrated in this figure, there are two main absorption regions at 350–450 nm and 900–1,000 nm with high and small values of oscillator strengths, respectively.

From the quantitative details reported elsewhere [45–48], we consider that highest energy band is ascribed to a π→π* electronic transition. However, lowest energy band is an ICT from electron donor to acceptor

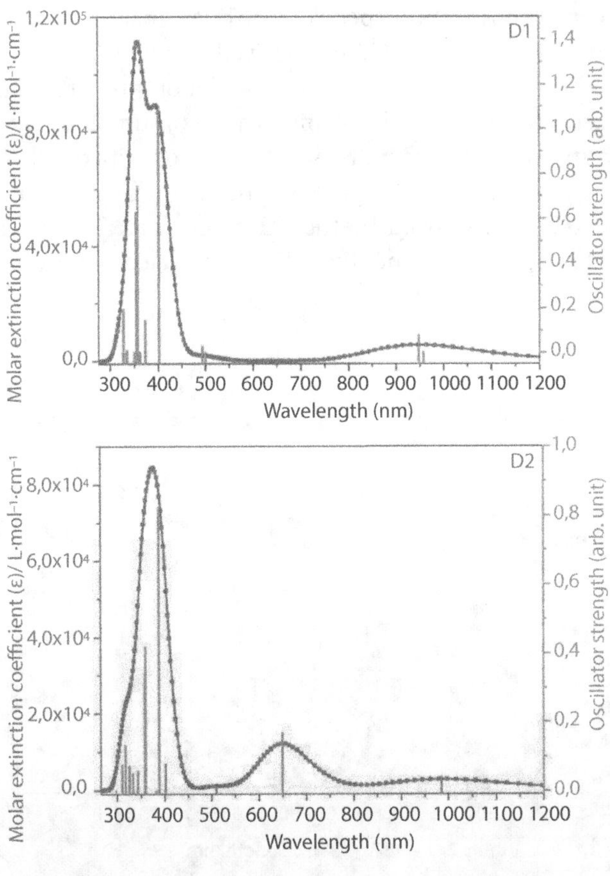

Figure 3.7 Simulated UV-Vis-NIR optical absorption spectra with oscillator strengths (vertical lines) of D1 and D2.

groups [49, 50]. In Figure 3.8, we have attributed the main absorption bands to the absorption maxima of the investigated molecules.

A first intense absorption band observed in the region from 350 nm to 450 nm is attributed to $\pi \rightarrow \pi^*$ transition originating from $S_0 \rightarrow S_1$ electronic transition. As expected, the optical absorption spectrum shows a large absorption band related to π-conjugated framework, highly planar characteristic and intense ICT generated by the photo-induced electron transfer in the D2 excited state [51, 52]. For D1, the band centered at 947 defines the ICT from the electron rich sulfur atom to electron-deficient $>C=C-(CN)_2$ group [53, 54]. This second band is located at around 948 nm in the case of D2. Moreover, D2 has a third band located at the region of 649 nm referring to the charge transfer from main framework to the carbonyl group.

The maximum absorption spectrum of D1 is made up of two absorption peaks placed at 357 and 402 nm derived mainly from H−2→L+1 and H→L+2 electronic transition with high oscillator strength values of 0.74 and 1.09, respectively. While D2 exhibits a maximum absorption band at around 387 nm originating from H→L+2 transition with oscillator strength of 0.82. As can be seen from Figure 3.8, there is a hypsochromic shift of UV-Vis spectrum introduced by the D2 compound. This hypsochromic effect could be explained by the slight different of electron properties of D2 that contains two different bridged groups.

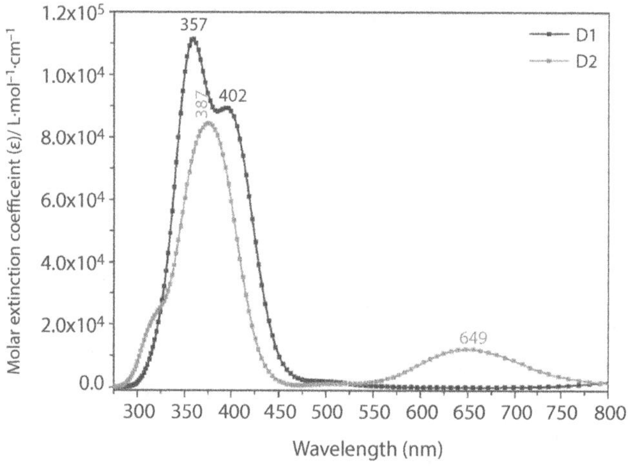

Figure 3.8 Compared UV-Vis-NIR optical absorption spectra of investigated compounds.

3.6.2 Emission Properties

To get deeper insights into the emission properties, the CIS method was used to optimize the excited states geometries of the investigated compounds [55–58]. The emission behavior of D1 and D2 was studied by means of TD-DFT//B3LYP/6-311g(d,p) level of theory. The emission (Em) spectra together with the high oscillator strengths are depicted in Figure 3.9. As a first observation, the obtained spectra show an accumulation of the highest values of the oscillator strengths in the region of 400–500 nm attributed to the range of the blue emission color.

The maximum emission wavelengths as well as the emission properties of D1 and D2 are tabulated in Table 3.4. The main emission bands are assigned to electron transition from the first excited state (S_1) to the ground state (S_0).

As illustrated in Figure 3.10, the maximum emission wavelengths are located at around 451 nm and 441 nm for D1 and D2, respectively, generating a blue emission. From another point of view, we found that there is a decrease of emission intensity in the main emission band of D2 compared to that of D1. This hypochromic or quenching effect could be due to the carbonyl group as a group quencher [59, 60]. Thus, it can be concluded that the D1 has better emission properties than D2.

The Stokes shifted between maximum absorption and maximum emission at about 49 nm for D1 and 54 nm for D2, indicating a little loss in energy during the relaxation process and efficient fluorescence. In addition, these small Stokes shifts are attributed to the high rigidity of the studied compounds.

To better understand the optical responses of our compounds (D1 and D2), the emission spectra were fitted by means of Gaussian peaks for D1 and using Lorentz peaks for D2 regarding the difference in curves' shape. As illustrated in Figure 3.11, the PL spectrum of D1 displays a FWHM of about 53 nm (λ = 451 nm) slightly smaller than D2 which depicts a broad FWHM at about 61 nm (λ = 441 nm). From these results, it seems clear that the investigated compounds can be used as active layers in OLEDs [61, 62].

The radiative lifetimes were calculated for a spontaneous emission using Einstein transition probabilities and approximated by [63]:

$$\tau = \frac{C^3}{2(E_{flu})^2 f}$$

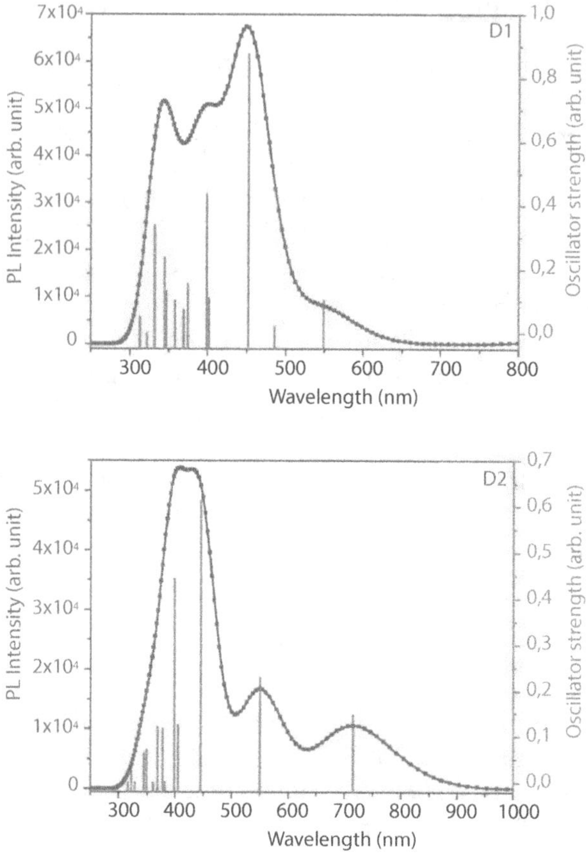

Figure 3.9 Emission spectra with oscillator strength (vertical lines) of the investigated compounds.

Where C, E_{flu}, and f represent the speed of light, excitation energy, and oscillator strength, respectively. The obtained values are 3.51 and 4.84 ns for D1 and D2, respectively. The obtained results reveal that compound D1 exhibits a lower radiative life time than D2, which indicates that D1 is more efficient in the emission of photons.

3.7 ICT Properties

As mentioned in the introduction part, the reorganization energy could be an important factor that governs the mobility of charge carriers. Then,

Table 3.4 The vertical transition energies (nm) and their oscillator strengths of emission from the first excited to ground ($S_1 \rightarrow S_0$) states of CPDT dimers calculated by TD-DFT//B3LYP/6-311g(d,p) level of theory.

Compounds	λ_{max}^{Em} (nm)	E_{flu} (eV)	f (a.u)	Electronic transition	τ (ns)
D1	339	3.5987	0.2351	L+1 → H−3 (51%)	
	389	3.1061	0.4341	L → H−2 (42%)	
	451	2.7426	0.8754	L+2 → H (79%)	3.51
	523	2.2583	0.1009	L→ H−1 (92%)	
D2	495	3.0609	0.1234	L+1 → H−1 (63%)	
	441	2.7867	0.6139	L+2 → H (73%)	4.84
	558	2.2550	0.2293	L → H−1 (92%)	
	722	1.7323	0.1449	L+1 → H (95%)	

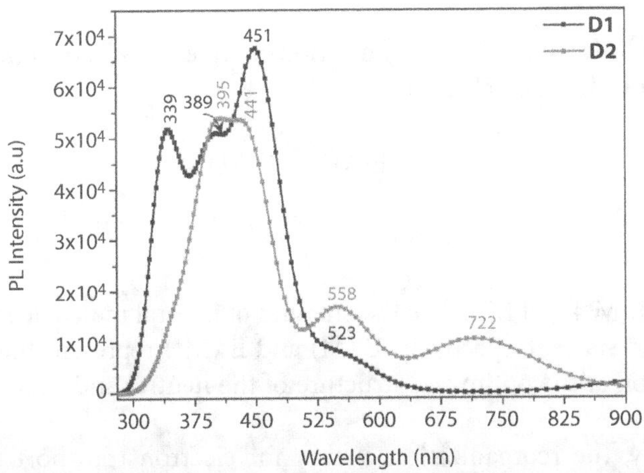

Figure 3.10 The simulated emission spectra at TD-DFT//B3LYP/6-311g(d,p) of D1 and D2.

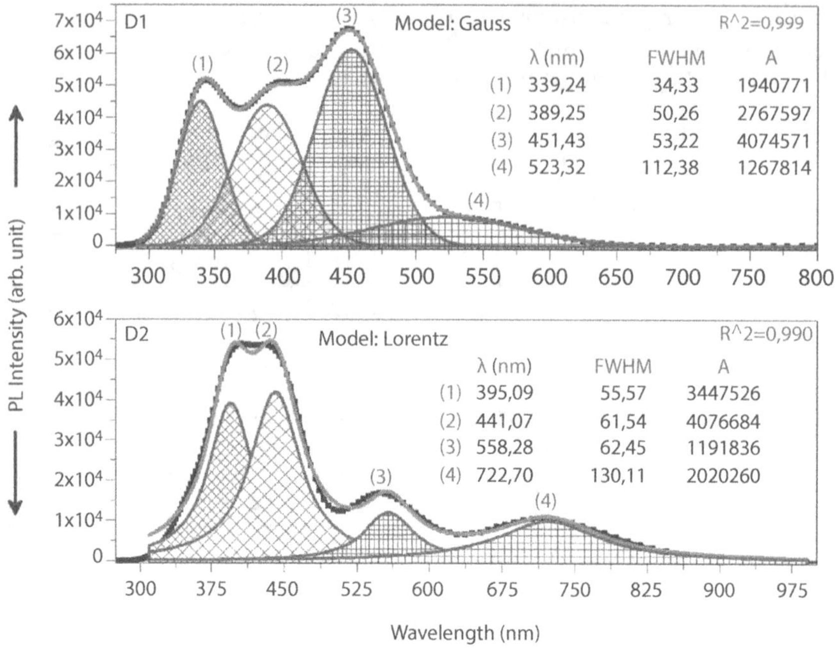

Figure 3.11 Fitting Gaussian and Lorentz peaks of emission spectra of CPDT dimers. The FWHM (full width at half maximum) and A (integrated area) are included in the graph.

as referred to [64–67], the reorganization energies for electron (λ_e) and for hole (λ_h) transfer can be defined as:

$$\lambda_1 = E(M^+) - E(M)$$

$$\lambda_2 = E^+(M) - E^+(M^+)$$

Where $E(M^+)$ and $E(M)$ are the energies of neutral molecule at cationic and ground states, respectively. $E^+(M)$ and $E^+(M^+)$ represent the energies of the cation at the optimized structure of the neutral and cationic states, respectively.

Similarly, the reorganization energy for electron transport ($\lambda_{electron} = \lambda_3 + \lambda_4$) is defined as:

$$\lambda_3 = E(M^-) - E(M)$$

$$\lambda_4 = E^-(M) - E^-(M^-)$$

Where $E(M^-)$ is the energy of neutral molecule at anionic structure. $E^-(M)$ and $E^-(M^-)$ correspond to the energies of the anion at the optimized structure of the neutral and anionic states, respectively. A schematic plot is shown in Figure 3.12.

On the basis of neutral and charged states of dimer optimized structures, we have calculated the reorganization energies and data reported in Table 3.5. According to Table 3.5, D1 plays an important part for the charge transfer ability with low reorganization energy for hole (λ_{hole} = 0.309) and electron ($\lambda_{electron}$ = 0.305). As for D2, we found the same reorganization energies' value of 0.310 for hole and electron, which ensures a great balance of charge injection within the active layer.

It is well know that electronic parameters including the adiabatic ionization potential (IPa) and adiabatic electron affinity (EAa) determination are important task to tune the carriers injection efficiency for OLEDs [68, 69]. The low IP (5.21 eV for D1 and 5.15 eV for D2) and the high EA (3.87 eV for D1 and 3.85 eV for D2) values signify that dimer molecules exhibit good performance to grasp hole and electron for favorable emission and efficient electroluminescent devices.

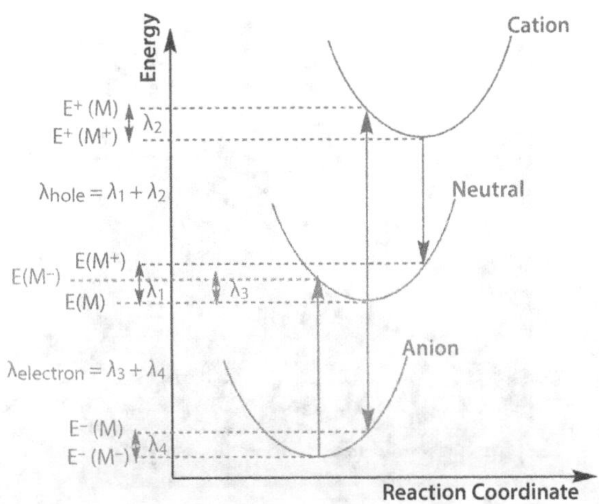

Figure 3.12 Schematic plot of reorganization energy.

Table 3.5 Reorganization energies λ(eV) for hole (λ_{hole}) and electron ($\lambda_{electron}$) transport calculated at DFT//B3LYP/6-311g(d,p) level of theory.

Compounds	λ_1	λ_2	λ_3	λ_4	λ_{hole}	$\lambda_{electron}$	IP	EA
D1	0.154	0.156	0.156	0.149	0.309	0.305	5.21	3.87
D2	0.155	0.155	0.154	0.156	0.310	0.310	5.15	3.85

3.8 OLEDs Modulation

It is well known that the improvement in the performance in OLEDs can be achieved by balanced charge injection and charge transport [70, 71]. This later is strictly related to the drift mobility of charge carriers. From electronic parameters including IPa, EAa, and reorganization energy, high carrier's mobility that leads to better device performance was achieved. As illustrated in Figure 3.13, the simulated positions on CIE coordinates system were found to be (0.16, 0.09) for D1 and (0.20, 0.15) for D2, corresponding to blue light emission.

As already proposed for the appropriate layers (HIL and EIL with optimal values of the thickness) to improve the efficiency of carriers injection [33] based on their energy levels diagram (HOMO and LUMO) [62–74], typical OLED device architecture is depicted in Figure 3.14.

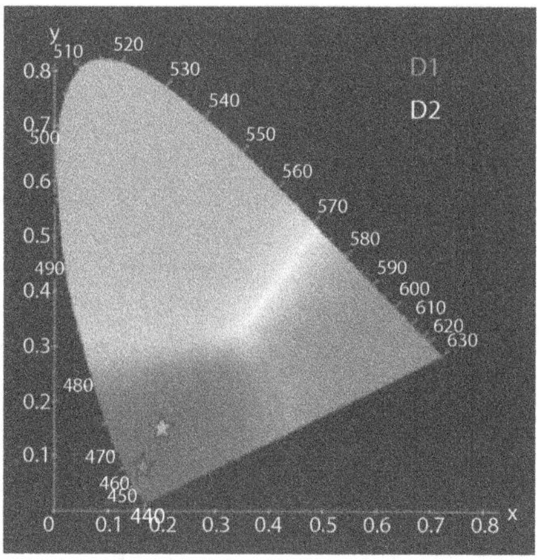

Figure 3.13 The representative CIE color coordinates of D1 and D2.

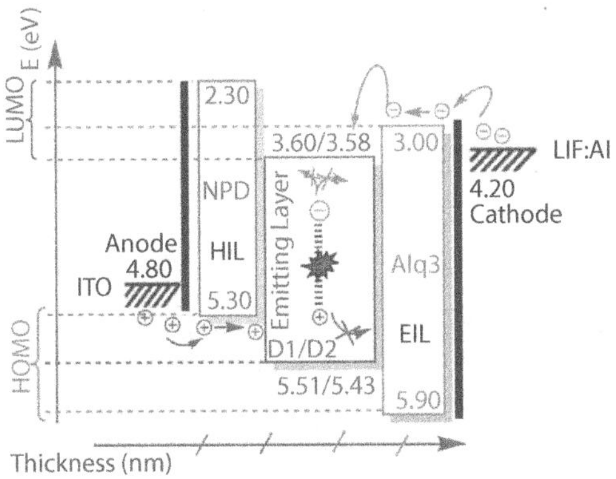

Figure 3.14 Typical OLED architecture including D1 or D2 as emitting layer.

In the three layers based OLED, the insertion of HIL and EIl layers of charge transport materials in addition to the emitter layer (CPDT derivatives) provides a powerful means to control charge injection, transport, and recombination in OLEDs.

It is worth reminding, that a large barrier for charge injection results in a high driving voltage, and that unipolar charge-transport capability leads to unbalanced charge transport and, consequently, low recombination efficiency. The deep HOMO level (5.9 eV) prevents efficiently the hole leakage from the emitting layer. In addition, the slightly small HOMO barrier between NPD (5.30 eV) and the dimer (5.43–5.51 eV) allows an efficient hole injection into the emission layer.

The electrical responses of the two kinds of devices, simulated using SILVACO software package, were determined based on the current-voltage (I-V) characteristics. The results were plotted in Figure 3.15. Typical voltages required to turn OLED devices were closed to 7.6 V for D1 and 6.5 V for D2. It is obviously that the relatively lower operating voltage was related to the efficient electron injection within device based D2 emitting layer.

For practical application, we believe that CPDT derivatives with highly planar structures that possess designed supra-molecular interactions, such as $\pi \rightarrow \pi$ stacking, can greatly enhance the efficiency and stability of electroluminescence (EL).

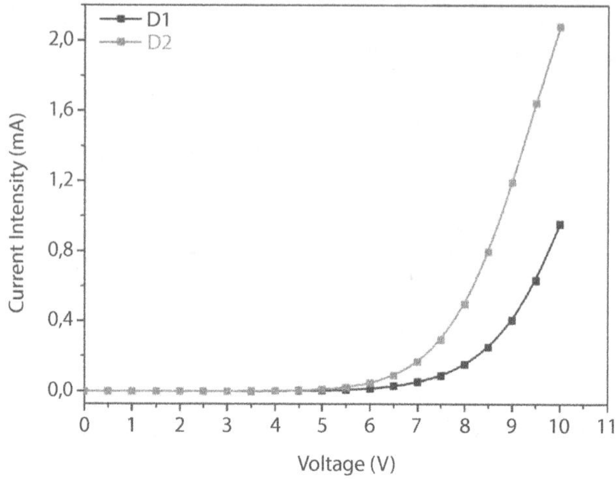

Figure 3.15 Current-voltage (I-V) characteristics of OLED devices based on CPDT dimers.

3.9 Conclusion

Examples of two types of π-conjugated CPDT derivatives with small band gaps, regarding their electronic structures, were investigated through theoretical calculations based on the DFT and TD-DFT at B3LYP functional with 6-311g(d,p) basis set in acetonitrile. The optical properties of the D2 are considerably affected due to the carbonyl bridge group. Then, the D2 compound showed strong ICT induced fluorescence quenching. According to the analysis of PL spectra, the studied compounds exhibit an emission in the blue region (451 nm for D1 and 441 nm for D2). The calculated reorganization energies reveal high performance for the charge mobility of D1 and D2. CIE coordinates are determined to be (0.16, 0.09) for D1 and (0.20, 0.15) for D2 located in the blue region. Further, current-voltage characteristic was also simulated. The threshold voltages were found to be 7.6 V for D1 and 6.5 V for D2. Accordingly, both dimers of molecule have excellent optical properties and electrical characteristics that make them attractive candidates for designing functional materials to be used as active layers in OLED devices.

References

1. Tang, C.W. and VanSlyke, S.A., Organic electroluminescent diodes. *Appl. Phys. Lett.*, 51, 913, 1987.

2. Burroughes, J.H., Bradley, D.D.C., Brown, A.R., Marks, R.N., Mackay, K., Friend, R.H., Burns, P.L., Holmes, A.B., Light-emitting diodes based on conjugated polymers. *Nature*, 347, 539, 1990.
3. Friend, R.H., Gymer, R.W., Holmes, A.B., Burroughes, J.H., Marks, R.N., Taliani, C., Bradley, D.D.C., Dos Santos, D.A., Brédas, J.L., Lögdlund, M., Salaneck, W.R., Electroluminescence in conjugated polymers. *Nature*, 397, 121, 1999.
4. Sekine, C., Tsubata, Y., Yamada, T., Kitano, M., Doi, S., Recent progress of high performance polymer OLED and OPV materials for organic printed electronics. *Sci. Technol. Adv. Mater.*, 15, 034203, 2014.
5. Jou, J.H., Kumar, S., Agrawal, A., Li, T.H., Sahoo, S., Approaches for fabricating high efficiency organic light emitting diodes. *J. Mater. Chem.*, C3, 2974, 2015.
6. Felscia, U.R., Rajkumar, B.J., Mary, M.B., Theoretical investigations on nonlinear fused 4-ring systems: Application to OLED and NLO devices. *Synth. Met.*, 246, 31, 2018.
7. Wong, M.Y. and Zysman-Colman, E., Purely Organic Thermally Activated Delayed Fluorescence Materials for Organic Light-Emitting Diodes. *Adv. Mater.*, 29, 1605444, 2017.
8. Im, Y., Byun, S.Y., Kim, J.H., Lee, D.R., Oh, C.S., Yook, K.S., Lee, J.Y., Recent Progress in High-Efficiency Blue-Light-Emitting Materials for Organic Light-Emitting Diodes. *Adv. Funct. Mater.*, 27, 1603007, 2017.
9. Chen, W.C., Lee, C.S., Tong, Q.X., Blue-emitting organic electrofluorescence materials: Progress and prospective. *J. Mater. Chem. C*, 3, 10957, 2015.
10. Chen, W.C., Yuan, Y., Zhu, Z.L., Jiang, Z.Q., Liao, L.S., Lee, C.S., Polyphenylnaphthalene as a Novel Building Block for High-Performance Deep-Blue Organic Light-Emitting Devices. *Adv. Opt. Mater.*, 6, 1700855, 2018.
11. Li, Z., Gan, G., Ling, Z., Guo, K., Si, C., Lv, X., Wang, H., Wei, B., Hao, Y., Easily available, low-cost 9, 9′-bianthracene derivatives as efficient blue hosts and deep-blue emitters in OLEDs. *Org. Electron.*, 66, 24, 2019.
12. Chalke, R.M. and Patil, V.R., Novel methoxy spirobifluorene and alkyl substituted diphenylacene based organic blue light emitting polymers for application in organic electronics. *Polym.*, 123, 355, 2017.
13. Qu, B., Feng, L., Yang, H., Gao, Z., Gao, C., Chen, Z., Xiao, L., Gong, Q., Color-stable deep red-emitting OLEDs based on a soluble terpolymer containing fluorene, thiophene and benzothiadiazole units. *Synth. Met.*, 162, 1587, 2012.
14. Kim, B., Park, Y., Lee, J., Yokoyama, D., Lee, J.H., Kido, J., Park, J., Synthesis and electroluminescence properties of highly efficient blue fluorescence emitters using dual core chromophores. *J. Mater. Chem. C*, 1, 432, 2013.
15. Fan, C., Wang, X., Luo, J., Blue organic light-emitting diodes based on terpyridine-substituted triphenylamine chromophores. *Opt. Mater.*, 64, 489, 2017.

16. Lee, K.H., Kwon, Y.S., Kang, L.K., Kim, G.Y., Seo, J.H., Kim, Y.K., Yoon, S.S., Blue organic light-emitting materials based on diphenylaminofluorene and *N*-phenylcarbazole derivatives. *Synth. Met.*, 159, 2603, 2009.
17. Zhu, Z.L., Chen, M., Chen, W.C., Ni, S.F., Peng, Y.Y., Zhang, C., Tong, Q.X., Lu, F., Lee, C.S., Removing shortcomings of linear molecules to develop high efficiencies deep-blue organic electroluminescent materials. *Org. Electron.*, 38, 323, 2016.
18. Yang, J., Huang, J., Li, Q., Li, Z., Blue AIEgens: Approaches to control the intramolecular conjugation and the optimized performance of OLED devices. *J. Mater. Chem. C*, 4, 2663, 2016.
19. Park, H., Lee, J., Kang, I., Chu, H.Y., Lee, J.I., Kwon, S.K., Kim, Y.H., Highly rigid and twisted anthracene derivatives: A strategy for deep blue OLED materials with theoretical limit efficiency. *J. Mater. Chem.*, 22, 2695, 2012.
20. Park, J.Y., Jung, S.Y., Lee, J.Y., Baek, Y.G., High efficiency in blue organic light-emitting diodes using an anthracene-based emitting material. *Thin Solid Films*, 516, 2917, 2008.
21. Agarwal, N., Nayak, P.K., Ali, F., Patankar, M.P., Narasimhan, K., Periasamy, N., Tuning of HOMO levels of carbazole derivatives: New molecules for blue OLED. *Synth. Met.*, 161, 466, 2011.
22. Zanoni, K. and Iha, N.M., Sky-blue OLED through PVK:Ir(Fppy)2(Mepic). active layer. *Synth. Met.*, 222, 393, 2016.
23. Goel, A., Chaurasia, S., Dixit, M., Kumar, V., Prakash, S., Jena, B., Verma, J.K., Jain, M., Anand, R., Manoharan, S.S., Donor-Acceptor 9-Uncapped Fluorenes and Fluorenones as Stable Blue Light Emitters. *Org. Lett.*, 11, 1289, 2009.
24. Vandana, T., Karuppusamy, A., Kannan, P., Polythiophenylpyrazoline containing fluorene and benzothiadiazole moieties as blue and white light emitting materials. *Polym.*, 124, 88, 2017.
25. Jung, M., Lee, J., Jung, H., Kang, S., Wakamiya, A., Park, J., Highly efficient pyrene blue emitters for OLEDs based on substitution position effect. *Dyes Pigm.*, 158, 42, 2018.
26. Lai, S.L., Tong, Q.X., Chan, M.Y., Ng, T.W., Lo, M.F., Lee, S.T., Lee, C.S., Distinct electroluminescent properties of triphenylamine derivatives in blue organic light-emitting devices. *J. Mater. Chem.*, 21, 1206, 2011.
27. Wang, Z., Liu, W., Xu, C., Ji, B., Zheng, C., Zhang, X., Excellent deep-blue emitting materials based on anthracene derivatives for non-doped organic light-emitting diodes. *Opt. Mater.*, 58, 260, 2016.
28. Zhu, M. and Yang, C., Blue fluorescent emitters: Design tactics and applications in organic light-emitting diodes. *Chem. Sci. Rev.*, 42, 4963, 2013.
29. Reddy, S.S., Sree, V.G., Park, H.Y., Maheshwaran, A., Song, M., Jin, S.H., A rational design strategy for an extremely deep-blue fluorescent emitter with a small CIE *y* value for solution processable, high efficiency, organic light-emitting diodes. *Dyes Pigm.*, 145, 63, 2017.

30. Xu, Z., Yu, D., Yu, M., The synthesis and photoluminescence characteristics of novel 4-aryl substituted thiophene derivatives with bis-diarylacrylonitrile unit. *Dyes Pigm.*, 95, 358, 2012.
31. Cinar, M.E. and Ozturk, T., Thienothiophenes, Dithienothiophenes, and Thienoacenes: Syntheses, Oligomers, Polymers, and Properties. *Chem. Rev.*, 115, 3036, 2015.
32. Matsushima, T. and Adachi, C., Highly Efficient Organic Light-Emitting Diodes Doped with Thiophene/Phenylene Co-Oligomer. *Chem. Mater.*, 20, 2881, 2008.
33. Zaier, R., Hajaji, S., Kozaki, M., S. Ayachi., D.F.T., and TD-DFT studies on the electronic and optical properties of linear π-conjugated cyclopentadithiophene (CPDT) dimer for efficient blue OLED. *Opt. Mater.*, 91, 108, 2019.
34. Frisch, M.J., Trucks, G.W., Schlegel, H.B., Scuseria, G.E., Robb, M.A., Cheeseman, J.R., Scalmani, G., Barone, V., Mennucci, B., Petersson, G.A., Nakatsuji, H., Caricato, M., Li, X., Hratchian, H.P., Izmaylov, A.F., Bloino, J., Zheng, G., Sonnenberg, J.L., Hada, M., Ehara, M., Toyota, K., Fukuda, R., Hasegawa, J., Ishida, M., Nakajima, T., Honda, Y., Kitao, O., Nakai, H., Vreven, T., Montgomery, J.A., Jr., Peralta, J.E., Ogliaro, F., Bearpark, M., Heyd, J.J., Brothers, E., Kudin, K.N., Staroverov, V.N., Kobayashi, R., Normand, J., Raghavachari, K., Rendell, A., Burant, J.C., Iyengar, S.S., Tomasi, J., Cossi, M., Rega, N., Millam, N.J., Klene, M., Knox, J.E., Cross, J.B., Bakken, V., Adamo, C., Jaramillo, J., Gomperts, R., Stratmann, R.E., Yazyev, O., Austin, A.J., Cammi, R., Pomelli, C., Ochterski, J.W., Martin, R.L., Morokuma, K., Zakrzewski, V.G., Voth, G.A., Salvador, P., Dannenberg, J.J., Dapprich, S., Daniels, A.D., Farkas, Ö., Foresman, J.B., Ortiz, J.V., Cioslowski, J., Fox, D.J., *Gaussian 09, Revision C.01*, Gaussian, Inc., Wallingford CT, 2009.
35. Silvaco Int., Atlas User's Manual online. Available: www.silvaco.com.
36. Hoang, M.H., Nguyen, D.N., Ngo, T.T., Um, H.A., Cho, M.J., Choi, D.H., Molecular-weight engineering of high-performing diketopyrrolopyrrole-based copolymer bearing high π-extended long donating units. *Polym.*, 83, 77, 2016.
37. Schiemenz, G.P., The Sum of van der Waals Radii - A Pitfall in the Search for Bonding. *Z. Naturforsch.*, 62b, 235, 2007.
38. Jackson, N.E., Savoie, B.M., Kohlstedt, K.L., Olvera de la Cruz, M., Schatz, G.C., Chen, L.X., Ratner, M.A., Controlling Conformations of Conjugated Polymers and Small Molecules: The Role of Nonbonding Interactions. *J. Am. Chem. Soc.*, 135, 10475, 2013.
39. Wang, X., Tang, A., Chen, F., Zhou, E., Side-chain effect in ethenylene fused thiophene-vinylene-thiophene (ETVT) based photovoltaic polymers. *Polym.*, 167, 31, 2019.
40. Zhou, Y., Zhang, M., Ye, J., Liu, H., Wang, K., Yuan, Y., Du, Y.Q., Zhang, C., Zheng, C.J., Zhang, X.H., Efficient solution-processed red organic light-emitting diode based on an electron-donating building block of pyrrolo [3,2-*b*]pyrrole. *Org. Electron.*, 65, 110, 2019.

41. Ran, X.Q., Feng, J.K., Ren, A.M., Tian, W.Q., Zou, L.Y., Liu, Y.L., Sun, C.C., Theoretical studies on the electronic and optical properties of arene- versus fluoroarene-thiophene co-oligomer. *J. Phys. Org. Chem.*, 22, 680, 2009.
42. Xu, Z., Li, Y., Zhang, W., Yuan, S., Hao, L., Xu, T., Lu, X., DFT/TD-DFT study of novel T shaped phenothiazine-based organic dyes for dye-sensitized solar cells applications. *Spectrochim. Acta A: Mol. Biomol. Spectrosc.*, 212, 272, 2019.
43. Ostovan, A., Mahdavifar, Z., Bamdad, M., Evaluation of photovoltaic properties and effective conjugated length of DTTTD-based polymers as donor in BHJ solar cells; quantum chemical approach. *Polym.*, 126, 162, 2017.
44. Wanwong, S., Poe, A., Balaji, G., Thayumanavan, S., The effect of heteroatom conformation on optoelectronic properties of cyclopentadithiophene derivatives. *Org. Biomol. Chem.*, 12, 2474, 2014.
45. Chitpakdee, C., Namuangruk, S., Khongpracha, P., Jungsuttiwong, S., Tarsang, R., Sudyoadsuk, T., Promarak, V., Theoretical studies on electronic structures and photophysical properties of anthracene derivatives as hole-transporting materials for OLEDs. *Spectrochim. Acta A: Mol. Biomol. Spectrosc.*, 125, 36, 2014.
46. Min, J., Luponosov, Y.N., Khanin, D.A., Dmitryakov, P.V., Svidchenko, E.A., Peregudova, S.M., Grodd, L., Grigorian, S., Chvalun, S.N., Ponomarenko, S.A., Effects of bridging atom in donor units and nature of acceptor groups on physical and photovoltaic properties of A-π-D-π-A oligomers. *Org. Electron.*, 55, 42, 2018.
47. Ajayakumar, M.R. and Mukhopadhyay, P., Naphthalene-bis-hydrazimide: Radical anions and ICT as new bimodal probes for differential sensing of a library of amines. *Chem. Commun.*, 25, 3702, 2009.
48. Zhang, T., Wang, R., Ren, H., Chen, Z., Li, J., Deep blue light-emitting polymers with fluorinated backbone for enhanced color purity and efficiency. *Polym.*, 53, 1529, 2012.
49. Ramkumar, V. and Kannan, P., Novel heterocyclic based blue and green emissive materials for opto-electronics. *Opt. Mater.*, 46, 314, 2015.
50. Solodukhin, A.N., Luponosov, Y.N., Buzin, M.I., Peregudova, S.M., Svidchenko, E.A., Ponomarenkoa, S.A., Unsymmetrical donor–acceptor oligothiophenes end-capped with triphenylamine and phenyldicyanovinyl units. *Mendeleev Commun.*, 28, 415, 2018.
51. Tseng, C.Y., Taufany, F., Nachimuthu, S., Jiang, J.C., Liaw, D.J., Design strategies of metal free-organic sensitizers for dye sensitized solar cells: Role of donor and acceptor monomers. *Org. Electron.*, 15, 1205, 2014.
52. Willot, P., De Cremer, L., Koeckelberghs, G., The Use of Cyclopenta [2,1-*b*;3,4-*b′*]dithiophene Analogues for the Development of Low-Bandgap Materials. *Macromol. Chem. Phys.*, 213, 1216, 2012.
53. Chang, Y.C., Chen, Y.D., Chen, C.H., Wen, Y.S., Lin, J.T., Chen, H.Y., Kuo, M.Y., Chao, I., Crystal Engineering for π-π Stacking via Interaction between

Electron-Rich and Electron-Deficient Heteroaromatics. *J. Org. Chem.*, 73, 4608, 2008.

54. Riul, A., Jr., Taylor, D., Mills, C., Murphy, P., Langmuir and Langmuir-Blodgett (LB) films of 4-dicyanomethylene,4H-cyclopenta2,1–b,3,4–b′] dithiophene. *Thin Solid Films*, 366, 249, 2000.
55. Yang, L., Feng, J.K., Ren, A.M., Theoretical studies on the electronic and optical properties of two thiophene–fluorene based π-conjugated copolymers. *Polym.*, 46, 10970, 2005.
56. Ayachi, S., Bergaoui, S., Ben Khalifa, I., Haj Said, A., Chemek, M., Massuyeau, F., Wéry, J., Faulques, E., Alimi, K., On the photo-physical properties of soluble oligomer from anodic oxidation of chlorine substituted anisole (OPClAn). *Synth. Met.*, 166, 22, 2013.
57. Ayachi, S., Ghomrasni, S., Bouachrine, M., Hamidi, M., Alimi, K., Structure-property relationships of soluble poly(2,5-dibutoxyethoxy-1,4-phenylene-alt-2,5-thienylene) (PBuPT) for organic-optoelectronic devices. *J. Mol. Struct.*, 1036, 7, 2013.
58. Ayachi, S., Ghomrasni, S., Alimi, K., A combined experimental and theoretical study on vibrational and optical properties of copolymer incorporating thienylene-dioctyloxyphenyle-thienylene and bipyridine units. *J. Appl. Polym. Sci.*, 123, 5, 2684, 2012.
59. Svagan, A.J., Busko, D., Avlasevich, Y., Glasser, G., Baluschev, S., Landfester, K., Photon Energy Upconverting Nanopaper: A Bioinspired Oxygen Protection Strategy. *ACS Nano*, 8, 8198, 2014.
60. Wu, J.H., Chen, W.C., Liou, G.S., Triphenylamine-based luminogens and fluorescent polyimides: Effects of functional groups and substituents on photophysical behaviors. *Polym. Chem.*, 7, 1569, 2016.
61. Ulla, H., Garudachari, B., Satyanarayan, M., Umesh, G., Isloor, A., Blue organic light emitting materials: Synthesis and characterization of novel 1,8-naphthalimide derivatives. *Opt. Mater.*, 36, 704, 2014.
62. Li, W., Yao, L., Liu, H., Wang, Z., Zhang, S., Xiao, R., Zhang, H., Lu, P., Yang, B., Ma, Y., Highly efficient deep-blue OLED with an extraordinarily narrow FHWM of 35 nm and a *y* coordinate <0.05 based on a fully twisting donor-acceptor molecule. *J. Mater. Chem. C*, 2, 4733, 2014.
63. Jungsuttiwong, S., Tarsang, R., Surakhot, Y., Khunchalee, J., Sudyoadsuk, T., Promarak, V., Namuangruk, S., Theoretical study of α-fluorenyl oligothiophenes as color tunable emissive materials for highly efficient electroluminescent device. *Org. Electron.*, 13, 1836, 2012.
64. Ran, X.Q., Feng, J.K., Ren, A.M., Li, W.C., Zou, L.Y., Sun, C.C., Theoretical Study on Photophysical Properties of Ambipolar Spirobifluorene Derivatives as Efficient Blue-Light-Emitting Materials. *J. Phys. Chem. A*, 113, 7933, 2009.
65. Koese, M.E., Mitchell, W.J., Kopidakis, N., Chang, C.H., Shaheen, S.E., Kim, K., Rumbles, G., Theoretical studies on conjugated phenyl-cored thiophene

dendrimers for photovoltaic applications. *J. Am. Chem. Soc.*, 129, 14257, 2007.
66. Deng, W.Q., Sun, L., Huang, J.D., Chai, S., Wen, S.H., Han, K.L., Quantitative prediction of charge mobilities of π-stacked systems by first-principles simulation. *Nat. Protoc.*, 10, 632, 2015.
67. Shi, X., Yang, Y., Wang, L., Li, Y., Introducing asymmetry induced by benzene substitution in a rigid fused π spacer of D-π-A-type solar cells; A computational investigation. *J. Phys. Chem. C*, 123, 4007, 2019.
68. Kamath, A., Raghu, S., Devendrappa, H., Methyl blue dyed polyethylene oxide films: Optical and electrochemical characterization and application as a single layer organic device. *Opt. Mater.*, 51, 213, 2016.
69. Ayachi, S., Mabrouk, A., Bouachrine, M., Alimi, K., Photophysical Properties of Two New Donor-Acceptor Conjugated Copolymers and Their Model Compounds: Applications in Polymer Light Emitting Diodes (PLEDs) and Polymer Photovoltaic Cells (PPCs), Singh, J. (Ed.) (2012). Organic Light Emitting Devices. (1 ed.) Croatia: In-Tech. in: *Organic Light Emitting Devices (1 ed.)*, J. Singh (Ed.), 2012. In-Tech, Croatia. https://doi.org/10.5772/2589.
70. Benor, A., Takizawa, S., Pérez-Bolivar, C., A. Jr., P., Energy barrier, charge carrier balance, and performance improvement in organic light-emitting diodes. *Appl. Phys. Lett.*, 96, 243310, 2010.
71. Kim, S.Y., Hong, K., Choi, H.W., Kim, K.Y., Tak, Y.H., Lee, J.L., Correlation Between Charge Injection and Charge Balance in Organic Light Emitting Diodes Using LiF and IrOx Interlayers. *J. Electrochem. Soc.*, 156, 4, J57, 2009.
72. Zhou, N., Shao, X., Wang, S., Xiao, Y., Li, X., Two trans-1-(9-anthryl)-2-phenylethene derivatives as blue-green emitting materials for highly bright organic light-emitting diodes application. *Org. Electron.*, 50, 228, 2017.
73. Yang, B., Kim, S.K., Xu, H., Park, Y.I., Zhang, H., Gu, C., Shen, F., Wang, C., Liu, D., Liu, X., The Origin of the Improved Efficiency and Stability of Triphenylamine-Substituted Anthracene Derivatives for OLEDs: A Theoretical Investigation. *ChemPhysChem*, 9, 2601, 2008.
74. Zhao, Z., Nie, H., Ge, C., Cai, Y., Xiong, Y., Qi, J., Wu, W., Kwok, R.T.K., Gao, X., Qin, A., Lam, J.W.Y., Tang, B.Z., Furan Is Superior to Thiophene: A Furan-Cored AIEgen with Remarkable Chromism and OLED Performance. *Adv. Sci.*, 4, 1700005, 2017.

4

Conjugated Polymer Light-Emitting Diodes

Sapana Jadoun[1,2*] and Ufana Riaz[2†]

[1]School of Basic & Applied Sciences, Department of Chemistry, Lingaya's Vidyapeeth, Faridabad, Haryana, India
[2]Material Research Laboratory, Department of Chemistry, Jamia Millia Islamia, New Delhi, India

Abstract

The development of organic light-emitting diodes (OLEDs) has attracted considerable interest in innovations for our daily life and future. They have promising applications in flat panel displays, energy-saving, eco-friendly, thinner and smaller in size, lightweight, and cost-effective fabrication process. For the development of OLEDs, numerous conjugated polymers have been studied due to their semiconductor nature, which is being associated with pi bond delocalization along the backbone of the polymer chain. Conjugated polymers are highly recommended for electroluminescent devices such as OLEDs. The chapter comprises basic knowledge of polymer light-emitting diodes, their construction, device function, and use of conjugated polymers in blue, red, green, and multicolored light-emitting diodes along with challenges and their future perspectives.

Keywords: Conjugated polymers, organic electronics, OLEDs, PLEDs, band gap, blue, green, and red emission region

4.1 Introduction

Light-emitting diode (LED) is a semiconductor light source that emits light in response to an electric current [1]. In semiconductors, electron recombine with holes, releasing energy in the form of photons [2]. The color of the light (corresponding to the energy of the photons) is determined by the

**Corresponding author*: sjadoun022@gmail.com; sapana@lingaysuniversity.edu.in
†*Corresponding author*: uriaz@jmi.ac.in

Inamuddin, Rajender Boddula, Mohd Imran Ahamed and Abdullah M. Asiri (eds.) Polymers for Light-Emitting Devices and Displays, (77–98) © 2020 Scrivener Publishing LLC

energy required for electrons to cross the band gap of the semiconductor [3]. Nowadays, organic light-emitting diodes (OLEDs) have attracted considerable attention of researchers, scientists, and industrial aspirants and bring innovation to our lives due to its unique properties such as low-cost, wide viewing angle, flexibility, lightweight, low power consumption, and thin panel thickness [4–8]. OLED device architecture includes the use of an organic carbon-based film which is made up of some conjugated polymers, sandwiched between two charged electrodes. These electrodes are the transparent anode (generally glass) and the other is a metallic cathode [9]. OLEDs have smart features of ultra-thin and fine image-quality along with lightweight as self-emitting devices [10]. Nowadays, they are very familiar to general consumers and civilians due to their uses in smartphones like Samsung's Galaxy series as well as in the manufacturing of large TV displays [11]. OLED can be fabricated by small molecules or polymers. The former can be fabricated by vacuum deposition method and polymer organic light-emitting diodes (PLEDs) are generally developed by solution deposition method [12, 13]. The latter method has attracted wide attention due to its ease of preparation and low cost [14]. The process of fabrication of PLEDs can be carried out at room temperature so that organic flexible substrate can be a platform for device fabrication [15]. This technology allows polymers to fabricate full color PLED pixels by various printing methods or techniques like dye diffusion [16], laser-induced thermal transfer [17], screen printing [18], inkjet printing [19, 20], and patterning with the photolithographic process [21, 22]. In the comparison of small molecules OLEDs, these patterning methods make PLEDs more useful in the larger size and high-resolution display [23]. For PLEDs, three colors, i.e., red [24], green [25], and blue [26] used for full-color display for which various types of conjugated polymers have been proposed and synthesized.

Conjugated polymers are commonly referred as "Organic Macromolecules" having conjugation of the single and double bond in their polymer backbone chain that makes them intrinsically conducting in nature [27]. In these organic macromolecules, the overlapping of p-orbitals takes place resulting delocalized π-electrons system which is responsible for unique optical and electronic properties [28]. In conjugated polymeric chain, injection of hole and electron *via* π (bonding) and π^* (antibonding) can create a self-localized excited state by delocalization of charge in valence and conduction bands which on emission can decay radiatively and revealed outstanding potential in electroluminescent devices [29]. The generation of charge carriers such as polarons, bipolarons, and solitons are responsible for the conduction mechanism [30].

These conjugated macromolecules show changes in electrical and optical properties when doped or functionalized by chemical species [31]. A numerous variety of conjugated polymers have been developed and evaluated since 1977, since the discovery of electrical conductivity in polyacetylene discovered [32]. These materials include polyanilines [31, 33] and their derivatives such as poly(o-phenylenediamine) [34, 35], polyluminol [36, 37], and some other polymers like polypyrroles [38], polynaphthylamines [39–41], polythiophenes [42], polyanisidine [43], polycarbazoles [44–46], polyphenylenevinylene (PPV) [47], and their derivatives [48], polyphenylenes [49], polyfluorenes [50], and polyaryleneethynylenes [51]. These have attracted numerous industries and academics for a different type of optoelectronic applications. Conjugated polymers have drawn considerable attention due to their strong fluorescence emission [52], excellent hole transport ability [53], flexibility [54], and solution processability [55–57]. Numerous reports have been published regarding enhancement of processability of conjugated polymers by blending or composite formation [58–61].

Conjugated polymers which are processable by solubility are the most promising candidate for optoelectronic devices such as LEDs [62]. Enhanced solubility and processability of conjugated polymers help in preparing an emissive electroluminescent thin film of organic compound which emits light in response to an electric current in optoelectronic devices [63, 64]. Chemical structure and extended conjugation in such type of polymers provides controlled morphology, enhanced solubility and processability for using them in OLEDs *via* preparing an emissive electroluminescent thin film of organic compound which emits light in response to an electric current [65]. In addition, extended conjugation also provides a low band gap as a functional property for these applications [66].

4.2 History, Classification, and Characteristics of Polymer OLED Material

In Cambridge University, a group worked on conjugated polymers and found electroluminescence in these materials in 1989 [67]. In this finding, the device had a very short lifetime of some minute and very weak external quantum efficiency (EQE) of 0.1%. Some of the companies worked on the progress of polymer OLEDs material and optimization of these devices fastly such as Sumitomo Chemical Co. Ltd, Cambridge Display Technology (CDT), Dow Chemical and Covion, etc., and the result of the research was achieved as several tens of thousands of hours long lifetime and high EQE of about 5%~10%.

OLEDs material can be broadly classified into two groups on the basis of emissive materials which includes small molecules based OLEDs and polymer-based OLEDs [68, 69]. Small molecules-based OLED can be fabricated by the vacuum deposition method [70] and polymer-based OLED can be developed by the solution deposition method, Table 4.1 [71]. The polymer material can be conjugated or non-conjugated further for the fabrication of OLED while dendrimers play the role of intermediate of polymers and small molecules, Figure 4.1. The multilayer device structure is used in small molecule OLEDs, and so there is a requirement of more amendments in chemical structures in addition to solubility while polymeric material to form OLEDs is willingly soluble by ink solvents and makes it a useful material for wet process or printing process discussed earlier in Section 4.1.

Copolymerized or modified conjugated polymeric materials work without a multilayer structure in OLEDs [72]. In these devices, one of the most important parameters in emission color and this is controlled by incorporation of emissive moieties into the polymeric backbone, which is generally a wide band gap structure [73].

Table 4.1 Comparison of small molecule LED and polymer LED *(Reprinted from Chizu Sekine et al., 2014 Sci. Technol. Adv. Mater. 15, 034203).*

	Small Molecule	Polymer
	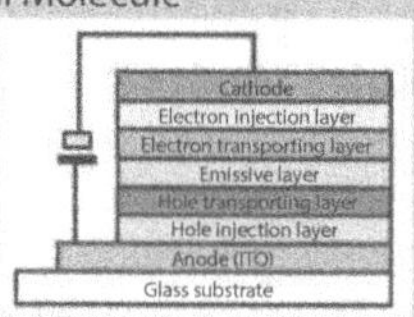	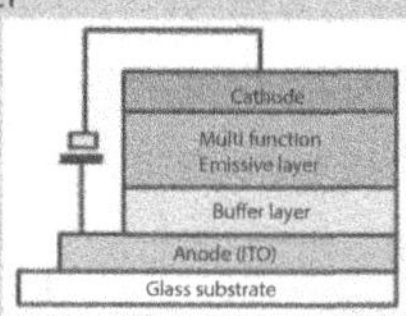
Process	Dry process (Vacuum evaporation) 	Wet process 
Patterning	Shadow mask	Printing (IJ etc.)
Structure	Complex layer structure (5-6) →Complex process	Simple layer structure (2-3) →Simple process, scalable
Material	Separated function	Integrated function
Issue	Layer structure complexity Difficulty in mask patterning	Performance (esp. LT) Patterning technology

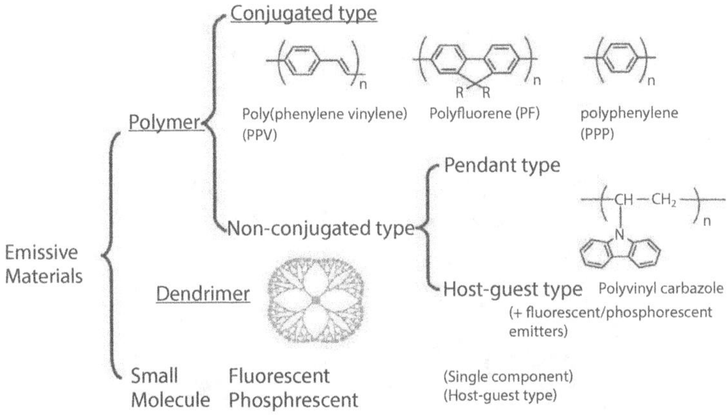

Figure 4.1 Classification of polymers according to emissive materials. (Reprinted from Chizu Sekine *et al.*, 2014 *Sci. Technol. Adv. Mater.* 15, 034203).

4.3 Polymer OLED Device Construction and Working

An OLED is a 100- to 500-nm-thick solid-state semiconductor device and about 200 times smaller to human hair [74]. The structure of OLED comprises a cathode, an emissive layer, a hole injection layer and an anode. In OLED there is an interlayer sandwiched in between cathode and anode electrodes [75]. The interlayer is composed of an organic layer in which the delocalization of pi electrons takes place and make this layer conductive in nature [76]. These materials are known as organic semiconductors due to the conductivity between conductors and insulators [77]. In organic semiconductors, their HOMO (highest occupied molecular orbitals) and LUMO (lowest unoccupied molecular orbitals) are considered as valence and conduction band present in the inorganic semiconductor [78]. An interlayer between hole injection and emissive layer makes an improvement in the device structure as well as in its performance due to cross-liking properties of the interlayer [79]. In polymer OLEDs, an interlayer is responsible for the enhancement in the emission efficiency. This layer is placed in between the hole injection layer and the emission layer. The function of this layer is hole-transporting as well as electron and exciton blocking. In summary, this layer is responsible for the separation of the hole injection layer and emission zone, as well as due to its electron blocking property, it accumulates electrons at the interface of the interlayer [80]. However, modern

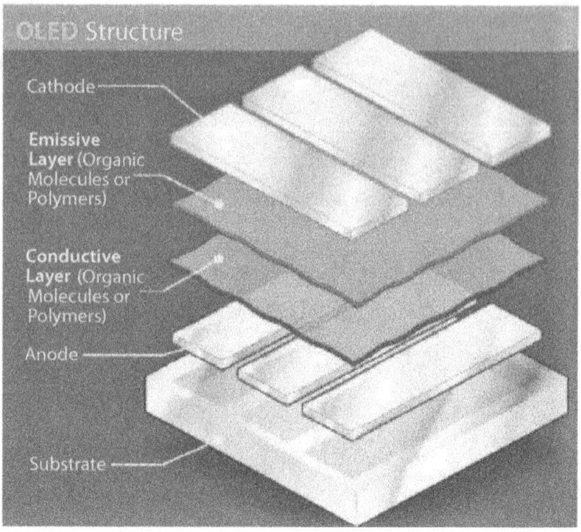

Figure 4.2 The basic OLED device architecture. Source: https://electronics.howstuffworks.com/oled1.htm.

OLEDs are fabricated by a very simple bilayer structure having a layer of emissive material and other is conductive material [9]. The basic OLED structure is shown in Figure 4.2.

Polymer OLEDs device performance is based on its emission efficiency and its lifetime. Its emission efficiency was discussed by Tokito and his labmates [81] and expressed as:

$$\Psi = \Upsilon.n_{e-h.}\Psi_{ph}.(1 - Q) \tag{4.1}$$

In the above equation, Ψ stands for electroluminescence efficiency, Υ stands for carrier balance for electron and hole, Ψ_{ph} stands for photoluminescence efficiency, and Q stands for quenching factor. It can be stated by the equation that by improving carrier balance for electron and hole, by enhancing photoluminescence efficiency and recombination rate or by suppressing cathode quenching, device performance can be improved in terms of electroluminescence efficiency.

4.4 Blue Light-Emitting Diodes

Various types of conjugated polymers have been proposed for blue light-emitting diodes which were highly soluble and reveled strong blue fluorescence or emit blue light in 350–440 nm range without any excimer formation on longer wavelength [82–85]. For blue OLED, some phenylene chains have been polymerized by meta linkage and displayed 4.69 cd A^{-1} efficiency [86].

In this regard, carbazole is a rigid plane biphenyl monomer having a wide band gap along with its high flexibility and high luminescent efficiency which helps in modification of the molecule skeleton for its use as a key chromophore in OLEDs [87, 88]. These are extensively used in the development of highly competent blue light-emitting diodes as a key role of host and charge-transporting materials [89]. By derivatives of carbazole, Morin [90] and his coworkers developed light-emitting diodes by electron and hole transport molecules, i.e., 2-(4-biphenylyl)-5-(4-tert-butylphenyl)-1,3,4-oxadiazole and N,Ń-diphenyl-N,Ń-bis(3-methylphenyl)-1,1′-biphenyl-4,4′-diamine, respectively, which showed electroluminescence in 424–432 nm wavelength (blue region) with Al and indium tin oxide as the electrodes. They also developed 2,7 carbazole based blue light-emitting copolymers with some other highly aromatic comonomers by Yamamoto or Suzuki cross-coupling reactions. These high molar mass modified with substituent species were highly water-soluble and amorphous in nature as well as these showed good redox properties and thermal stability for their application in layer blue light-emitting diodes [91].

Numerous works have also been reported on polyfluorenes that is also a blue light-emitting material [92]. Although, polyfluorenes exhibited promising features, literature suggested the tendency of forming excimers and aggregation which caused shifting of emission spectra towards longer wavelengths or showed bathochromic shift and decreases fluorescence quantum yields [93]. To remove such type of problems, scientists and research have done a number of studies and done functionalization by bulky side substituents or copolymerization with other monomers or by introduction of lithe spacers amid the chromophores [94–98]. Oligo(9, 9-dihexyl-2,7-fluorene ethynylene)s were synthesized by Tsutsui *et al.* [99] for extraordinary photoluminescence having high quantum yields as well as high oxidative and thermal stability for their use in OLED application. Chan *et al.* [100] developed poly(9,9-dihexyl-2,7-dibenzosilole) by copolymerization of monomers dibromo and bis(boronate) which showed better efficiency in a single layer light-emitting device in comparison of polyfluorene.

4.5 Green Light-Emitting Diodes

Seino and his labmates developed a green organic light-emitting device (OLED) whose power efficiency was too high (100 lm W^{-1}). This was achieved *via* energy transfer from an exciplex and it was 1.6-times higher than green thermally activated delayed fluorescence (TADF) OLEDs [101]. Tanaka *et al.* [102] developed highly efficient green organic light-emitting diodes (OLEDs) using a conjugated polymer dipyridylphenyl moieties and fac tris(2-phenylpyridine)iridium, $Ir(ppy)_3$ as a phosphorent emitter and yield 29% at 100 cd/m^2 and 26% at 1000 cd/m^2 external quantum efficiencies which were too high and lead to 133 lm/W and 107 lm/W ultra high power efficiencies at 100 cd/m^2 and 1000 cd/m^2 respectively. A highly efficient green OLED was fabricated by using 9,9-diarylfluorene-terminated 2,1,3-benzothiadiazole derivative in Ku's lab [103] which revealed EQE (ηext) of 3.7% and maximum brightness at 168,000 cd m^{-2}. Zhu and his coworkers [104] proposed green and blue-green phosphorescent OLEDs based on tetraphenylimidodiphosphinate and iridium complex. These were highly efficient, due to shorter excited stated lifetime and better carrier transport ability along with good electroluminescence performance. 3′,5′-di(carbazol-9-yl)-[1,1′-biphenyl]-3,5-dicarbonitrile was used as a host material for green thermally activated delayed phosphorescent and fluorescent OLEDs by Cho *et al.* [105]. This materials showed high quantum efficiency close to 25%. Chen [106] and his coworkers achieved highly efficient green phosphorescent OLEDs by 1,3,5-Triazine derivatives as new electron transport-type host materials. A dicarbazole-triazine hybrid bipolar host material was developed in which dicarbazole moiety is electron donor and triazene is electron acceptor. This device was highly efficient green phosphorescent OLEDs having maximum efficiency of up to 20.0% [107].

4.6 Red Light-Emitting Diodes

Red OLEDs have an emitting layer comprising of one material during their initial development. Okada *et al.* [108] developed a highly efficient complex for red OLED by using 1-phenylisoquinoline. The complex revealed emission peak in the red region, i.e., 598–635 nm, the quantum yield of 0.17–0.32 and lifetime of those complexes ranged 1.07–2.34 μs, respectively. Jung and his labmates [109] reported red organic light-emitting devices (OLEDs) by incorporating various donor-acceptor moieties in polymeric backbone which showed emission at 637–677 nm and highest brightness of 8,300 cd m^{-2} with 4.46% maximum EQE at 7 V, the current efficiency and the power efficiency was found 3.43 cd A^{-1} and 1.64 lm W^{-1}, respectively. 3-2-(3,3-dicyanomethylene-5,5-dimethyl-1-cyclohexylidene) vinyl-N-naphthyl-carbazole(NCz-2CN)and3,6-bis(2-(3,3-dicyanomethylene-5,5-dimethyl-1-cyclohexylidene)vinyl-N-phenyl-carbazole (PCz-4CN) which are derivatives of carbazole were used as donor-Л-acceptor for obtaining red OLEDs by Fu *et al.* [110]. They emit in red region 630–666 nm and maximum luminance of 4,110 cd/m^2 attained at 15 V. The current efficiency was found 2.09 cd/A while maximum luminous efficiency was found at 0.49 lm/W.

4.7 Multicolor Light-Emitting Diodes

Bouillud *et al.* [111] presented conjugated polymers based on fluorene with the incorporation of thiophene and phenylene moieties in their copolymer having outstanding properties of tunability of the electroluminescent properties. By changing the composition of comonomers, it has the ability to change their emission from blue to green or yellow. They also suggested hole injection and hole transport phenomena by the addition of an insulating buffer layer as well as the hole transporting material into the material. This electroluminescence efficiency was increased from 4.5 to 125 cd/m^2 by incorporation of N,N'-diphenyl-N,N'-bis(3-methylphenyl)-1,1-biphenyl-4,4'-diamine as the hole transporting layer and LiF as an insulating buffer layer.

Wong and his coworkers [112] studied the use of conjugated and non-conjugated polymeric backbone in terms of synthesis, physical properties, sample quality, and device performances for blue, green, and red OLED materials (Figure 4.3), while Chou *et al.* [113] prepared universal bipolar host by phosphine oxide and two carbazole groups for blue, green, and red phosphorescent OLEDs which was highly efficient.

4.8 Advantages of OLEDs over Other Liquid Crystal Display

OLEDs revealed many advantages with other conventional liquid crystal displays (LCDs) such as polymer OLEDs that show different colors such as red, green, blue (RGB) as well as white too which is important to many applications [114, 115]. OLEDs are biodegradable. An OLED display can be many times lighter, more flexible, and thinner compared to LCD, as well as it can display deep black levels because it works without a backlight [116]. These are much brighter as compared to LCD.

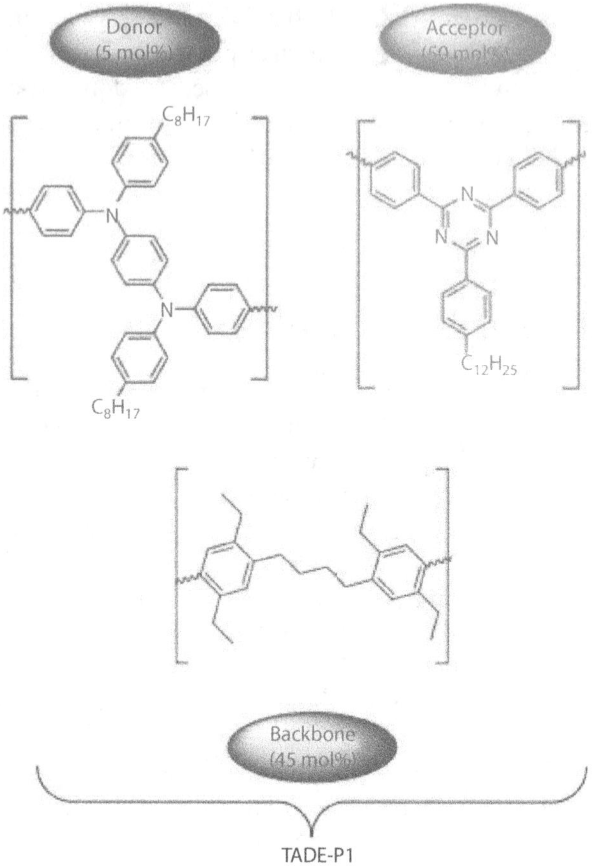

Figure 4.3 Functionalized conjugated polymer having donor (5 mol.% triphenylamine), acceptor (50 mol.% 1,3,5-triazine) and backbone (45 mol.%) with an insulating n-butyl link (*Reprinted from Michael Y. Wong, 2017, Journal of Electronic Materials. 46, 6246–6281).*

These have the active light-emitting capability as well as the OLEDs are driven by the direct current with 10-V approximate voltage [117]. They are also highly efficient in a light color so that full-color display can be realized easily and in place of early vacuum display devices, these are all-solid-state devices [118]. In addition, they have a lithe information display technology with low energy consumption and high light-emitting efficiency [119]. These devices have outstanding light-emitting properties and temperature features which least affected by temperature fluctuations and faster response time. They are easy to prepare at a very low cost [120].

4.9 Applications of OLEDs

In 2003, Kodak has used OLED in the fabrication of its first digital camera [121]. OLEDs are widely being used in the latest smartphones, DVD players, digital cameras, digital watches, etc., due to its high flexibility and foldability which helps in saving space and weight [122]. After this, numerous companies have used this technology in their products including Nokia [123], Samsung [124], etc. Since the year 2019, OLED screen is readily used in many mobile phones including Honor view 10, Nokia 6.1 plus, Samsung Galaxy A8+, Vivo V15 Pro, Gionee S11S, OnePlus 7 Pro, etc., and in TV such as Sony Bravia Android Smart OLED TV KD-55A8G, LG Smart TV OLED55C8PTA, METZ Android OLED TV M55S9A, etc. OLEDs have also applicable in high-end television systems, flat-panel displays, computer monitors, some pocket-size systems like digital cameras, portable gaming consoles, media players, android phones, and mini-screens [125–127].

These are also widely used in various multiple-input output wireless optical channels [128]. Some other applications of OLED are shown in Figure 4.4.

4.10 Challenges and Future Possibilities

OLED has a wide area of applications in different areas, but still, there are many challenges involved in this technology such as its is costly in

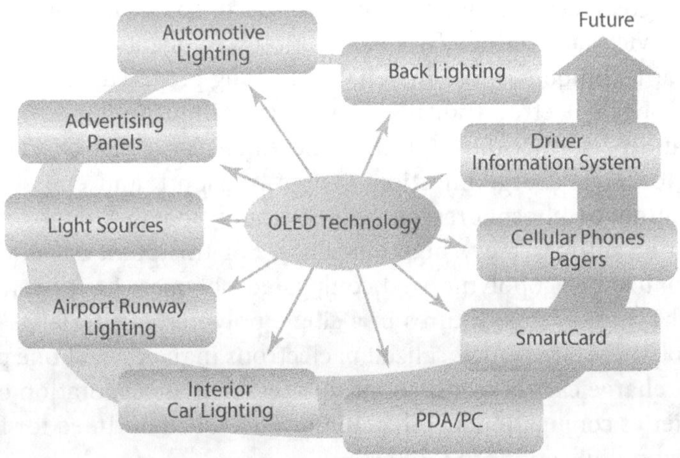

Figure 4.4 Applications of OLED. Source: https://www.elprocus.com/oled-display-technology-architecture-applications/.

comparison to LCD or LEDs. There is still a lack of a broad range of products incorporating OLEDs. Light efficiency is still low and compared to other display devices, their lifetime is shorter. Red, green, and white LEDs give longer lifetime but the blue OLED revealed a limited lifetime of 1.6 years. Technology is highly sensitive to water so can be damaged by moisture. When OLEDs compared with LCD in direct sunlight, these show worse scenario. In the future, these problems can be resolved by using various types of technologies and conjugated polymers. OLEDs can be used for curved display in the future as well as transparent displays entrenched in windows. The property of flexibility of OLEDs maybe uses in roll to roll manufacturing process which allows flexible display architecture. Flexible, stretching, and self-healing materials need to incorporate for OLED device fabrication that can also improve barrier layers to protect OLEDs from moisture and oxygen. By removal of triplet exciton from conjugated polymer backbone could expressively enhance the device stability, specifically at the initial stage of pouring, i.e., *via* the introduction of additive which quenches lead the high efficiency and device lifetime. For this, various features of polymers could be introduced in single polymer *via* its copolymerization, doping, medication, or *via* functionalization of monomers then copolymerization.

4.11 Conclusion

Great progress has been done in the synthesis of conjugated polymers for OLED device fabrication. By modification, copolymerization, doping, and other techniques, those materials are being prepared which helps in applications of electron transport materials to improve the device fabrication and performance of OLEDs. The efficiency of the OLED device is dependent on charge injection, charge transport, and emission. The combination of phosphorescent emitters and conjugated polymers is very important to achieve high efficiency. For this point of view, many functionalized or copolymerized conjugated polymers have been investigated having various features in a single polymeric material as conjugated polymers having delocalized pi electrons in their backbone possess brilliant charge carrier transport properties so by incorporation of various moieties conjugation length could be controlled or tuned for further improved modifications in OLEDs.

References

1. Lin, Y.-S., Liu, R.-S., Su, H.-Y., White light-emitting device, United States Lite-On Technology Corporation (Taipei, TW), 7026656, 2006, http://www.freepatentsonline.com/y2005/0247951.html
2. Sundaram, S.K. and Mazur, E., Inducing and Probing Non-Thermal Transitions in Semiconductors Using Femtosecond Laser Pulses. *Nat. Mater.*, 1, 4, 217, 2002.
3. Jang, J.S., Kim, H.G., Lee, J.S., Heterojunction Semiconductors: A Strategy to Develop Efficient Photocatalytic Materials for Visible Light Water Splitting. *Catal. Today*, 185, 1, 270–77, 2012.
4. Kalyani, N.T., Swart, H.C., Dhoble, S.J., *Principles and Applications of Organic Light Emitting Diodes (OLEDs)*, Woodhead Publishing, Sawston, Cambridge, 2017.
5. Chatterjee, T. and Wong, K.-T., Perspective on Host Materials for Thermally Activated Delayed Fluorescence Organic Light Emitting Diodes. *Adv. Opt. Mater.*, 7, 1, 1800565, 2019.
6. Ashok Kumar, S., Shankar, J.S., Periyasamy, B.K., Nayak, S.K., Device Engineering Aspects of Organic Light-Emitting Diodes (OLEDs). *Polym.-Plast. Technol. Mater.*, 58, 1–28, 2019.
7. Zhong, C., Duan, C., Huang, F., Wu, H., Cao, Y., Materials and Devices toward Fully Solution Processable Organic Light-Emitting Diodes. *Chem. Mater.*, 23, 3, 326–40, 2010.
8. Zhou, G., Wong, W.-Y., Suo, S., Recent Progress and Current Challenges in Phosphorescent White Organic Light-Emitting Diodes (WOLEDs). *J. Photochem. Photobiol. C: Photochem. Rev.*, 11, 4, 133–56, 2010.
9. Geffroy, B., Le Roy, P., Prat, C., Organic Light-emitting Diode (OLED) Technology: Materials, Devices and Display Technologies. *Polym. Int.*, 55, 6, 572–82, 2006.
10. Templier, F., *OLED Microdisplays: Technology and Applications*, Wiley Online Library, Hoboken, New Jersey, 2014.
11. Kopola P., Tuomikoski M.,Suhonen R., Maaninen A., Gravure printed organic light emitting diodes for lighting applications, *Thin Solid Films*, 517, 5757–5762, 2009.
12. Lee, S.Y., Yasuda, T., Komiyama, H., Lee, J., Adachi, C., Thermally Activated Delayed Fluorescence Polymers for Efficient Solution-Processed Organic Light-Emitting Diodes. *Adv. Mater.*, 28, 21, 4019–24, 2016.
13. Liu, Y., Li, C., Ren, Z., Yan, S., Bryce, M.R., All-Organic Thermally Activated Delayed Fluorescence Materials for Organic Light-Emitting Diodes. *Nat. Rev. Mater.*, 3, 4, 18020, 2018.
14. Lakshmanan, R. and Kiruba Daniel, S.C.G., Engineered Nanomaterials for Organic Light-Emitting Diodes (OLEDs), in: *Handbook of Nanomaterials for Industrial Applications*, pp. 312–23, Elsevier, Oxford, United Kingdom, 2018.

15. Forrest, S.R., Lee, K., Devices combining thin film inorganic LEDs with organic LEDs and fabrication thereof, United States, The Regents of the University of Michigan (Ann Arbor, MI, US) 10229958, 2019, http://www.freepatentsonline.com/10229958.html
16. Wang, R., Zhang, D., Xiong, Y., Zhou, X., Liu, C., Chen, W., Wu, W., Zhou, L., Xu, M., Wang, L., TFT-Directed Electroplating of RGB Luminescent Films without a Vacuum or Mask toward a Full-Color AMOLED Pixel Matrix. *ACS Appl. Mater. Interfaces*, 10, 21, 17519–25, 2018.
17. Delaporte, P., Alloncle, A.-P., Lippert, T., Laser Printing of Electronic Materials, in: *Laser Printing of Functional Materials: 3D Microfabrication, Electronics and Biomedicine*, pp. 271–97, 2018.
18. Niu, Q., Shao, Y., Xu, W., Wang, L., Han, S., Liu, N., Peng, J., Cao, Y., Wang, J., Full Color and Monochrome Passive-Matrix Polymer Light-Emitting Diodes Flat Panel Displays Made with Solution Processes. *Org. Electron.*, 9, 1, 95–100, 2008.
19. MacPherson, C., Anzlowar, M., Innocenzo, J., Kolosov, D., Lehr, W., O'Regan, M., Sant, P., Stainer, M., Sysavat, S., Venkatesh, S., 39.4: Development of Full Color Passive PLED Displays by Inkjet Printing, in: *SID Symposium Digest of Technical Papers*, vol. 34, pp. 1191–93, 2003, Wiley Online Library.
20. De Gans, B.-J., Duineveld, P.C., Schubert, U.S., Inkjet Printing of Polymers: State of the Art and Future Developments. *Adv. Mater.*, 16, 3, 203–13, 2004.
21. Lee, S.T., Lee, J.Y., Kim, M.H., Suh, M.C., Kang, T.M., Choi, Y.J., Park, J.Y., Kwon, J.H., Chung, H.K., Baetzold, J., 21.3: A New Patterning Method for Full-Color Polymer Light-Emitting Devices: Laser Induced Thermal Imaging (LITI), in: *SID Symposium Digest of Technical Papers*, vol. 33, pp. 784–87, 2002, Wiley Online Library.
22. Tachikawa, T., Itoh, N., Handa, S., Aoki, D., Miyake, T., 34.3: High-Resolution Full-Color Polymer Light-Emitting Devices Using Photolithography, in: *SID Symposium Digest of Technical Papers*, vol. 36, pp. 1280–83, 2005, Wiley Online Library.
23. Sekine, C., Tsubata, Y., Yamada, T., Kitano, M., Doi, S., Recent Progress of High Performance Polymer OLED and OPV Materials for Organic Printed Electronics. *Sci. Technol. Adv. Mater.*, 15, 3, 34203, 2014.
24. Cao, D., Liu, Q., Zeng, W., Han, S., Peng, J., Liu, S., Synthesis and Characterization of Novel Red-emitting Alternating Copolymers Based on Fluorene and Diketopyrrolopyrrole Derivatives. *J. Polym. Sci. Part A: Polym. Chem.*, 44, 8, 2395–2405, 2006.
25. Montilla, F. and Mallavia, R., On the Origin of Green Emission Bands in Fluorene-Based Conjugated Polymers. *Adv. Funct. Mater.*, 17, 1, 71–78, 2007.
26. Li, Z., Ye, S., Liu, Y., Yu, G., Wu, W., Qin, J., Li, Z., New Hyperbranched Conjugated Polymers Containing Hexaphenylbenzene and Oxadiazole Units: Convenient Synthesis and Efficient Deep Blue Emitters for PLEDs Application. *J. Phys. Chem. B*, 114, 28, 9101–8, 2010.

27. Pecher, J. and Mecking, S., Nanoparticles of Conjugated Polymers. *Chem. Rev.*, 110, 10, 6260–79, 2010.
28. Petty, M., *Molecular Electronics*, Springer, Salmon Tower Building, New York City, 2007.
29. Brédas, J.-L., Beljonne, D., Coropceanu, V., Cornil, J., Charge-Transfer and Energy-Transfer Processes in π-Conjugated Oligomers and Polymers: A Molecular Picture. *Chem. Rev.*, 104, 11, 4971–5004, 2004.
30. Lu, Y., Wang, J.-Y., Pei, J., Strategies To Enhance the Conductivity of N-Type Polymer Thermoelectric Materials. *Chem. Mater.*, 31, 17, 6412–23, 2019.
31. Liao, G., Li, Q., Xu, Z., The Chemical Modification of Polyaniline with Enhanced Properties: A Review. *Prog. Org. Coat.*, 126, 35–43, 2019.
32. Harun, M.H., Saion, E., Kassim, A., Yahya, N., Mahmud, E., Conjugated Conducting Polymers: A Brief Overview. *UCSI Acad. J.: Journal for the Advancement of Science & Arts*, 2, 63–68, 2007.
33. Jadoun, S. and Riaz, U., A Review on the Chemical and Electrochemical Copolymerization of Conducting Monomers: Recent Advancements and Future Prospects. *Polym.-Plast. Technol. Mater.*, 59, 1–21, 2019.
34. Riaz, U., Ashraf, S.M., Jadoun, S., Budhiraja, V., Kumar, P., Spectroscopic and Biophysical Interaction Studies of Water-Soluble Dye Modified Poly (o-Phenylenediamine) for Its Potential Application in BSA Detection and Bioimaging. *Sci. Rep.*, 9, 1, 8544, 2019.
35. Riaz, U., Jadoun, S., Kumar, P., Arish, M., Rub, A., Ashraf, S.M., Influence of Luminol Doping of Poly(o-Phenylenediamine) on the Spectral, Morphological and Fluorescent Properties: A Potential Biomarker for Leishmania Parasite. *ACS Appl. Mater. Interfaces*, 9, 33159–33168, 2017, acsami.7b10325, https://doi.org/10.1021/acsami.7b10325.
36. Riaz, U., Jadoun, S., Kumar, P., Kumar, R., Yadav, N., Microwave-Assisted Facile Synthesis of Poly (Luminol-Co-Phenylenediamine) Copolymers and Their Potential Application in Biomedical Imaging. *RSC Adv.*, 8, 65, 37165–75, 2018.
37. Jadoun, S., Ashraf, S.M., Riaz, U., Microwave-assisted Synthesis of Copolymers of Luminol with Anisidine: Effect on Spectral, Thermal and Fluorescence Characteristics. *Polym. Adv. Technol.*, 29, 2, 1007–17, 2018.
38. Jadoun, S., Biswal, L., Riaz, U., Tuning the Optical Properties of Poly(o-Phenylenediamine-Co-Pyrrole) *via* Template Mediated Copolymerization. *Des. Monomers Polym.*, 21, 1, 75–81, 2018, https://doi.org/10.1080/15685551.2018.1459078.
39. Jadoun, S., Verma, A., Ashraf, S.M., Riaz, U., A Short Review on the Synthesis, Characterization, and Application Studies of Poly(1-Naphthylamine): A Seldom Explored Polyaniline Derivative. *Colloid Polym. Sci.*, 295, 1443–1453, 2017, https://doi.org/10.1007/s00396-017-4129-2.
40. Riaz, U., Ashraf, S.M., Aleem, S., Budhiraja, V., Jadoun, S., Microwave-Assisted Green Synthesis of Some Nanoconjugated Copolymers: Characterisation and Fluorescence Quenching Studies with Bovine Serum Albumin. *New J. Chem.*, 40, 5, 4643–4653, 2016, https://doi.org/10.1039/C5NJ02513C.

41. Jadoun, S., Sharma, V., Ashraf, S.M., Riaz, U., Sonolytic Doping of Poly(1-Naphthylamine) with Luminol: Influence on Spectral, Morphological and Fluorescent Characteristics. *Colloid Polym. Sci.*, 295, 4, 715–724, 2017, https://doi.org/10.1007/s00396-017-4055-3.
42. Perepichka, I.F., Perepichka, D.F., Meng, H., Wudl, F., Light-emitting Polythiophenes. *Adv. Mater.*, 17, 19, 2281–2305, 2005.
43. Jadoun, S., Verma, A., Riaz, U., Luminol Modified Polycarbazole and Poly (o-Anisidine): Theoretical Insights Compared with Experimental Data. *Spectrochim. Acta Part A: Mol. Biomol. Spectrosc.*, 204, 64–72, 2018.
44. Riaz, U., Ashraf, S.M., Fatima, T., Jadoun, S., Tuning the Spectral, Morphological and Photophysical Properties of Sonochemically Synthesized Poly(Carbazole) Using Acid Orange, Fluorescein and Rhodamine 6G. *Spectrochim. Acta - Part A: Mol. Biomol. Spectrosc.*, 173, 986–993, 2017, https://doi.org/10.1016/j.saa.2016.11.003.
45. Jadoun, S., Ashraf, S.M., Riaz, U., Tuning the Spectral, Thermal and Fluorescent Properties of Conjugated Polymers: Via Random Copolymerization of Hole Transporting Monomers. *RSC Adv.*, 7, 52, 32757–32768, 2017, https://doi.org/10.1039/c7ra04662f.
46. Riaz, U., Ashraf, S.M., Kumar Saroj, S., Zeeshan, M., Jadoun, S., Microwave-Assisted Solid State Intercalation of Rhodamine B and Polycarbazole in Bentonite Clay Interlayer Space: Structural Characterization and Photophysics of Double Intercalation. *RSC Adv.*, 6, 41, 2016, https://doi.org/10.1039/c5ra27387k.
47. Pande, N., Chauhan, N.P.S., Pande-Shrama, D., Mozafari, M., Chapter 4 - Conjugated Polymers Having Semiconducting Properties, in: B T - *Advanced Functional Polymers for Biomedical Applications*, M. Mozafari and N.P.S. Chauhan (Eds.), pp. 65–82, Elsevier, Oxford, United Kingdom, 2019, https://doi.org/https://doi.org/10.1016/B978-0-12-816349-8.00004-7.
48. Li, J., Hou, Y.-x., Wang, Y.-y., Ye, F., Zhong, G.-y., The Piezoresistance of a Device with Polyphenylenevinylene Derivative PSS-PPV Film. *Microsyst. Technol.*, 25, 2, 423–30, 2019, https://doi.org/10.1007/s00542-018-3973-4.
49. Zhang, X., Higashihara, T., Ueda, M., Wang, L., Polyphenylenes and the Related Copolymer Membranes for Electrochemical Device Applications. *Polym. Chem.*, 5, 21, 6121–41, 2014, https://doi.org/10.1039/C4PY00898G.
50. Javier, A.E., Varshney, S.R., McCullough, R.D., Chain-Growth Synthesis of Polyfluorenes with Low Polydispersities, Block Copolymers of Fluorene, and End-Capped Polyfluorenes: Toward New Optoelectronic Materials. *Macromolecules*, 43, 7, 3233–37, 2010, https://doi.org/10.1021/ma100519g.
51. Wu, W., Xin, S., Xu, Z., Ye, C., Qin, J., Li, Z., Main-Chain Second-Order Nonlinear Optical Polyaryleneethynylenes Containing Isolation Chromophores: Enhanced Nonlinear Optical Properties, Improved Optical Transparency and Stability. *Polym. Chem.*, 4, 11, 3196–3203, 2013, https://doi.org/10.1039/C3PY00257H.

52. Jin, Y.-J., Bae, J.-E., Cho, K.-S., Lee, W.-E., Hwang, D.-Y., Kwak, G., Room Temperature Fluorescent Conjugated Polymer Gums. *Adv. Funct. Mater.*, 24, 13, 1928–37, 2014, https://doi.org/10.1002/adfm.201302829.
53. Xu, Y., Bu, T., Li, M., Qin, T., Yin, C., Wang, N., Li, R. *et al.*, Non-Conjugated Polymer as an Efficient Dopant-Free Hole-Transporting Material for Perovskite Solar Cells. *ChemSusChem*, 10, 12, 2578–84, 2017, https://doi.org/10.1002/cssc.201700584.
54. Han, A.-R., Dutta, G.K., Lee, J., Lee, H.R., Lee, S.M., Ahn, H., Shin, T.J., Oh, J.H., Yang, C., ε-Branched Flexible Side Chain Substituted Diketopyrrolopyrrole-Containing Polymers Designed for High Hole and Electron Mobilities. *Adv. Funct. Mater.*, 25, 2, 247–54, 2015, https://doi.org/10.1002/adfm.201403020.
55. Li, G., Chang, W.-H., Yang, Y., Low-Bandgap Conjugated Polymers Enabling Solution-Processable Tandem Solar Cells. *Nat. Rev. Mater.*, 2, 8, 17043, 2017, https://doi.org/10.1038/natrevmats.2017.43.
56. Mei, J. and Bao, Z., Side Chain Engineering in Solution-Processable Conjugated Polymers. *Chem. Mater.*, 26, 1, 604–15, 2014, https://doi.org/10.1021/cm4020805.
57. Vohra, V., Giovanella, U., Tubino, R., Murata, H., Botta, C., Electroluminescence from Conjugated Polymer Electrospun Nanofibers in Solution Processable Organic Light-Emitting Diodes. *ACS Nano*, 5, 7, 5572–78, 2011, https://doi.org/10.1021/nn201029c.
58. Wong, W.-Y., Wang, X.-Z., He, Z., Djuri, A.B., Yip, C.-T., Cheung, K.-Y., Wang, H., Mak, C.S.K., Chan, W.-K., Metallated Conjugated Polymers as a New Avenue towards High-Efficiency Polymer Solar Cells, in: *Materials for Sustainable Energy*, pp. 51–57, Co-Published with Macmillan Publishers Ltd, UK, 2010, https://doi.org/doi:10.1142/9789814317665_0007.
59. Li, G., Shrotriya, V., Huang, J., Yao, Y., Moriarty, T., Emery, K., Yang, Y., High-Efficiency Solution Processable Polymer Photovoltaic Cells by Self-Organization of Polymer Blends, in: *Materials for Sustainable Energy*, pp. 80–84, Co-Published with Macmillan Publishers Ltd, UK, 2010, https://doi.org/doi:10.1142/9789814317665_0012.
60. Fu, B., Baltazar, J., Sankar, A.R., Chu, P.-H., Zhang, S., Collard, D.M., Reichmanis, E., Enhancing Field-Effect Mobility of Conjugated Polymers Through Rational Design of Branched Side Chains. *Adv. Funct. Mater.*, 24, 24, 3734–44, 2014, https://doi.org/10.1002/adfm.201304231.
61. Mike, J.F. and Lutkenhaus, J.L., Recent Advances in Conjugated Polymer Energy Storage. *J. Polym. Sci. Part B: Polym. Phys.*, 51, 7, 468–80, 2013, https://doi.org/10.1002/polb.23256.
62. Huang, F., Wu, H., Cao, Y., Water/Alcohol Soluble Conjugated Polymers as Highly Efficient Electron Transporting/Injection Layer in Optoelectronic Devices. *Chem. Soc. Rev.*, 39, 7, 2500–2521, 2010.
63. Zhu, X.-H., Peng, J., Cao, Y., Roncali, J., Solution-Processable Single-Material Molecular Emitters for Organic Light-Emitting Devices. *Chem. Soc. Rev.*, 40, 7, 3509–24, 2011.

64. Duan, L., Hou, L., Lee, T.-W., Qiao, J., Zhang, D., Dong, G., Wang, L., Qiu, Y., Solution Processable Small Molecules for Organic Light-Emitting Diodes. *J. Mater. Chem.*, 20, 31, 6392–6407, 2010.
65. Zhao, X. and Zhan, X., Electron Transporting Semiconducting Polymers in Organic Electronics. *Chem. Soc. Rev.*, 40, 7, 3728–43, 2011.
66. Liu, X., Sun, Y., Perez, L.A., Wen, W., Toney, M.F., Heeger, A.J., Bazan, G.C., Narrow-Band-Gap Conjugated Chromophores with Extended Molecular Lengths. *J. Am. Chem. Soc.*, 134, 51, 20609–12, 2012.
67. Burroughes, J.H., Bradley, D.D.C., Brown, A.R., Marks, R.N., Mackay, K., Friend, R.H., Burns, P.L., Holmes, A.B., Light-Emitting Diodes Based on Conjugated Polymers. *Nature*, 347, 6293, 539, 1990.
68. Martens, H.C.F., Huiberts, J.N., Blom, P.W.M., Simultaneous Measurement of Electron and Hole Mobilities in Polymer Light-Emitting Diodes. *Appl. Phys. Lett.*, 77, 12, 1852–54, 2000.
69. Rehmann, N., Hertel, D., Meerholz, K., Becker, H., Heun, S., Highly Efficient Solution-Processed Phosphorescent Multilayer Organic Light-Emitting Diodes Based on Small-Molecule Hosts. *Appl. Phys. Lett.*, 91, 10, 103507, 2007.
70. Lee, T.-W., Noh, T., Shin, H.-W., Kwon, O., Park, J.-J., Choi, B.-K., Kim, M.-S., Shin, D.W., Kim, Y.-R., Characteristics of Solution-processed Small-molecule Organic Films and Light-emitting Diodes Compared with Their Vacuum-deposited Counterparts. *Adv. Funct. Mater.*, 19, 10, 1625–30, 2009.
71. Gilissen, K., Stryckers, J., Verstappen, P., Drijkoningen, J., Heintges, G.H.L., Lutsen, L., Manca, J., Maes, W., Deferme, W., Ultrasonic Spray Coating as Deposition Technique for the Light-Emitting Layer in Polymer LEDs. *Org. Electron.*, 20, 31–35, 2015.
72. Choi, M.-C., Kim, Y., Ha, C.-S., Polymers for Flexible Displays: From Material Selection to Device Applications. *Prog. Polym. Sci.*, 33, 6, 581–630, 2008.
73. Purohit, V., Banu, T., Daiya, K., AMOLED: an emerging trends in LED, *International Journal of Scientific & Engineering Research*, 3, 1–4, 2012.
74. Zakhidov, A.A., Papadimitratos, A., Monolithic Parallel Multijunction Oled with Independent Tunable Color Emission, *Google patents, United States Solarno, Inc. (Coppell, TX, US),The Board of Regents of the University of Texas System (Austin, TX, US)* 20130240847, 2013, http://www.freepatentsonline.com/y2013/0240847.html
75. Zakhidov, A.A. and Papadimitratos, A., Monolithic Parallel Multijunction Oled with Independent Tunable Color Emission, 2013. Google Patents.
76. Huang, X., Sheng, P., Tu, Z., Zhang, F., Wang, J., Geng, H., Zou, Y., Di, C.-a., Yi, Y., Sun, Y., A Two-Dimensional π–d Conjugated Coordination Polymer with Extremely High Electrical Conductivity and Ambipolar Transport Behaviour. *Nat. Commun.*, 6, 7408, 2015.
77. Molapo, K.M., Ndangili, P.M., Ajayi, R.F., Mbambisa, G., Mailu, S.M., Njomo, N., Masikini, M., Baker, P., Iwuoha, E.I., Electronics of Conjugated Polymers (I): Polyaniline. *Int. J. Electrochem. Sci.*, 7, 12, 11859–75, 2012.

78. Zahn, D.R.T., Gavrila, G.N., Gorgoi, M., The Transport Gap of Organic Semiconductors Studied Using the Combination of Direct and Inverse Photoemission. *Chem. Phys.*, 325, 1, 99–112, 2006.
79. Yan, H., Lee, P., Armstrong, N.R., Graham, A., Evmenenko, G.A., Dutta, P., Marks, T.J., High-Performance Hole-Transport Layers for Polymer Light-Emitting Diodes. Implementation of Organosiloxane Cross-Linking Chemistry in Polymeric Electroluminescent Devices. *J. Am. Chem. Soc.*, 127, 9, 3172–83, 2005.
80. Small, C.E., Tsang, S.-W., Kido, J., So, S.K., So, F., Origin of Enhanced Hole Injection in Inverted Organic Devices with Electron Accepting Interlayer. *Adv. Funct. Mater.*, 22, 15, 3261–66, 2012.
81. Tokito, S., Adachi, C., Murata, H., *Organic EL Display*, pp. 111–14, Ohmsha, Ltd, 2004.
82. Zhang, Z., Guo, K., Li, Y., Li, X., Guan, G., Li, H., Luo, Y., Zhao, F., Zhang, Q., Wei, B., A Colour-Tunable, Weavable Fibre-Shaped Polymer Light-Emitting Electrochemical Cell. *Nat. Photonics*, 9, 4, 233, 2015.
83. Lin, Y., Chen, Y., Chen, Z., Ma, D., Zhang, B., Ye, T., Dai, Y., Triphenylamine and Quinoline-Containing Polyfluorene for Blue Light-Emitting Diodes. *Eur. Polym. J.*, 46, 5, 997–1003, 2010.
84. Wetzelaer, G.A.H., Kuik, M., Nicolai, H.T., Blom, P.W.M., Trap-Assisted and Langevin-Type Recombination in Organic Light-Emitting Diodes. *Phys. Rev. B*, 83, 16, 165204, 2011.
85. Zheng, H., Zheng, Y., Liu, N., Ai, N., Wang, Q., Wu, S., Zhou, J., Hu, D., Yu, S., Han, S., All-Solution Processed Polymer Light-Emitting Diode Displays. *Nat. Commun.*, 4, 1971, 2013.
86. Liu, J. and Pei, Q., Poly (M-Phenylene): Conjugated Polymer Host with High Triplet Energy for Efficient Blue Electrophosphorescence. *Macromolecules*, 43, 23, 9608–12, 2010.
87. Jiang, H., Hosts for High-Performance Phosphorescent Organic Light-Emitting Diodes Based on Carbazole Derivatives. *Asian J. Org. Chem.*, 3, 2, 102–12, 2014.
88. Zhu, M. and Yang, C., Blue Fluorescent Emitters: Design Tactics and Applications in Organic Light-Emitting Diodes. *Chem. Soc. Rev.*, 42, 12, 4963–76, 2013.
89. Jiang, H., Sun, J., Zhang, J., A Review on Synthesis of Carbazole-Based Chromophores as Organic Light-Emitting Materials. *Curr. Org. Chem.*, 16, 17, 2014–25, 2012.
90. Morin, J.-F., Beaupré, S., Leclerc, M., Lévesque, I., D'Iorio, M., Blue Light-Emitting Devices from New Conjugated Poly (N-Substituted-2, 7-Carbazole) Derivatives. *Appl. Phys. Lett.*, 80, 3, 341–43, 2002.
91. Morin, J.-F., Boudreault, P.-L., Leclerc, M., Blue-Light-Emitting Conjugated Polymers Derived From 2,7-Carbazoles. *Macromol. Rapid Commun.*, 23, 17, 1032–36, 2002, https://doi.org/10.1002/marc.200290000.

92. Shu, C.-F., Dodda, R., Wu, F.-I., Liu, M.S., Jen, A.K.-Y., Highly Efficient Blue-Light-Emitting Diodes from Polyfluorene Containing Bipolar Pendant Groups. *Macromolecules*, 36, 18, 6698–6703, 2003.
93. Akcelrud, L., Electroluminescent Polymers. *Prog. Polym. Sci.*, 28, 6, 875–962, 2003.
94. Ego, C., Grimsdale, A.C., Uckert, F., Yu, G., Srdanov, G., Müllen, K., Triphenylamine-Substituted Polyfluorene—A Stable Blue-Emitter with Improved Charge Injection for Light-Emitting Diodes. *Adv. Mater.*, 14, 11, 809–11, 2002.
95. Jiang, X., Liu, S., Ma, H., Jen, A.K.-Y., High-Performance Blue Light-Emitting Diode Based on a Binaphthyl-Containing Polyfluorene. *Appl. Phys. Lett.*, 76, 14, 1813–15, 2000.
96. Tu, G.L., Mei, C.Y., Zhou, Q.G., Cheng, Y.X., Geng, Y.H., Wang, L.X., Ma, D.G., Jing, X.B., Wang, F.S., Highly Efficient Pure-White-Light-Emitting Diodes from a Single Polymer: Polyfluorene with Naphthalimide Moieties. *Adv. Funct. Mater.*, 16, 1, 101–6, 2006.
97. Khodabakhshi, E., Blom, P.W.M., Michels, J.J., Efficiency Enhancement of Polyfluorene: Polystyrene Blend Light-Emitting Diodes by Simultaneous Trap Dilution and β-Phase Formation. *Appl. Phys. Lett.*, 114, 9, 93301, 2019.
98. Ma, Y., Peng, F., Guo, T., Jiang, C., Zhong, Z., Ying, L., Wang, J., Yang, W., Peng, J., Cao, Y., Semi-Orthogonal Solution-Processed Polyfluorene Derivative for Multilayer Blue Polymer Light-Emitting Diodes. *Org. Electron.*, 54, 133–39, 2018.
99. Lee, S.H., Nakamura, T., Tsutsui, T., Synthesis and Characterization of Oligo(9,9-Dihexyl-2,7-Fluorene Ethynylene)s: For Application as Blue Light-Emitting Diode. *Org. Lett.*, 3, 13, 2005–7, 2001, https://doi.org/10.1021/ol010069r.
100. Chan, K.L., McKiernan, M.J., Towns, C.R., Holmes, A.B., Poly(2,7-Dibenzosilole): A Blue Light Emitting Polymer. *J. Am. Chem. Soc.*, 127, 21, 7662–63, 2005, https://doi.org/10.1021/ja0508065.
101. Seino, Y., Inomata, S., Sasabe, H., Pu, Y.-J., Kido, J., High-Performance Green OLEDs Using Thermally Activated Delayed Fluorescence with a Power Efficiency of over 100 Lm W−1. *Adv. Mater.*, 28, 13, 2638–43, 2016, https://doi.org/10.1002/adma.201503782.
102. Tanaka, D., Sasabe, H., Li, Y.-J., Su, S.-J., Takeda, T., Kido, J., Ultra High Efficiency Green Organic Light-Emitting Devices. *Jpn. J. Appl. Phys.*, 46, 1, L10–12, 2006, https://doi.org/10.1143/jjap.46.l10.
103. Ku, S.-Y., Chi, L.-C., Hung, W.-Y., Yang, S.-W., Tsai, T.-C., Wong, K.-T., Chen, Y.-H., Wu, C.-I., High-Luminescence Non-Doped Green OLEDs Based on a 9,9-Diarylfluorene-Terminated 2,1,3-Benzothiadiazole Derivative. *J. Mater. Chem.*, 19, 6, 773–80, 2009, https://doi.org/10.1039/B814082K.
104. Zhu, Y.-C., Zhou, L., Li, H.-Y., Xu, Q.-L., Teng, M.-Y., Zheng, Y.-X., Zuo, J.-L., Zhang, H.-J., You, X.-Z., Highly Efficient Green and

Blue-Green Phosphorescent OLEDs Based on Iridium Complexes with the Tetraphenylimidodiphosphinate Ligand. *Adv. Mater.*, 23, 35, 4041–46, 2011, https://doi.org/10.1002/adma.201101792.

105. Cho, Y.J., Yook, K.S., Lee, J.Y., A Universal Host Material for High External Quantum Efficiency Close to 25% and Long Lifetime in Green Fluorescent and Phosphorescent OLEDs. *Adv. Mater.*, 26, 24, 4050–55, 2014, https://doi.org/10.1002/adma.201400347.
106. Chen, H.-F., Yang, S.-J., Tsai, Z.-H., Hung, W.-Y., Wang, T.-C., Wong, K.-T., 1,3,5-Triazine Derivatives as New Electron Transport–Type Host Materials for Highly Efficient Green Phosphorescent OLEDs. *J. Mater. Chem.*, 19, 43, 8112–18, 2009, https://doi.org/10.1039/B913423A.
107. Chang, C.-H., Kuo, M.-C., Lin, W.-C., Chen, Y.-T., Wong, K.-T., Chou, S.-H., Mondal, E. *et al.*, A Dicarbazole–Triazine Hybrid Bipolar Host Material for Highly Efficient Green Phosphorescent OLEDs. *J. Mater. Chem.*, 22, 9, 3832–38, 2012, https://doi.org/10.1039/C2JM14686J.
108. Okada, S., Okinaka, K., Iwawaki, H., Furugori, M., Hashimoto, M., Mukaide, T., Kamatani, J., Igawa, S., Tsuboyama, A., Takiguchi, T., Substituent Effects of Iridium Complexes for Highly Efficient Red OLEDs. *Dalton Trans.*, 9, 1583–90, 2005.
109. Jung, B.-J., Lee, J.-I., Chu, H.Y., Do, L.-M., Lee, J., Shim, H.-K., A New Family of Bis-DCM Based Dopants for Red OLEDs. *J. Mater. Chem.*, 15, 25, 2470–75, 2005, https://doi.org/10.1039/B419408J.
110. Fu, H.-y, Wu, H.-r., Hou, X.-y., Xiao, F., Shao, B.-x., N-Aryl Carbazole Derivatives for Non-Doped Red OLEDs. *Synth. Met.*, 156, 11, 809–14, 2006, https://doi.org/https://doi.org/10.1016/j.synthmet.2006.04.013.
111. Donat-Bouillud, A., Lévesque, I., Tao, Y., D'Iorio, M., Serge, B., Blondin, P., Ranger, M., Bouchard, J., Leclerc, M., Light-Emitting Diodes from Fluorene-Based π-Conjugated Polymers. *Chem. Mater.*, 12, 7, 1931–36, 2000, https://doi.org/10.1021/cm0001298.
112. Wong, M.Y., Recent Advances in Polymer Organic Light-Emitting Diodes (PLED) Using Non-Conjugated Polymers as the Emitting Layer and Contrasting Them with Conjugated Counterparts. *J. Electron. Mater.*, 46, 11, 6246–81, 2017, https://doi.org/10.1007/s11664-017-5702-7.
113. Chou, H.-H. and Cheng, C.-H., A Highly Efficient Universal Bipolar Host for Blue, Green, and Red Phosphorescent OLEDs. *Adv. Mater.*, 22, 22, 2468–71, 2010, https://doi.org/10.1002/adma.201000061.
114. Kunić, S. and Šego, Z., OLED Technology and Displays, in: *Proceedings ELMAR-2012*, pp. 31–35, IEEE, 2012.
115. Thejokalyani, N. and Dhoble, S.J., Novel Approaches for Energy Efficient Solid State Lighting by RGB Organic Light Emitting Diodes–A Review. *Renewable Sustainable Energy Rev.*, 32, 448–67, 2014.
116. Chen, H.-W., Lee, J.-H., Lin, B.-Y., Chen, S., Wu, S.-T., Liquid Crystal Display and Organic Light-Emitting Diode Display: Present Status and Future Perspectives. *Light: Sci. Appl.*, 7, 3, 17168, 2018.

117. Li, Z., Li, Z.R., Meng, H., *Organic Light-Emitting Materials and Devices*, CRC Press, 2006.
118. Goushi, K., Yoshida, K., Sato, K., Adachi, C., Organic Light-Emitting Diodes Employing Efficient Reverse Intersystem Crossing for Triplet-to-Singlet State Conversion. *Nat. Photonics*, 6, 4, 253, 2012.
119. Odaki, T., Tasaki, M., Ichikawa, A., Takagi, K., Hashimoto, K., Light-Emitting Diode Device, *Google Patents, United States Asahi Rubber Inc. (Saitama, JP)*, 6521915, 2006, http://www.freepatentsonline.com/6521915.html
120. Freudenrich, C., How OLEDs Work, How Stuff Works, 2005, http://electronics.howstuffworks.com/oled.htm
121. Freudenrich, C., How Oleds Work, 2010. Retrieved November 10: 2010.
122. Buckley, A., *Organic Light-Emitting Diodes (OLEDs): Materials, Devices and Applications*, Elsevier, 2013.
123. Poor, A., Display Week 2014 Review: OLEDs. *Inf. Disp.*, 30, 5, 10–13, 2014.
124. Semenza, P., OLEDs in Transition. *Inf. Disp.*, 27, 10, 14–16, 2011.
125. Young, R., OLEDs Expected to Gain Ground as LCD Investment Slows. *Inf. Disp.*, 35, 3, 24–29, 2019.
126. Mizusaki, M., Shibasaki, M., Tsuchiya, H., Tsukamoto, Y., Umeda, T., Nakamura, K., Ohshita, T., Shimada, S., P-44: Long Lifetime and High Performance OLED Display with Wide Temperature Range for Automotive Application, in: *SID Symposium Digest of Technical Papers*, vol. 50, pp. 1400–1403, Wiley Online Library, North Carolina, 2019.
127. Mertens, R., *The OLED Handbook (2019 Edition)*, 2019, Lulu. com.
128. Chavez-Burbano, P., Vitek, S., Teli, S.R., Guerra, V., Rabadan, J., Perez-Jimenez, R., Zvanovec, S., Optical Camera Communication System for Internet of Things Based on Organic Light Emitting Diodes. *Electron. Lett.*, 55, 6, 334–36, 2019.

5

Application of Electrospun Materials in LEDs

Subhash B. Kondawar*, Mahelaqua A. Haque and Chaitali N. Pangul

Department of Physics, Rashtrasant Tukadoji Maharaj Nagpur University, Nagpur, India

Abstract

An era for the upgradation of powerful info-communication tool had commanded the research to realize the unknown. Integration of electronics and the textile had made it possible for humans to achieve the new human machine interfaces. Light-emitting fibers, the very core element in the wearable light-emitting devices, are the chief fundamental of the research areas. These light-emitting fibers, when constrained into nano size dimensions, a plentiful of diverse properties are attainable. Engineering the light-emitting nanofibers via assortment of techniques has been carried out but electrospinning had stunned the ways and evidenced to be the very simple, efficient, and robust method to fabricate nanofibers. Utilization of such electrospun nanofibers in LEDs has accomplished a significant interest. These electrospun light-emitting fibers overcome many of the hardships faced by solid lighting devices. Amalgamation of various elements with the polymers have sightseen the very desirable features from the light-emitting nanofibers. In this chapter, we present the very novel and noteworthy work carried out on electrospun nanofibers for usage in LEDs.

Keywords: Electrospun nanofibers, polymer, light-emitting diodes (LEDs), smart textile

5.1 Introduction

Nanotechnology is a growing interdisciplinary technology which offers fabrication techniques for the conversion of optoelectronics into textiles.

**Corresponding author*: sbkondawar@yahoo.co.in

Inamuddin, Rajender Boddula, Mohd Imran Ahamed and Abdullah M. Asiri (eds.) Polymers for Light-Emitting Devices and Displays, (99–124) © 2020 Scrivener Publishing LLC

The introduction of nanotechnology in textile industry has occurred due to the raising needs of classy fabrics integrated with distinct features. Cumulative clients mandate to have robust and purposeful attire whose manufacturing is also done in maintainable way had opened up the ways for integration of nanotechnology into the garment fields. One of the enthralling and attainable applications of nanotechnology is nano-textile. Nanofibers are so fabricated that the optimum of the properties could be tweaked. The inheritance of the textiles including their flexibility, ease of wearing, and wettability are maintained, such a manner of fabrication is chosen [1]. The innovation in wearable technologies has transformed the concept of clothing. Fashion industry is straining efforts so that the amalgamation of the textiles along with the communicating devices and flexible optoelectronics is conceivable one day at a very easy manner. A costume named Bubelle has been designed by Philips, which owes the asset of tuning its colors reliant on the customers' mood. Organic light-emitting diode had been utilized by Black Eyed Peas on the platform in the clothing for their acts [2]. Plentiful innovative attires have been efficaciously fabricated with the use of nanotechnology in particular manufacture of fibers, yarns, and finishing techniques. Along the decades, the application of nanotechnology in textile industry had remarked numerous references in the literatures. Optical fibers, light-emitting films, and nano additives have been also considered as a participant in the manufacture of the light-emitting textiles and garments in order to enhance the performance and supplementary functionalities.

Clothing that can respond to the logic of ecological circumstances in the form of power-driven, thermal, magnetic, chemical, and electrical sources are usually referred as smart textiles. When the textiles are gifted to respond or perform in the diverse way, then the regular products are counted as the smart textiles. The science boundaries are pushed forward through the newer discoveries in the field of optoelectronics and textiles which represents the functioning of smart textiles. Functionalizing large area surfaces at very high speed is the driving force for smart textiles research [3]. Owing to the flexible light-emitting materials, the electronic devices are becoming portable. With the hidden technology beneath the coating and fibers, the functionality may be tuned as such the total control of the circuitries can allow the textile to produce astounding effects (Figure 5.1). Fashion design walk ways are becoming the profound platform for the innovative upcoming of the assortment of technical garments [4].

LEDs definitely outstand the class of luminous species. They possess long lifetimes and low energy consumptions and are physically more compact, thinner, and lighter that enable them to be more eco-friendly devices.

Figure 5.1 Light-emitting smart textile (Reproduced with the permission from Ref. [4], Smart Textiles and Nano-Technology: A General Overview, *J. Textile. Sci. Eng.*, 2015).

They form a major component in the sensors owing to their fast reaction and interaction. Other beneficiaries of LEDs lie in their variation of colors, illumination and numerous patterns, and adaptability into countless devices. Light-emitting fibers occupied smart textiles and their future is emerging in a vast manner and the field is captivating the researcher for the further expansion.

5.2 Electrospun Nanofibers Technology

Assortment of properties is owned by electrospun nanofibers as in when the fibers of polymers are shortened from micrometers to submicrons or nanometers. Very large surface area to volume ratio, flexibility in surface functionalities, and superior mechanical strength are some of the assets of the electrospun fibers when compared to the any other known materials. Following these amazing properties, the nanofibers are proven to be the apt candidate for variety of practical applications. Electrospinning has emerged out be and very competent, humble, robust, and operative technique to fabricate nanofibers overcoming the variety of preparation techniques including template synthesis, drawing, phase separation, self-assembly [5, 6]. The fundamental code of this electrospinning technology is "electrostatic interactions". A very high voltage is applied to needle of the peristaltic pump, which is forced to drop down the solution in a controlled manner. The voltage is applied within the syringe needle and the ground collector which is usually of the metal collector type. The thesis says the formation of fibers is the resultant of the electrostatic interactions. Figure 5.2 represents the illustration of the electrospinning instrument [7].

Electrospinning setup comprises of important three constituents. The high voltage supply is the very first need of the electrospinning setup, the syringe needle is the secondary required thing, whereas last but not the least is the ground collector for the collection of the nanofibrous mat. Different types of the spinneret and the collector facilitate the variation in the electrospin instrument which are very readily available in the market [8]. Along with the structure of the instrument, there are many more factors influencing the electrospinning of fibers.

A variety of polymer nanofibers could be electrospun by varying the precursor solutions. Organic polymer nanofibers, organic inorganic hybrid polymer nanofibers, polymer blends nanofibers, composite nanofibers, etc., could be easily fabricated via electrospinning technique. The precursors for the electrospun nanofibers could be of vast variety and include metal salt solutions, rare earth ions solution, perovskite solutions, also various complexes incorporated solutions homogenated with polymer solutions. The viscosity of the solution has to be such that the nanofibers are electrospun while surface tension of the solution and the conductivity also play a vital role in defining the structures of the nanofibers. Multi-structured nanofibers having beaded morphology, ribbon-like morphology, and core-shell

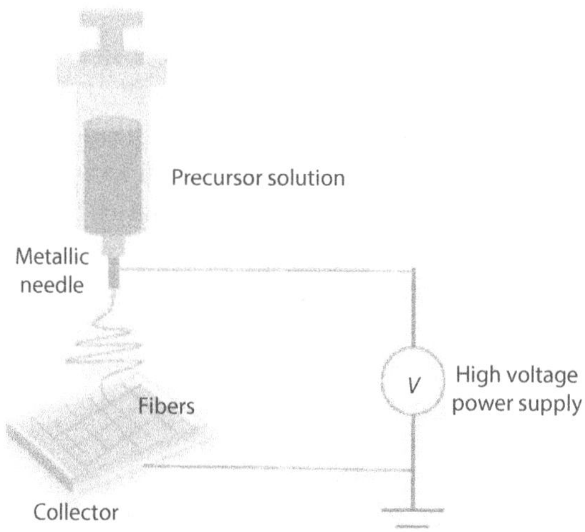

Figure 5.2 Illustration of electrospinning instrument (Reproduced with permission from Ref. [7], Synthesis and Characterization of Structure-Controlled Micro-/Nanocomposite TiO_2 Fibers with Enhanced Photocatalytic Activity, *Journal of Nanomaterials,* 2014).

Figure 5.3 Various applications of electrospun nanofibers (Reproduced with permission from Ref. [9], Electrospun nanofibers: New generation materials for advanced applications, *Materials Science and Engineering B*, 2017).

structured nanofibers could be attained by varying the viscosity and conductivity of the precursor solutions. Along with the precursor's properties, the applied voltage also plays a key role as well in the formation of nanofibers. The appiled voltage must be that suffcient high, the surface tension of the solution should overcome and the formation of Taylor cone happens. The formation of the Taylor cone is the combination of the various factors. Conductivity, viscosity, surface tension, and applied high voltage are the crucial factors for Taylor cone formation and so is the flow rate. Flow rate has to be defined in such a manner that the polarization of the polymeric solution is sufficiently done. After the formation of the Taylor cone, the solution jet gets attracted towards the collector, so it becomes essential to give plenty of time to the force driven jet towards the collector to vaporize and solidifies into a fiber. Once the fiber approaches the collector, further, the collector alignment comes in picture. The orientation of the collector also influences the collection pattern of the nanofibers. Wheels, drum rotate collector, disk, etc., are the different types of collectors which can define the morphology of the nanofibers. Apart from these factors, the environmental parameters too affect the morphology of the electrospun nanofibers. Humidity and temperature are the ambient parameters which hampers the formation of the nanofibers. Considering these many parameters, the various morphological structured nanofibers could be electrospun which finds their application in various fields such as sensors, filters, energy storage devices, cosmetology, biomedical fields, protective clothing, catalysts, etc. Figure 5.3 illustrates the various applications of electrospun nanofibers [9].

5.3 Electrospun Materials for LEDs

GaN chips-based light-emitting diodes have gained noteworthy consideration lately in the fields of general lighting, displays, indicators, and communication [10]. Electrospinning had turned out to be the advantageous technology which yet again finds its place even in the light-emitting fields. Beneficial light scattering and reflectance ability of the nanofibers allow them to be the optimum materials for the light emission devices. Diameter of the electrospun nanofibers and film thickness are linearly related to the ability of scattering. There is much room to study the utilization of superior scattering and reflectance ability of the electrospun nanofibers to influence the luminous efficacy of the light-emitting diodes. Higher luminous efficacy is achieved by arrangement of multiple LEDs closely packed together with white LED and blue LED with remote phosphor layer. In order to meet the requirements of the day-to-day lighting arrangements, luminous efficacy, color rendering index, scattering and reflectance ability, and diffusion ability are the major factors whose close study is to be done and integrate to achieve an optimum result. Light-emitting nanofibers are one such class of materials which are found to bear all these properties and with foresight of light-emitting applications could be utilized. Light-emitting nanofibers possess superior properties owned by the class of the nanofibers. Accompanied by the various assets, the photophysical behavior of the light-emitting nanofibers outclass them.

There is a gigantic scope in the field of electrospun nanofibers that are proficient of light emission. Large variety of materials could be easily manipulated to achieve the desired precursor solution which is utilized for the electrospinning. Polymer are one of the core constituents in the process of electrospinning. The high molecular weight of the polymer enables the alignment of the nanostructured nanofibers even from the lower molecular weight precursors. Optical fibers obtained from polymers termed as plastic optical fibers are booming the research fields due to easy handling, long wavelength transparency, superior flexibility, and excellent compatibility. When these polymers are combined with various other constituents like metal ions, rare earth ions, complexes, perovskite, etc., tend to achieve the supreme light-emitting properties even pursuing the superior surface functionalities.

In this chapter, the current progress on the applications of light-emitting electrospun nanofibers, with distinct prominence on polymer nanofibers functionalized with rare earth ion dopants, metal oxide semiconductors, and perovskite materials are focussed.

5.3.1 Metal Oxide Semiconducting Electrospun Nanofibers

Semiconductor nanomaterials displays size controlled spectral tunability and chemical flexibility which allows them to be the efficient materials for use in novel and evolving applications, such as nanoelectronics, lasing, fluorescent tagging, and light-emitting diodes [11]. Semiconducting materials have established an imperative dwelling in various fields of research owing to their high binding energies, wide band gap, tailoring parameters, etc. [12]. Explicit optical, electrical, and electronic properties facilitate metal oxide semiconductors as the beneficial basic building blocks in the photoelectric devices. Low-dimensional nanostructures with appropriate bandgap, exclusive conduction properties, and confined carrier conduction tracks made them aptaspirants in photoelectric devices [13]. High abundance, electrochemical stability, favorable surface kinetics, and long-lived charge transfer states have enabled metal oxides to be an extensively photocatalysts for water oxidation [14]. Metal oxide semiconductors are amazingly diverse when derives to electronic structure, materials design concepts, defect states, charge transport mechanisms, thin-film processing, and optoelectronic properties, thereby empowering equally conformist and wholly innovative functions [15]. These metal oxide semiconducting nanomaterials when constrained in one-dimensional nanomaterials, the properties tend to enhance drastically. The interest on exploring new one-dimensional nanostructured luminescent materials is growing in recent years as they possess specific and fascinating luminescent properties like various emission lifetimes and increased luminescence efficiencies, which proves to be different from their bulk equals [16]. Semiconducting materials include a wide variety of materials, namely, ZnO, TiO_2, SnO_2, CdS, ZnS, etc. Nanostructure of these wonder materials enhances further properties following quantum confinement effect [17]. The electronic arrangement of positive hole and negative electron pairs, referred hereon exciton, results into quantum confinement in semiconductors. An exciton of Bohr radius which is regular size of electron and hole pair in a substance crystal, offers estimated measurement for the inception of effects belonging to quantum confinement. Semiconductor's operative bandgap intensifies when the exciton is enfolded into a nanocrystal whose dimensions are about to reach Bohr radius. The size of the nanocrystal varies inversely to the effective bandgap and light emission energy consequential from exciton recombination [18].

Electrospun semiconducting nanomaterials possess advantage of large surface area to volume ratio which plays a vital role in defining the light-emitting properties. Cheng Wang and co-workers [19] fabricated remarkably photoluminescent electrospun TiO_2-PPV hybrid nanoparticlepolymer fibers. A close comparative study between pure PPV nanofibers and TiO_2/PPV nanofibers synthesized via *in situ* method explores the various facts. The morphological studies of the nanofibers were observed using FESEM and fluorescence microscopy. This indicates that TiO_2-PPV fibers exhibited robust photoluminescence when compared to pure PPV fibers. The same is also confirmed upon the photoluminescence studies of prepared nanofibers. Yellow-green light emission was obtained from PPV fibers. However, with incorporation of TiO_2 into PPV fibers, the photoluminescence spectrum displays two emission peaks. The change in the PL spectrum is devoted to the interaction between the PPV precursor and TiO_2 sol-gel by thermolytic development of the final TiO_2-PPV hybrid electrospun nanofibers. This also widens the band gap and chain length of PPV is also abbreviated. Muzammil *et al.* [20] utilized electrospinning technique to fabricate GaN nanofibers. Gallium oxide nanofibers of various precursors solutions were first fabricated which then further undergo the process of ammonization which results in GaN nanofibers. These nanofibers were studied using HRTEM, FTIR, XRD, and UV-Vis spectroscopy. Characterizations revealed higher band gap energy for 0.5M GaN nanofibers. Photoluminescence studies ensures upon a broad spectrum having sub-spectrum representing near band gap emission is related with the direct band gap emission and defect emissions are documented to the defects and dislocations in the GaN fibers. These studies revealed the utility of highly crystalline GaN nanofibers in opto-electronic devices like LEDs.

Manoel and co-workers [21] studied silver (Ag)-zinc oxide boosted electrospun cellulose acetate fibrous mat for UV laser emission. Initially, Silver (Ag) nanoprisms in tri sodium citrate aqueous solution were prepared wherein the electrospun cellulose acetate (CA)- ZnO nanofibrous mat was immersed for adsorption process. The final product of (CA)-ZnO-Ag enriched electrospun nanofibers were studied through various characterizations. Laser studies were performed and it is noteworthy to mention that with the use of ZnO as the gain medium when compared with organic dyes, the interface within the gain source and pump demonstrates almost no degradation effect which is very common in dyes. Whereas, no plasmonic effect is observed for 150 nm and larger size of Silver (Ag) nanoprisms which would be probable in the nanoparticle of comparative size. Thus, incorporation of ZnO and Ag has turned out to be beneficial for enhancing the light-emitting properties of nanofibers. Kallol *et al.* [22] expressed light-dependent NDR (negative differential resistance) in MEH/PPV poly [2methoxy-5-(2-ethylhexyloxy)-1,4-phenylenevinylene] decorated electrospun TiO_2 mat. The negative differential resistance devices are of great importance in this era in the field of electronics. The purchased TiO_2 nanoparticles were dissolved in PVA solution in order to get homogenous solution and electropsinning was carried out to get TiO_2/PVA nanofibers. This nanofibrous mat was then immersed in MEH/PPV solution. The UV-Vis study and photoluminescence studies were carried out on the the prepared smaples. Photoluminescence spectra of pristine TiO_2 mat, TiO_2 waterlogged in MEH/PPV, and light exposed TiO_2 matrix water logged in MEH/PPV which revealed totally different light emission behavior. When light is exposed to TiO_2 mat water logged in MEH/PPV, the presence of a different spectrum surely determines the definite energy conjugation within the TiO_2 mat waterlogged in MEH/PPV. The energy transfer occurs form MEH/PPV to TiO_2 particles via excited electrons in the MEH/PPV.

Alberto and team [23] demonstrated MoS_2 nanoparticle doped light-emitting nanofibers fabricated through a simple and reasonable sonication process in N-methyl-2-pyrrolidone and effectively condense the electrospun nanofibers in polymer filament. Molybdenum disulphide possesses highly interesting semiconducting properties. Implanting such materials in polymer architecture such as nanofibers have offered supplementary benefits counting with the easy management of functional nanoparticles, their integration in flexible optoelectronic components, and amended stability. MoS_2 nanoparticles were prepared by chemical ultrasonication process which is driven by sonochemical cavitation phenomenon. These nanoparticles were studied in the powder form and then incorporated into the fiber matrix via electrospinning process. The comparative photoluminescence study was carried out and it was concluded that the emission from MoS_2 was well preserved upon electrospinning. Thus, these highly luminescent MoS_2 nanofibrous mat is a deserving material vastly motivating for incorporation in compound, system level platforms for sensing and in flexible optoelectronics. Tang *et al.* [24] proposed the enhancement of angular color uniformity of remote phosphor converted light-emitting diodes using electrospun nanofiber diffusing

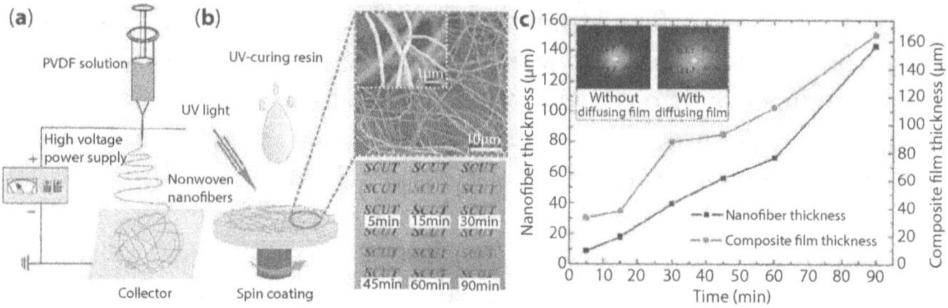

Figure 5.4 (a) Schematic of the electrospinning. (b) Composite-film preparation. (c) Nanofiber and composite-film thicknesses as a function of the time. The inset shows the green-laser-light spots. (Reproduced with permission from Ref. [24], Enhancement of angular color uniformity of remote-phosphor converted light-emitting diodes by electrospun nanofiber diffusing films, *Materials Letters*, 2018).

films. Figure 5.4 depicts the typical procedure of fabrication of nanofibrous films along with the diffusion technique with an ultraviolet curable resin $(C_3H_4O_2)_n$ via spin coating method. The alongside graph is the graph of thickness of films as function of time. The inset is the picture of the green laser light spots with and without diffusing films which clearly shows the enhanced diffusion ability of the composite films. For further studies, diffuse transmittance and haze were measured of these films.

5.3.2 Perovskite Electrospun Nanofibers

Metal halide perovskite materials are one more class of solution processable semiconductors seeking attention since they possess significant potential for use in optoelectronic device such as light-emitting devices, solar cells, etc. During the past few years, in precise, there has been striking advancement for the peripheral quantum efficiency of perovskite LEDs. The exceptional feature of hybrid perovskite materials is from the outstanding electro-optic properties, such as tunable band gap, strong absorption coefficient, high mobility, and long carrier diffusion distance [25–29]. Despite of these amazing phenomena, poor film morphology is one of the limitations in perovskite which restricts their use for high-efficiency LEDs. Strong leakage current is the probable outcome of such crystalline perovskite film which is rough or discontinuous. The performance of LEDs is hampered in a massive manner. Therefore, a dense and uniform perovskite film is desirable for the better performance of LEDs and numerous efforts are made that way [30, 31].

Ercan *et al.* [32] studied the synthesis of multicolor luminescent nanofibers having core-shell structure and their vast scale electrospun fibrous matrix of perovskite-thermoplastic polyurethane. Optical and mechanical property studies were carried out for the synthesized films. High mechanical endurance was seen in fabricated core-shell nanofibers which are contributed towards healthy flexible assets of thermoplastic polyurethane which maintain the intensity of the luminescence even in the stretched state. The stretching test was carried out by following thousand stretching relaxing sequences. Additionally, white light emission from fibers was achieved by incorporating MEH/PPV which emits orange light into thermoplastic polyurethane and at the core site turquoise blue light-emitting perovskite nanoparticles were coupled. Finally, a LED covering anemissive nanofibrous layer of orange light-emitting polyfluorene was fabricated whose working of the perovskite-based thermoplastic polyurethane fibers was understood. Effort made in this way verified an operative way to synthesize steady and light-emitting stretchable electrospun nanofibers. Incorporation of these light-emitting nanofibers into light-emitting devices could affluence the forthcoming expansion of smart textiles.

Han *et al.* [33] revealed the synthesis process of perovskite unified alumina-lanthana (AL) nanofibrous matrix. The embellishment of cesium lead bromide (CPB) perovskite crystals on electrospun AL fibers was the key process in the synthesis. Initially, these AL fibers were prepared by the process of electrospinning utilizing PVP as the carrier medium. These electrospun nanofibers were then calcined to get rid of the organic components. It is noteworthy to mention that even after calcinations, the flexibility was maintained of the AL fibers which was confirmed upon swirling AL fiber which remain intact indicating that excellent flexibility was even obtained in the single fiber. The behavior of flexible polymer nanofibers was ensured upon abundant resistance to torsion from microscopic. Immersion process of AL fibers into perovskite nanocrystals precursor solution was carried out, and hence, by a simple supersaturated recrystallization process, CPB crystal accumulation was carried out on the exterior of nanofibers. These final products were named as CCAL membranes. In histrionic contrast to the fragile property of inorganic materials, high flexibility of CCAL nanofibrous matrix [avg. thickness of CCAL (1) is 63 μm] was observed. Figure 5.5 illustrates the further light emission studies of the CCAL membranes. It is observed that the PL intensity shows a remarkable increase in the CCAL (1) membrane. This extremely high luminescence has been credited to best passivation of the CPB nanocrystals.

Thus, CCAL (1) membrane is derived to be the most efficient green light-emitting membrane, and thus, fabrication of model light-emitting diodes by means of optimum CCAL matrix and ultra violet light-emitting chip by remote packing way was carried out. Figure 5.6 shows the

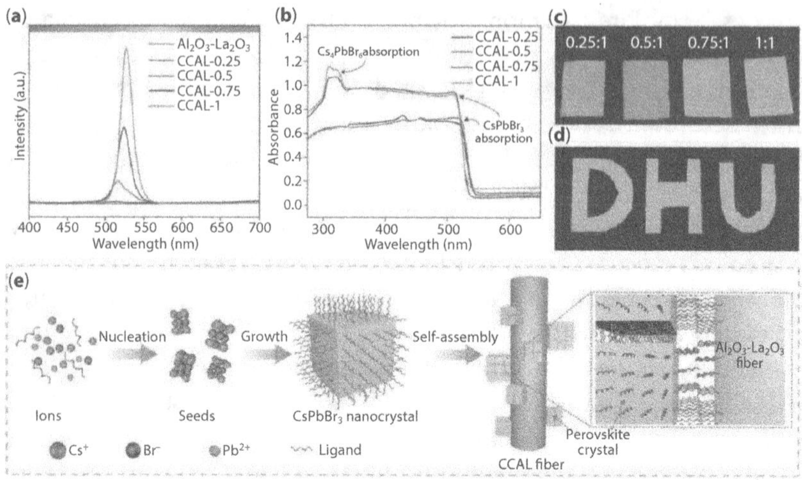

Figure 5.5 (a) PL spectrum, and (b) UV-Vis absorption of CCAL membranes with different ratios of $CsBr/PbBr_2$. (c) Photograph of different samples under UV excitation. (d) CCAL-1 membranes showing excellent flexibility and ease of cutting the "DHU" pattern. (e) The proposed formation process of CCAL membranes. (Reproduced with permission from Ref. [33], Fluorescent Nanofibrous Membranes: Self-Assembly of Perovskite Crystals Anchored Al_2O_3-La_2O_3 Nanofibrous Membranes with Robust Flexibility and Luminescence. *Small*, 2018).

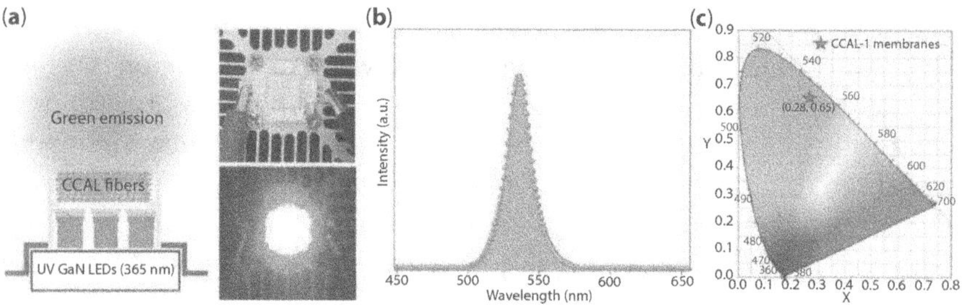

Figure 5.6 (a) Diagram (left) and photographs (right) of the configuration of the prototype LEDs device combining UV LED chips with CCAL-1 membrane. (b) The electroluminescence spectrum of a CCAL-1 membrane LEDs. (c) CIE chromaticity diagram of CCAL-1 membrane. (Reproduced with permission from Ref. [33], Fluorescent Nanofibrous Membranes: Self-Assembly of Perovskite Crystals Anchored Al_2O_3-La_2O_3 Nanofibrous Membranes with Robust Flexibility and Luminescence. *Small*, 2018).

fabrication of LEDs. The electroluminescence spectra were recorded for the model light-emitting diode which displays the efficient emission intensity. The chromaticity coordinates for the model light-emitting diodes were calculated and it could be seen that coordinates reach green region (0.28, 0.65)

(Figure 5.6c). Over all, these CPB ornamented AL fibers constitute to be promising candidate for LEDs.

Paris and co-workers [34] in order to fabricate robust hydrophilic and hydrophobic bright polymer fiber emitters with capable applications in textile-based light-emitting devices utilized electrospinning setup. Hydrophobic robust nanocrystals-polymer fiber emitters were fabricated using electrospinning process to encapsulate the perovskite nanocrystal into polymethyl methacrylate fibers, and a simple immersion process was followed for nanocrystal coating on hydrophilic cross-linked polyvinylpyrrolidone fibrous membranes. A very detailed method was carried out to synthesize caesium lead bromide (CPB) and formamidinium lead bromide (FPB) perovskite nanocrystals. Lead Halide Perovskite (LHP) solution in toluene was added to the solution of poly methyl methacrylate in order to fabricate electrospun poly methyl methacrylate/LHP nanofibers. Another set of nanofibers of polyvinylpyrrolidone/LHP was obtained by first fabricating electrospun nanofibers and then simple immersion into LHP nanocrystal solution into toluene. Another method utilized to obtain polyvinylpyrrolidone/LHP nanofibers was to spin coat. The prepared samples were studied through various methods. Figure 5.7 represents the actual photograph, SEM image, and fluorescence microscopy images for FPB nanocrystals: poly methyl methacrylate electrospun fibers.

SEM analysis estimated the average diameters of 0.1% wt FPB and CPB nanocrystals fibers equal to 3.3 ± 0.3 μm and 5.5 ± 1.3 μm, respectively. Uniform encapsulation of FPB nanocrystals within the poly methyl methacrylate could be seen in fluorescence microscopy images. Air and water stability studies were carried out for the recognition of practical use of perovskite nanocrystals in optoelectronics. Poly methyl methacrylate/LHP and polyvinylpyrrolidone/LHP nanofibers and pristine LHP were warehoused for about a year where the temperature was maintained at 21°C and the humidity levels were varied throughout the year. Quantum yield was recorded at the onset of manufacture and a year later, whereas the atmospheric parameters were maintained constant for poly methyl methacrylate/LHP and polyvinylpyrrolidone/LHP and pristine nanocrystals. The photoluminescence intensity showed a judicious red shift after a year. Quantum yield quenching was speckled broadly. The CPB nanocrystal system displays reduced emission quantum yield, whereas both the nanocrystals pristine films also reveal a noteworthy decline in emission efficiency. In divergence, the quantum yield of the PMMA fibers containing nanocrystals was found to reduce to <10% compared to its real quantum yield. This is devoted to the suppression of oxygen and moisture

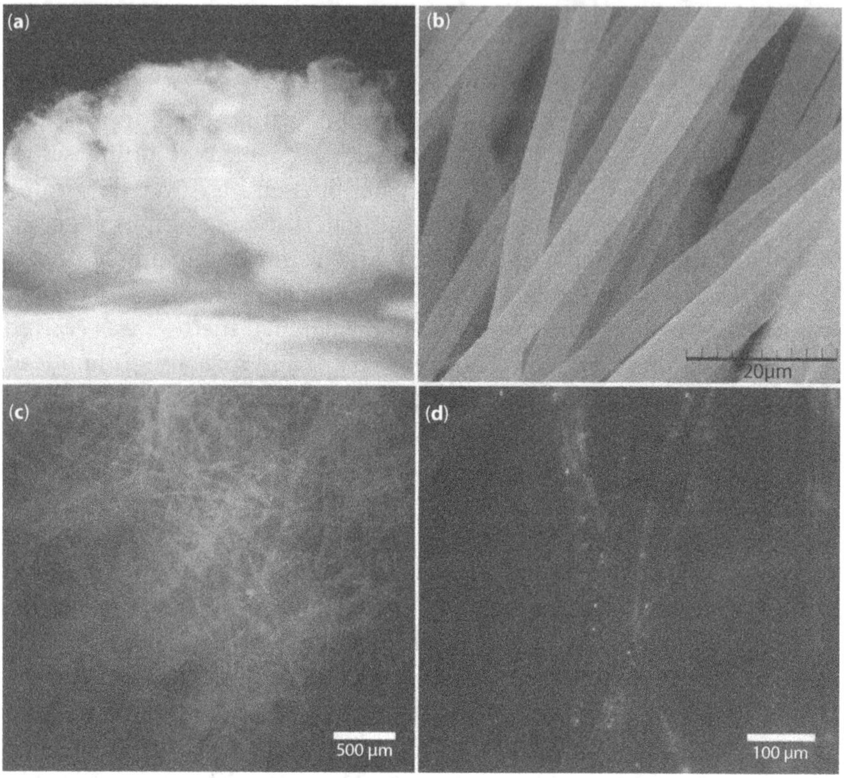

Figure 5.7 (a) Photograph, (b) SEM image, (c, d) Fluorescence microscopy images at different magnifications for electrospun $FAPbBr_3$ NC: PMMA fibers with polymer: NC mass ratio of 100:1 (1%wt). (Reproduced with permission from Ref. [34], Robust hydrophobic and hydrophilic polymer fibers sensitized by inorganic and hybrid lead halide perovskite nanocrystal emitter, *Front. Chem.* 2019).

transmission to the nanocrystals by the polymer matrix. Figure 5.8 shows the graphical studies of combined luminescence studies for water-soaked samples. The utmost optical stability in water surrounding is obtained by nanocrystals encapsulated poly methyl methacrylate fibers during electrospinning whose emission intensity is reduced for less than 15% after a week of water absorption. Nanocrystal encapsulated polyvinylpyrrolidone fibers possess steady emission in water immersion for the first 10–30 min, while the emission quenches completely after a period of a day, whereas, within 100 min of water immersion, the emission from the both pristine nanocrystals system instantly water suppressed. These studies revealed the utilization of electrospun LHP/polymer nanofibers to be the resourceful candidates for smart clothing.

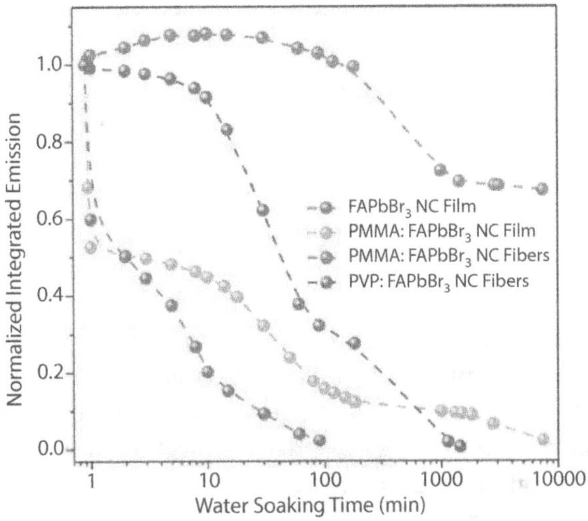

Figure 5.8 Integrated emission vs. water soaking time for NCs embedded in films and fibers. (Reproduced with permission from Ref. [34], Robust hydrophobic and hydrophilic polymer fibers sensitized by inorganic and hybrid lead halide perovskite nanocrystal emitter, *Front. Chem.* 2019).

5.3.3 Rare Earth Ion Doped Electrospun Nanofibers

Color tunability could be easily achieved by changing compositions, dopants, and host matrices. Rare earth ions are one such potential dopants used for light-emitting applications. Rare earth elements are the set of 17 elements in the periodic table, starting from Lanthanum (La) to Lutetium (Lu) altogether with Yttrium (Y) and Scandium (S). Rare earth elements are often termed as "Lanthanides". Lanthanide ions own enthralling optical properties and extremely been employed in solid-state lasers, luminescent lamps, optical fibers, flat displays, and other photonic devices for decades [35]. The lanthanides possess amazing properties due to their partially filled 4f orbitals which are shielded by the chemically stable and filled 5s, 5p orbitals [36]. Most of the lanthanides displays f-f transitions which are quite narrow and the chemical environment of the ion does not affect their optical properties. Therefore, the transitions from lanthanides are straightforwardly detectable, making them superlative applicants for optical probes [37].

Pangul and co-workers [38] fabricated electrospun Dy^{3+} doped ZnO nanofibers. *In situ* approach was followed to fabricate Dy^{3+} doped ZnO nanofibers. The precursor solution of zinc acetate and dysprosium nitrate was mixed to the polymeric solution of polyvinyl alcohol with additional acetic acid. This resultant solution was subjected to electrospinning procedure, and thereafter, the electrospun nanofibers were calcined to achieve Dy^{3+} doped ZnO nanofibers. The fibers were examined through various characterization techniques. The average diameter of the nanofibers was found in the range of 250 nm to 650 nm. Details of photoluminescence studies revealed the energy transfer mechanism within the host ZnO and dopant Dy, which enhances the emission from ZnO nanofibers in the visible region. ZnO following its characteristic emission emits in UV region and visible region emission is related to defect states. Upon addition of dopant into ZnO, the photoluminescence properties of ZnO changes drastically. The UV emission form ZnO is overcome by the characteristic emission of Dy. The emission in blue and yellow region corresponds to the magnetically allowed ${}^4F_{9/2} \rightarrow {}^6H_{15/2}$ transition and forced electrical ${}^4F_{9/2} \rightarrow {}^6H_{13/2}$ transition. Figure 5.9 is the Commission Internationale de l'Eclairage (CIE) diagram for ZnO nanofibers and Dy^{3+} doped ZnO nanofibers. The color transverse is very evident from the figure. The study of

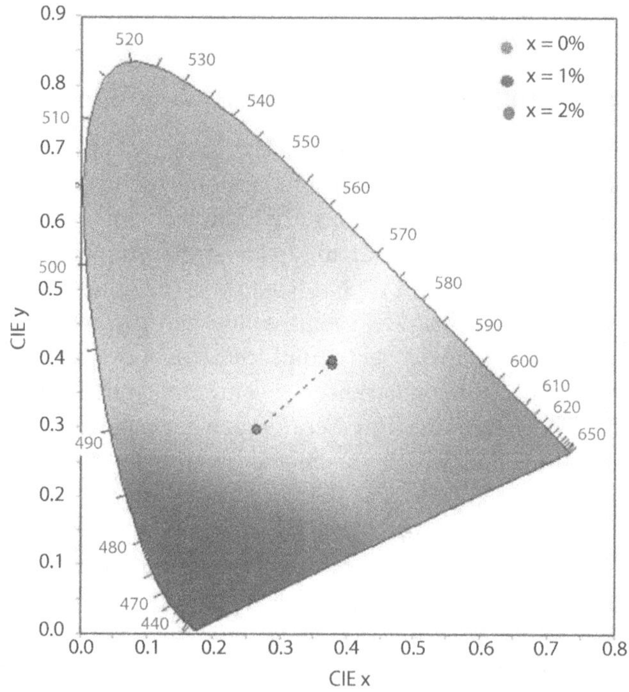

Figure 5.9 CIE diagram of emission spectra of x% Dy^{3+}-doped ZnO nanofibers (x = 0, 1, 2) (Reproduced with permission from Ref. [35], Enhanced photoluminescence properties of electrospun Dy^{3+} doped ZnO nanofibers for white lighting devices. *Luminescence*, 2018).

CIE parameters ensures the use of such kind of electrospun materials in LEDs, smart clothing, etc.

Chen *et al.* [39] studied Gd_2O_2S/Dy^{3+}, Eu^{3+} nanofibers through electrospinning united with dual boat sulfurization process. The average diameters of polyvinylpyrrolidone/($Gd(NO_3)^{3+}Dy(NO_3)_3$+$Eu(NO_3)_3$) electrospun nanofibers and Gd_2O_2S/Dy^{3+}, Eu^{3+} nanofibers are 454.10 ± 4.97 nm and 162.93 ± 0.75 nm respectively. The comparative change in diameter is observed due to removal of organic components during the process of calcinations. Fluorescence characterizations revealed the effective energy transfer mechanisms within the dopants. Inokuti-Hirayama (I-H) model was adopted to study the Dy^{3+} and Eu^{3+} energy transfer process and confirms the doublet-doublet interaction mechanism for energy transfer between Dy^{3+} and Eu^{3+}. This energy transfer mechanisms within the dopants enables the Gd_2O_2S/Dy^{3+}, Eu^{3+} nanofibers to emit white light which is the resultant of yellow, blue, and red light from dopant Dy^{3+} and Eu^{3+}. Such intelligent materials present a good platform for the utilization of electrospun nanomaterials in efficient LEDs.

Shiwei *et al.* [40] elaborated the thermal, structural, and optical properties of electrospun Europium/Terbium benzoic acid phen complex-polyacrylonitrile fibers. Initially, Terbium benzoic acid phen complex and Europium benzoic acid phen complex powders were prepared by solvo-thermal reactions which were then utilized along with polyacrylonitrile solution for electrospinning. In the fabrication process, three kinds of electrospun nanofibers are prepared, namely, Eu complex PAN fibers, Tb complex PAN fibers, and Eu/Tb complex PAN fibers. The diameters of the nanofibers were recorded from surface morphological studies and the diameter size was increased upon addition of the dopants. This is contributed to the increased viscosity of the solution upon addition of the complex dopant into the PAN solution. Fluorescence intensities for 2.5 wt % Eu complex PAN fibers and 2.0 wt % for Tb complex PAN fibers was bizarrely enhanced with the raising ratio of complexes into PAN. Electrospun Eu-Tb complex PAN fibers definitely marked this way in the field of LEDs.

Yu and co-workers [41] fabricated the core-shell wire designed flexible nanofibers via electrospinning. A heterogenous material was fabricated for studying the differentiation of photon generation which could be used for LEDs displays. Rare earth complexes, namely, Terbium 2, 4 pentane Dionte and Europium $(TTA)_3$phen, and polyacrylonitrile were utilized for electrospinning core-shell structure nanofibers and blended nanofibers. Figure 5.10a illustrates the detailed schematic method for fabrication of core-shell structure nanofibers. Figure 5.10b is the TEM image of the complex nanofibers presenting the core-shell structure of heterogenous coaxial

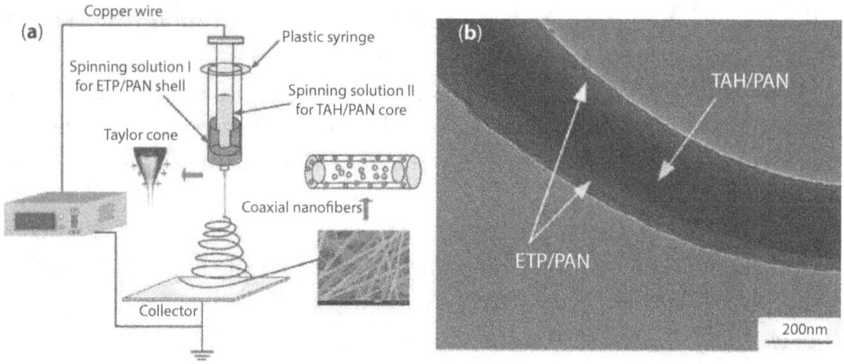

Figure 5.10 (a) Schematic representation of the electrospun setup for core-shell fiber structure and (b) TEM image of the core-shell nanofiber. (Reproduced with permission from Ref. [41], Differentiation of photon generation depended on electrospun configuration in Eu^{3+}/Tb^{3+} doped polyacrylonitrile nanofibers, *Journal of Alloys and Compounds*, 2019).

electrospun nanofiber. The comparative study of the blended and core-shell electrospun fibers presents probable applications from electrospun complexes of Tb^{3+}-Eu^{3+} doped poly acrylonitrile in light-emitting devices. Luminous flux and luminous intensity were calculated for both the variety of electrospun nanofibers. Both materials show intensifying tendency for luminous fluxes with the increase in europium complex concentration. The total luminous flux for blended electrospun fibers reaches to 558.3 mlm and for core-shell structured electrospun fibers reaches to 377.8 mlm. The luminous efficacies were found around 4.83 and 3.27 mlm/W, respectively, for fusion and core-shell nanofibers. The brightness parameter is enhanced for blended electrospun fibers than core-shell nanofibers which are devoted to the tougher interface within rare earth ions and ligands. Dimension expansion for color modification could be possibly achieved by strange spatial circulations of core-shell organized nanofibers but might have a minor variation in luminous intensity. The distinct spatial formation of coaxial fibers enables the shell to absorb more UV radiation which in turn allows less light to enter the core layer correspondingly. This results in reduction of characteristic peak intensity of Tb^{3+}a bit, whereas the intensity of Eu^{3+} is equivalently improved. As compared to blended fibers, the distinguished emission spectra of core-shell designed fibers show a remarkable deviation, which sidesteps the interface within the terbium and europium ions, but certainly holds the discrete fluorescence features of rare earth ions also. The terbium and europium ions co-doped complex fibers alter chromicity in reply to several peripheral spurs and constitute a striking alternate for color displays.

Dandekar *et al.* [42] presents the incorporation of rare earth ion complexes into polymer blends via facile method of electrospinning. Polymer blends of polyvinylidene fluoride (PVDF), polystyrene (PS), and polymethyl methacrylate (PMMA) were used to fabricate electrospun Europium complex PVDF/PS, Europium complex PVDF/PMMA, and Europium complex PS/PMMA nanofibers. The surface morphology seems to be randomly oriented nanofibers with variation of diameter. The diameter variation is contributed to the different polymeric solutions utilized for electrospinning. The photoluminescence spectra for Europium complex and Europium complex/polymer blends nanofibers were studied, and Europium complex/polymer blends show remarkable enhancement in the intensity. This improvement is due to operative ET (energy transfer) within organic ligands and central Eu^{3+} ions. The luminescence intensity is boosted in the rare earth ion complex merged polymers in comparison of $Eu(TTA)_3phen$ complex. Judd-Ofelt analysis was carried out considering hypersensitive $^5D_0 \rightarrow {}^7F_2$ shift and is extra subtle to the proportion and arrangement of ligand fields that denote that values of Ω_2 are higher for Europium complex/polymer blends nanofibers when compared to Europium complex. This suggests stronger hypersensitive behaviors of $^5D_0 \rightarrow {}^7F_2$ transition in the presence of polymer blend matrix [41].

Yu and co-workers [43] fabricated ultralong well-aligned TiO_2:Ln^{3+}[Ln = Eu, Sm, or Er] fibers arrays. Polyethylene oxide along with Tetra-N-Butyl Titanate ($Ti(OC_4H_9)_4$) and lanthanide nitrates were taken as precursors and electrospinning was performed. The ultrahigh molecular weight of polyethylene was used to fabricate highly aligned fibers. The high molecular weight helps to eliminate the bending instability of the electrospinning jet. The change in electrospinning parameter greatly influences the morphological properties of nanofibers. TiO_2 precursor fibers have random structures when collecting drum was allowed to rotate at speed of less than the 800 rpm, whereas, when the speed of collecting drum exceeded 1,100 rpm, aligned TiO_2 fibers were obtained. The alignment of the as-spun fibers found to advance when the speed of rotation of collecting drum was increased. Photoluminescence studies for Eu doped TiO_2, Sm doped TiO_2, and Er doped TiO_2 nanofibers illustrate the effective energy transfer mechanism between host TiO_2 and the respective dopants. Thus, variation of color range is achieved from variation of dopants which possess temperature dependent luminescence.

Another successful study of rare earth ion doped electrospun nanofibers was carried out by Shi and team [44]. The anhydrous SiO_2/$Tb(acac)_3phen$ (SiO_2:Tb^{3+}) nanoparticles were merged in polyvinylpyrrolidone (PVP) hybrid nanofibers via electrospinning. These electrospun nanofibers were used to examine the photoluminescence properties of SiO_2:Tb^{3+} nanoparticles under polymer environment. A strong green emission is seen from SiO_2:Tb^{3+}- PVP nanofibers, which is contributed to characteristic emission from Tb^{3+} ions. The life time of emission for Tb^{3+} doped SiO_2 nanoparticles is 1.54 ms and for SiO_2:Tb^{3+}-PVP nanofibers is 1.83 ms. The radiative and nonradiative transition rate of Tb^{3+} ions are the major factors affecting the lifetimes. The rate of radiative transition is effectively increased due to the presence of polymer matrix around the Tb^{3+} ions where the rare earth ions are located at an inversion symmetry. Thus, presence of polymer matrix has turned out to be advantageous for the long-term emission from SiO_2–Tb^{3+} ions.

5.3.4 Electrospun Coordination Polymeric Nanofibers

Coordination polymers are structures that could be extended to one, two, or three dimensions by mere linking metal centres to each other via organic ligands (linkers) through coordination bonds [45]. Group B elements are frequently employed in place of convenient jointers in case of coordination polymers. Linear, T or Y shaped, tetrahedral, square-planar, square-pyramidal, trigonal-bipyramidal, octahedral, trigonal-prismatic, pentagonal-bipyramidal, etc., geometries of coordination polymers could be framed which depends upon the metal and its oxidation state [46]. Considerable attention has been engrossed by luminescent lanthanide ions and organic joint ligands framing coordination polymer. Novel organic inorganic hybrid materials having variety of structures which possess unique physical properties are one of the characteristics exhibited by coordination polymers having alternating sequences of metal ions and organic ligands [47].

Batabyal and team [48] studied coordination polymeric nanofibers and their field-emission properties. Organic polymers exclusively possess 1-D structures while coordination polymers occur in crystal-like, 1-D, 2-D, and 3-D designs in the solid state. Coordination polymers in 1-D structure are apt for imitating the assets of organic polymers. Hence, electrospinning technique was utilized to fabricate nanofiber bundles from coordination polymer gels on the centimeter scale and in gram quantities. Inverted-test-tube method was utilized for the preparation of transparent greenish gel. Methanol was used to dilute the gel in the volume ratio of 1:1 followed by stirring. Viscosity, solvent evaporation rate, and composition of gel define the class of electrospun fibers. Electrospun coordination polymer nanofibers having Ni^{2+} ions in the presence of 4-trifluoromethylbenzoic acid were examined via field emission studies. Electrospun coordination polymer nanofibers are proved to bear outstanding photo emission and electron emission properties, upon which they are allowed to be a probable candidate for their utilization in light emitting devices and in optoelectronics [48].

Bai and co-workers [49] worked to synthesize novel material with introduction of a nitrile-CN group in a triazolepyridine supported copper iodide dimer for enhanced luminescence. The versatile copper catalyzedazide-alkyne cycloaddition reactions (CuAAC) reaction has been used to prepare pyridine triazole supported copper iodide complex (Complex 1) over which a nitrile (-CN) is replaced in benzyl hybridized pyridine triazole. The nitrile (-CN) substituted copper iodide complex (Complex 1) found to be compatible with PMMA for fabrication of electrospun nanofibers. Solution of Complex 1 was allowed to homogenize with PMMA solution via stirring for two days. The resultant solution was injected in syringe for electrospinning procedure. Photoluminescence studies explored the highly luminescence properties of Complex 1 which when incorporated in PMMA matrix enables the nanofibrous film to be luminescent. Such highly efficient materials open various ways for utilization of variety of elements for assortment of applications.

5.4 Conclusions

Even though a large number of nanofibers have been stated until now, here, only the tip of the iceberg with respect to the photonic application potential of electrospun nanofibers is reported. Electrospun nanofibers have no boundaries for the use of assortment of materials. Organic molecules can be used for producing OLEDs that can be incorporated into textiles, owing to their high luminous efficiency. Inorganic materials like semiconducting metal oxides possess high binding energies, wide band gap, and tailoring parameters which assures them to be the promising candidate in the field of light emission. With the variety of doping in the metal oxides semiconductors, the further properties could be enhanced and tailored very efficiently. Perovskite materials are another one such potential candidates for tuning the luminescent properties in the form of one-dimensional nanomaterials like nanofibers. Owing to the altering bandgap and huge mobility perovskite materials allows themselves as a contributor for the light-emitting devices. Rare earth ions undoubtedly will always be the integral part of the LEDs. The sharp spectral luminescence and participation in effective energy transfer mechanism labels rare earth ions to be the optimum dopants ever after. The rigidity, crystallinity, and poor luminous efficacies could be intelligently tackled with the combination of organic, inorganic, hybrid, and heterogenous materials which when takes the form of the electrospun nanomaterials enables their use in smart and technical textiles, LEDs. The forthcoming era is optimistic and extensively resourceful due to such electrospun nanomaterials.

References

1. Srinivasan, K., Rajanikumar, K., Sheetal Bhardwaj, K., Lalitha Kumari, B., Chavali, M., Nanotechnology Trends in Fashion and Textile Engineering. *Curr. Trends Fashion Technol. Textile Eng.*, 2, 3, 1–4, 2018. 10.19080/ctftte.2018.02.555590
2. Yetisen, A.K., Qu, H., Manbachi, A., Butt, H., Dokmeci, M.R., Hinestroza, J.P., Skorobogatiy, M., Khademhosseini, A., Yun, S.H., Nanotechnology in Textiles. *ACS Nano*, 10, 3, 3042–3068, 2016. 10.1021/acsnano.5b08176
3. Cherenack, K. and Pieterson, L., Smart textiles: Challenges and opportunities. *J. Appl. Phys.*, 112, 091301, 2012.
4. Syduzzaman, M., Patwary, S.U., Farhana, K., Ahmed, S., Smart Textiles and Nano-Technology: A General Overview. *J. Text. Sci. Eng.*, 5, 181, 2015. 10.4172/2165-8064.1000181
5. Huang, Z.M., Zhang, Y.Z., Kotaki, M., Ramakrishna, S., A review on polymer nanofibers by electrospinning and their applications in nanocomposites. *Compos. Sci. Technol.*, 63, 2223–2253, 2003.
6. Mirjalili, M. and Zohoori, S., Review for application of electrospinning and electrospun nanofibers technology in textile industry. *J. Nanostruct. Chem.*, 6, 207–213, 2016. 10.1007/s40097-016-0189-y
7. Liu, C., Ding, C., Zhao, T., Jiang, L., Synthesis and Characterization of Structure-Controlled Micro-Nanocomposite TiO_2 Fibers with Enhanced Photocatalytic Activity. *J. Nanomater.*, 2014 , ID 261034, 1–10, 2014. 10.1155/2014/261034
8. Zubia, J. and Arrue, J., Plastic optical fibers: An introduction to their technological processes and applications. *Opt. Fiber Technol.*, 7, 2, 101–140, 2001. 10.1006/ofte.2000.0355
9. Thenmozhi, S., Dharmaraj, N., Kadirvelu, K., Kim, H.Y., Electrospun nanofibers: New generation materials for advanced applications. *Mater. Sci. Eng. B*, 217, 36–48, 2017. 10.1016/j.mseb.2017.01.001
10. Tang, Y., Li, Z., Liang, G., Li, Z., Li, J., Yu, B., Enhancement of luminous efficacy for LED lamps by introducing polyacrylonitrile electrospinning nanofiber film. *Opt. Express*, 26, 21, 27716–27725, 2018. 10.1002/cphc.200500428
11. Wang, P., Wang, Y., Tong, L., Functionalized polymer nanofibers: A versatile platform for manipulating light at the nanoscale. *Light: Sci. Appl.*, 2, 102, 2013. 10.1038/lsa.2013.58
12. Bashouti, M., Salalha, W., Brumer, M., Zussman, E., Lifshitz, E., Alignment of Colloidal CdS Nanowires Embedded in Polymer Nanofibers by

Electrospinning. *ChemPhysChem*, 7(1), 102–106, 2006. 10.1002/cphc.200500428

13. Zheng, W.P., Zu, R.D., Zhong, L.W., Nanobelts of Semiconducting Oxides. *Science*, 291, 9, 2001. 10.1126/science.105 8120
14. Weixin, O., Feng, T., Jr-Hau, H., Xiaosheng, F., Enhancing the Photoelectric Performance of Photodetectors Based on Metal Oxide Semiconductors by Charge-Carrier Engineering. *Adv. Funct. Mater.*, 29(9), 1807672 (1-20) 2019. 10.1002/adfm.201807672
15. Somnath, B., Jakub, H., Stephen, L., Robert, B., Highly Localized Charge Transfer Excitons in Metal Oxide Semiconductors. *Nano Lett.*, 18, 2, 1228–1233, 2018. 10.1021/acs.nanolett.7b04818
16. Xinge, Y., Tobin, J.M., Antonio, F., Metal oxides for optoelectronic applications. *Nat. Mater.*, 15, 383, 2016. 10.1038/nmat4599
17. Guogang, L., Zhiyao, H., Chong, P., Wenxin, W., Ziyong, C., Chunxia, L., Hongzhou, L., Jun, L., Electrospinning Derived One-Dimensional LaOCl: Ln 3 + (Ln = Eu/Sm, Tb, Tm) Nanofibers, Nanotubes and Microbelts with Multicolor-Tunable Emission Properties. *Adv. Funct. Mater.*, 20, 3446–3456, 2010. 10.1002/adfm.201001114
18. Buhro, W.E., Semiconducting Nanocrystals Shape Matters. *Nat. Mater.*, 2, 138, 2003. 10.1038/nmat844
19. Wang, C., Yan, E., Huang, Z., Zhao, Q., Xin, Y., Fabrication of Highly Photoluminescent TiO_2/PPV Hybrid Nanoparticle-Polymer Fibers by Electrospinning. *Macromol. Rapid Commun.*, 28, 2, 205–209, 2007. 10.1002/marc.200600626
20. Muzammil, P., Munawar, B.S., Loganathan, R., Shakil, G., Electrospinning, Synthesis of GaN fibers by varying precursor concentrations using. *Optik - International Journal for Light and Electron Optics*, 184, 75–81, 2019. 10.1016/j.ijleo.2019.03.026
21. Manoel, S.N., Mário, C.A.O., Christian, T.D., Raquel, E.M.L., Nikifor, R., Cid, B.A., Leonardo, S.M., Helinando, P.O., Anderson, S.L.G., UV random laser emission from fexible ZnO-Ag-enriched electrospun cellulose acetate fiber matrix. *Sci. Rep.*, 9, 1765, 2019. 10.1038/s41598-019-48056-w
22. Kallol, M., Karthega, M., Batabyal, S.K., Light-dependent negative differential resistance in MEH-PPV decorated electrospun TiO_2 mat. *Appl. Phys. A*, 124, 4, 34, 2018. 10.1007/s00339-018-1758-3
23. Alberto, P., Luigi, R., Vito, F., Riccardo, C., Andrea, C., Filippo, F., Francesco, C., Dario, P., Luana, P., Low-defectiveness exfoliation of MoS_2 nanoparticles and their embedment in hybrid light-emitting polymer nanofibers. *Nanoscale*, 10, 21748, 2018. 10.1039/c8nr06294c
24. Tang, Y., Liang, G., Song, C., Li, Z., Yu, S., Ding, X., Enhancement of angular color uniformity of remote-phosphorconverted light-emitting diodes by electrospun-nanofiber diffusing films. *Mater. Lett.*, 227, 104–107, 2018. 10.1016/j.matlet.2018.05.058

25. Lin, K., Xing, J., Quan, L.N., Arguer, F.P.G., Gong, X., Lu, J., Xie, L., Zhao, W., Zhang, D., Yan, C., Li, W., Liu, X., Lu, Y., Kirman, J., Sargent, E., Xiong, Q., Wei, Z., Perovskite light-emitting diodes with external quantum efficiency exceeding 20 per cent. *Nature*, 562, 245–248, 2018. 10.1038/s41586-018-0575-3
26. Shen, X., Zhang, Y., Kershaw, S.V., Li, T., Wang, C., Zhang, X., Wang, W., Li, D., Wang, Y., Lu, M., Zhang, L., Sun, C., Zhao, D., Qin, G., Bai, X., Yu, W.W., Rogach, A.L., Zn-Alloyed CsPbI3 Nanocrystals for Highly Efficient Perovskite Light Emitting Devices. *Nano Lett.*, 19, 3, 1552–1559, 2019. 10.1021/acs.nanolett.8b04339
27. Sutherland, B.R. and Sargent, E.H., Perovskite photonic sources. *Nat. Photonics*, 10, 295–302, 2016. 10.1038/nphoton.2016.62
28. Qiu, X., Cao, B., Yuan, S., Chen, X., Qiu, Z., Jiang, Y., Ye, Q., Wang, H., Zheng, H., Liu, J., Kanatzidis, G., From unstable CsSnI3 to air-stable Cs2SnI6: A lead-free perovskite solar cell light absorber with bandgap of 1.48 eV and high absorption coefficient. *Sol. Energ. Mat. Sol. C.*, 159, 227–234, 2017. 10.1016/j.solmat.2016.09.022
29. Kim, H.J., Kim, U., Kim, H.M., Kim, T.H., Mun, H.S., Jeon, B.G., Hong, K.T., Lee, W.J., Ju, C., Kim, K.H., Char, K., High Mobility in a Stable Transparent Perovskite Oxide. *Appl. Phys. Express*, 5, 6, 061102, 2012. 29. 10.1143/apex.5.061102
30. Li, Y., Yan, W., Li, Y., Wang, S., Wang, W., Bian, Z., Xiao, L., Gong, Q., Direct Observation of Long Electron-Hole Diffusion Distance in CH3NH3PbI3 Perovskite Thin Film. *Sci. Rep.*, 5, 14485, 2015. 10.1038/srep14485
31. Shi, Y., Wu, W., Dong, H., Li, G., Xi, K., Divitini, G., Ran, C., Yuan, F., Zhang, M., Jiao, B., Hou, X., A Strategy for Architecture Design of Crystalline Perovskite Light-Emitting Diodes with High Performance. *Adv. Mater.*, 30, 25, 1800251, 2018. 31. 10.1002/adma.201800251
32. Ercan, E., Tsai, P.C., Chen, J.Y., Lam, J.Y., Hsu, L.C., Chueh, C.C., Chen, W.C., Stretchable and Ambient Stable Perovskite/Polymer Luminous Hybrid Nanofibers of Multicolor Fiber Mats and Their White LED Applications. *Appl. Mater. Interfaces*, 11, 26, 23605–23615, 2019. 10.1021/acsami.9b05527
33. Han, W., Cui, F., Si, Y., Mao, X., Ding, B., Fluorescent Nanofibrous Membranes: Self-Assembly of Perovskite Crystals Anchored Al_2O_3-La_2O_3 Nanofibrous Membranes with Robust Flexibility and Luminescence. *Small*, 14, 45, 1801963, 2018. 10.1002/smll.201801963
34. Paris, G.P., Andreas, M., Androniki, A., Petroula, K., Grigorios, I., Robust hydrophobic and hydrophilic polymer fibers sensitized by inorganic and hybrid lead halide perovskite nanocrystal emitter. *Front. Chem.*, 7, 87, 1–12, 2019. 10.3389/fchem.2019.00087
35. Luo, W., Liu, Y., Chen, X., Lanthanide-doped semiconductor nanocrystals: Electronic structures and optical properties. *Sci. China Mater.*, 58, 819–850, 2015. 10.1007/s40843-015-0091-9
36. Jacquier, Guokui, L., Bernard, Spectroscopic Properties of Rare Earths in Optical Materials, in: *Springer Series in Material Science, Springer-Verlag Berlin Heidelberg*, 83 (1), eISSN 978-3-540-28209-9, 2005. 10.1007/3-540-28209-2

37. Bunzli, J.C.G., Benefiting from the Unique Properties of Lanthanide Ions. *Acc. Chem. Res.*, 39, 53–61, 2006. 10.1021/ar0400894
38. Pangul, C.N., Anwane, S.W., Kondawar, S.B., Enhanced photoluminescence properties of electrospun Dy^{3+} doped ZnO nanofibres for white lighting devices. *Luminescence*, 33, 1087–1093, 2018. 10.1002/bio.3513
39. Chen, J., Song, Y., Li, D., Ma, Q., Dong, X., Yu, W., Wang, X., Yang, Y., Wang, J., Liu, G., Investigating Efficient Energy Transfer in Novel Strategy-obtained Gd_2O_2S:Dy^{3+} Eu^{3+} Nanofibers Endowed with White Emitting and Magnetic Dual-functionality. *J. Lumin.*, 206, 509–517, 2018. 10.1016/j.jlumin.2018.10.087
40. Shiwei, W., Guangbo, X., Jingjing, Z., Sen, Z., Tingju, L., Structure, thermal and luminescence properties of Eu/Tb(BA)3phen/PAN fibers fabricated by electrospinning. *Opt. Mater.*, 78, 445–451, 2018. 10.1016/j.optmat.2018.02.020
41. Yu, M., Shen, L.F., Li, D.S., Pun, E.Y.B., Guo, J., Lin, H., Differentiation of photon generation depended on electrospun configuration in Eu3þ/Tb3þ doped polyacrylonitrile nanofibers. *J. Alloys Compd.*, 786, 1040–1050, 2019. 10.1016/j.jallcom.2019.02.012
42. Dandekar, M.P., Itankar, S.G., Kondawar, S.B., Nandanwar, D.V., Koinkar, P., Photoluminescent electrospun europium complex $Eu(TTA)_3$phen embedded polymer blends nanofibers. *Opt. Mater.*, 85, 483–490, 2018. 10.1016/j.optmat.2018.09.019
43. Yu, H., Li, Y., Song, Y., Wu, Y., Lan, X., Liu, S., Tang, Y., Xu, S., Chen, B., Ultralong well-aligned TiO_2:Ln^{3+} (Ln = Eu, Sm, or Er) fibres prepared by modified electrospinning and their temperature-dependent luminescence. *Sci. Rep.*, 7, 44099, 2017. 10.1038/srep44099
44. Shi, J., Wang, Y., Huang, L., Lu, P., Sun, Q., Wang, Y., Tang, J., Belfiore, L.A., Kipper, M.J., Polyvinylpyrrolidone Nanofibers Encapsulating an Anhydrous Preparation of Fluorescent SiO_2–Tb^{3+} Nanoparticles. *Nanomaterials*, 9, 4, 510, 2019. 44. 10.3390/nano9040510
45. Loukopoulos, E. and Kostakis, G.E., Review: Recent advances of one-dimensional coordination polymers as catalysts. *J. Coord. Chem.*, 71, 3, 371–410, 2018. 10.1080/00958972.2018.1439163
46. Kitagawa, S., Kitaura, R., Noro, S., Functional Porous Coordination Polymers. *Angew. Chem. Int. Ed.*, 43, 2334–2375, 2004. 10.1002/anie.200300610
47. Hasegawa, Y. and Nakanishi, T., Luminescent lanthanide coordination polymers for photonic applications. *RSC Adv.*, 5, 338–353, 2015. 10.1039/c4ra09255d
48. Batabyal, S.K., Malik, A., Puthan, P., Ramakrishna, S., Haur, C.S., Vittal, J.J., Coordination-Polymeric Nanofibers and their Field-Emission Properties. *Macromol. Rapid Commun.*, 30, 1356–1361, 2009. 10.1002/marc.200900230
49. Bai, S.Q., Kai, D., Karen, L.K., Ming, L., Lu, J., Ying, J., David, J., Young, X., Jun, L., Xu, L., Andy, H., A Triazolyl-Pyridine-Supported CuI Dimer: Tunable Luminescence and Fabrication of Composite Fibers. *ChemPlusChem*, 80, 1235–1240, 2015. 10.1002/cplu.201500202

6

Luminescent Polymer Light-Emitting Devices and Displays

Nayan Ranjan Singha[1*], Pijush Kanti Chattopadhyay[2], Mousumi Deb[1], Mrinmoy Karmakar[1], Manas Mahapatra[1], Madhushree Mitra[2] and Arnab Dutta[1]

[1]Advanced Polymer Laboratory, Department of Polymer Science and Technology, Government College of Engineering and Leather Technology (Post Graduate), Salt Lake, Kolkata, India
[2]Advanced Polymer Laboratory, Department of Leather Technology, Government College of Engineering and Leather Technology (Post Graduate), Salt Lake, Kolkata, India

Abstract

The present century has observed rapid techno-commercial advancement in the field of luminescent polymer light-emitting devices and displays. Level of efficiencies as well as performance parameters, such as luminance or brightness and lifetime, have been improved remarkably, and such developments are still in progress. Enhanced efficiency, performance, longevity, and stability of polymeric light emitting diodes (PLEDs) can be attained achieving the optimum balance between hole and electron mobility, leading to balanced charge injection and transport. Nature and wavelength of the emitted light from PLEDs can be altered by structural tuning of the polymer chains *via* blending, doping, and chemical treatment through incorporation of spacer and/or narrow/wide band-gap moieties. Moreover, the elevated efficiency, longevity, and diversity in terms of color emission can be achieved by suitable modification of the hole/electron transporting/blocking nature of the adjacent layers and/or constitution of electrodes. Therefore, in this particular chapter, special attention has been paid to elucidate the roles of different injecting or blocking layers and their respective modifications in alterations of the efficiency and longevity of PLEDs, as well as the diversified color emanating from emitting layer(s). Finally, some recent developments have been highlighted

**Corresponding author*: drs.nrs@gmail.com

Inamuddin, Rajender Boddula, Mohd Imran Ahamed and Abdullah M. Asiri (eds.) Polymers for Light-Emitting Devices and Displays, (125–176) © 2020 Scrivener Publishing LLC

including quantum dot/pervoksite containing white light emitting PLED and PLEDs emitting thermally activated delayed fluorescence or near infrared light.

Keywords: Conventional/inverted top/bottom emitting diode, π-π stacking interactions, electron injection/transmission layer, hole injection/transmission layer, polaron exciton, thermally activated delayed fluorescence, quantum dot based composite, external quantum efficiency

Abbreviation

LED: light emitting diode, LD: laser diode, BDOH-PF: poly[9,9-bis(3,6-dioxaheptyl)-fluorene-2,7-diyl], LEC: light-emitting electrochemical cells, PEO: poly-ethylene oxide, PLED: polymer light emitting diode, EIL: electron-injection layer, ETL: electron-transport layer, EML: emitting layer, HTL: hole-transport layer, HIL: hole injection layer, HBL: hole blocking layer, EBL: electron blocking layer, ITO: indium tin oxide, TFB: poly(9,9-dioctylfluorene-alt-*N*-(4-sec-butylphenyl)-diphenylamine), TBT: 4,7-bis(3-hexylthien)-2,1,3-benzothiadiazole, PLQE: photoluminescence quantum efficiency, PFO: poly-(9,9-dioctylfluorene), TFB: poly(9,9-dioctylfluorene-alt-*N*-(4-sec-butylphenyl)-diphenylamine), PFB: poly 9,9'-dioctylfluorene-co-bis-*N*,*N*'-(4-butylphenyl) bis-*N*,*N*'-phenyl-1,4-phenylenediamine, MEH-PPV: poly(2-methoxy, 5-(2'-ethyl-hexoxy)-1,4-phenylenevinylene, PMMA: poly-(methyl methacrylate), TAZ: triazole, PVK: poly(*N*-vinylcarbazole), TQB: 1,3,5-tris(4-phenylquinolin-2-yl)benzene, TPBI: ,3,5-tris(1-phenyl-1H-benzimidazol-2-yl)benzene, PEDOT:PSS: poly(styrene sulfonate)-doped poly(3,4-ethylenedioxythiophene), GO: graphene oxide, MZO: magnesium-alloyed zinc oxide, IZO: indium zinc oxide, IO: indium oxide, EQE: external quantum efficiency, QD: quantum-dot, DPEPO: bis(2-(diphenylphosphino)phenyl) ether oxide, and CP: conducting polymer

6.1 Introduction

In general, a device is a piece of mechanical or electrical equipment fulfilling a specific purpose. Among the various kinds of devices, light-emitting devices or photonic devices play a pivotal role in a wide variety of diversified applications including optical communications, optical computing, and interconnects, data transmission and signal processing, optical storage, sensors and optical imaging, solid-state lamps, and displays. Mostly, light-emitting devices, such as LEDs, LDs, modulators, and photodetectors, are composed of semiconductor materials, and among the various types of semiconductor materials, III-V semiconductors are especially suitable for manufacturing light-emitting devices, as III-V semiconductors possess the basic advantages including high carrier mobilities and direct energy gaps, along with the availability of both n-type and p-type semiconductors. In addition to III-V semiconductors, II-VI semiconductors are also frequently used in light-emitting devices including LEDs. Technically, ZnTe is a kind of II-VI semiconductors, as Zn and Te are the member of group IIB and VI of the periodic table, whereas GaN is a type of III-V semiconductors, since Ga and N are positioned in the groups III and V of the periodic table. Moreover, both n-type and p-type semiconductors are extrinsic semiconductors obtained by doping a pure semiconductor crystal *via* addition of a trace element or chemical as doping agent. In this regard, an n-type semiconductor is a type of extrinsic semiconductor that has been doped with electron donor atoms. Accordingly, the majority of charge carriers in the n-type semiconductor crystal are electrons. In contrast, a p-type semiconductor is an extrinsic semiconductor that is doped with electron acceptor atoms, and thus, the majority of charge carriers in a p-type semiconductor crystal are positive holes or electron deficient centers.

Basically, polymer is a large molecule or macromolecule composed of many repeating subunits. Nowadays, large varieties of polymers are available worldwide, and these macromolecules are playing essential and ubiquitous roles in everyday life. In this regard, several conjugated polymeric materials are now used in making organic light-emitting devices. In fact, the development of polymeric semiconductors was rooted from the application of organic materials in the fabrication of practical electroluminescent devices. The basic advantage of organic materials is their extremely high fluorescence quantum efficiencies in the visible spectrum including the blue regions [1]. In 1987, for the first time, Tang and VanSlyke fabricated organic electroluminescent diodes *via* suitable assembly of aromatic diamine layer with another layer constituted of fluorescent metal-chelate complexes. Thus, organic materials including polymeric materials are ideally suited as devices for multicolor display applications. In this regard, an ideal polymer light-emitting device should envisage the following characteristics: low operating voltage, high electron/photon conversion efficiency, and high power efficiency [2]. Earlier, researchers encountered several problems, such as lack of long-term stability, weaker photoluminescence, and insufficient quantum yield, in developing the polymeric semiconductors. In this context, polyacetylene was the most widely studied light-emitting polymer, which showed very weak photoluminescence [3].

For the first time, Burroughes *et al.* introduced conjugated polymer-based semiconductors in fabricating organic light-emitting devices [3]. Herein, poly-*p* phenylene vinylene was used as an active element of large-area light-emitting diode. The basic advantage of a polymeric semiconductor is its ability to produce robust and useful structures based on the requirement and field of applications of the device. Thus, in contrast to an inorganic semiconductor, polymeric semiconductors can be tailor-made to produce devices for large-area display in a cost-effective manner. Though inorganic LEDs are currently entering the marketplace, because of suffering from a range of constraints, these are mainly suitable for point-source illumination rather than surface-source illumination.

Therefore, in this chapter, a detailed discussion has been furnished relating to history and chronological development of luminescent polymer as a component of efficient and multifunctional light-emitting devices and displays, followed by theoretical background of those photoluminescence phenomena. Finally, the most updated status of recently developed luminescent polymers has been discussed with special emphasis on the dependence of properties and performances on structure and constitution of luminescent polymers.

6.2 Chronological Development

Though the development of polymeric light-emitting devices was initiated in 1990 by Burroughes *et al.*, such development process was stimulated by the evolvement of first organic light-emitting diode fabricated in 1987 by Tang and VanSlyke (Table 6.1). However, the quantum efficiency of those polymeric light-emitting devices based on poly-p phenylene vinylene was about 0.05%, which was insufficient and non-practicable for large-scale usage. Nevertheless, such an effort gave birth to a new horizon in the field of applied materials science. Such poor quantum efficiency was explained by the failure at the polymer/thin metal interface because of excessive localized heating [3]. In this context, non-radiative processes at the defect sites reduced the quantum efficiency. To improve quantum efficiency and overall performance properties, same group developed copolymer-based light-emitting diode envisaging significantly higher quantum efficiency, which was 30-folds of the previously reported result [4]. Such elevated quantum efficiency was achieved changing the molecular and supramolecular architecture of the conjugated system via copolymerization of two different monomers, such as poly-*p* phenylene vinylene and poly-(2,5 dimethoxy *p* phenylene vinylene), wherein three different conjugated systems were generated via copolymerization and further treatments (Figure 6.1), which contributed in restricting the migration of excitons to the quenching center. Meanwhile, Braun and Heege confirmed the discovery of polymeric light-emitting devices by Burroughes *et al.* through fabrication of diodes based on semiconducting polymer, i.e., poly(2-methoxy, 5-(2′-ethyl-hexoxy)-1, 4-phenylenevinylene, via simplified casting process without post-processing steps or heating [5]. Later on, there was an increasing demand for polymeric diodes emitting lights of variable color, and in this context, Millard reported fabrication of polyfluorine-based diodes of higher efficiency, i.e., >20 lm W^{-1}, higher brightness, low operating bias, and excellent color stability [6]. In case of green-light-emitting polyfluorine-based polymeric structures having benzothiodiozyl functionalities (Figure 6.2), a consistent efficiency of more than 20 lm W^{-1} was repeatedly obtained.

In fact, the development of polyfluorine-based light-emitting diodes, a major class of polymeric emitters, was started with the successful synthesis of blue electroluminescent soluble poly(9,9-dialkylfluorene-2,7-diyl) in 1989 by Fukuda *et al.* [7]. For the first time, Ohmori *et al.* reported

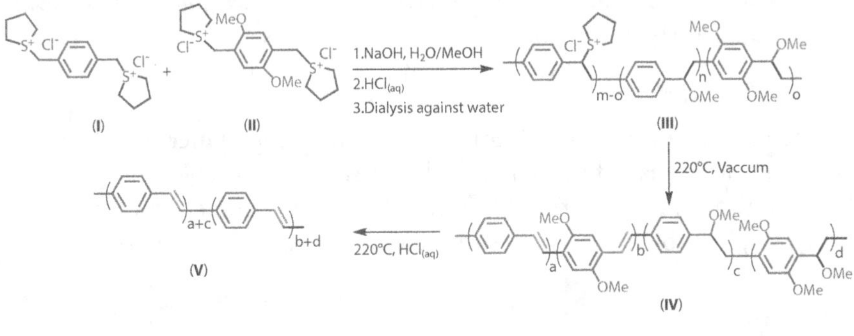

Figure 6.1 Copolymerization of (I) poly-p phenylene vinylene and (II) poly-(2,5 dimethoxy p phenylene vinylene) to produce three different copolymers having (III) non-conjugated, (IV) partially conjugated, and (V) fully conjugated moieties [4].

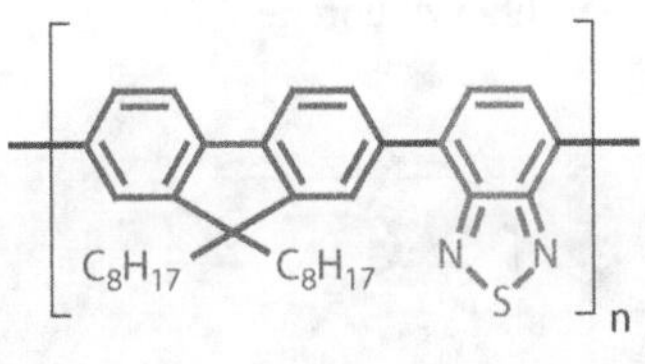

Figure 6.2 Structure of polyfluorine-based green light-emitting diode comprising of the benzothiodiozyl functionalities [6].

the electroluminescence of polyfluorine-based light-emitting diode [8], in which indium tin oxide, Mg:In, and poly(9,9-dihexylfluorene-2,7-diyl) were employed as anode, cathode, and light-emitting polyfluorine, respectively (Figure 6.3).

Later, Pei and Yang have studied photoluminescence and electroluminescence from BDOH-PF (Figure 6.4), synthesized substituting PF homopolymer with oligo(ethylene oxide) side chains [2, 9]. The main objective behind incorporation of the side chain was to promote the required ion transport for LECs. These LECs have performed better than the traditional light-emitting diodes and are capable to emit blue and/or green or white light with high quantum efficiency, high power efficiency, and high brightness at <4 V operating voltage [2] (Table 6.1). In this regard, the emission of blue and/or green or white light depends on the composition of polymeric component. For instance, if a blend of BDOH-PF, PEO, and $LiCF_3SO_3$ in 10:1:2 weight ratio of is incorporated within a LEC containing indium tin-oxide and aluminum as anode and cathode, respectively, LEC emits white light. However, in the absence of PEO, the similarly constructed LEC has been capable to emit only the blue light instead of white light. Because of the incorporation of PEO, the emission of white light with decreased quantum efficiency is associated with phase separation between PEO and BDOH-PF, leading to alterations in the stacking pattern of the BDOH-PF polymer chains. However, by that time, the exact details behind the origin of such color change have been unknown.

Thereafter, several polyfluorine and their copolymer-based light-emitting diodes have been fabricated by researchers (Table 6.1) [11–14]. By that time, LEDs fabricated from PF-homopolymers provided relatively brighter blue emission with higher efficiencies compared to the

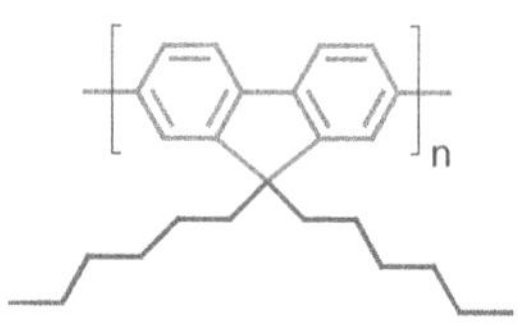

Figure 6.3 Structure of poly(9,9-dihexylfluorene-2,7-diyl).

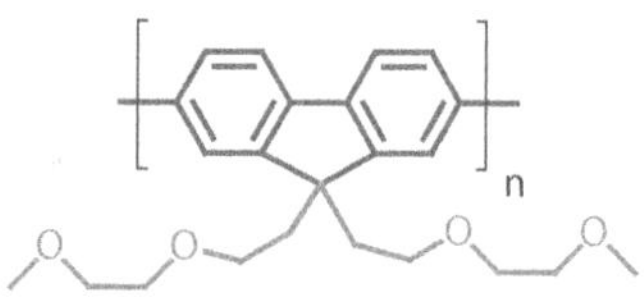

Figure 6.4 Structure of BDOH-PF [10].

Table 6.1 Chronological development and performances of luminescent polymer light-emitting devices and displays.

Year	Particulars	Device architecture	Properties								Remarks	Ref.
			EQE[a] (%, photons/ hole)	Luminous efficiency (lm W^{-1})	Brightness (cd m^{-2}) (operating voltage)	Emission wavelength (λ_{max}) (nm) (color)	Luminance efficiency (cd A^{-1})	Photoluminescence quantum efficiency (%)	Lifetime/ halflife (h)	Others		
1987	first organic LED[b]	ITO[c]//aromatic diamine+8-hydroxyquinoline aluminum//Mg:Ag	1	1.5	> 1000 (< 10 V)	550 (green)	–	–	–b	–	poor EQE[a] and luminous efficiency; emits green light	[1]
1989	electroluminescent PFs[d] synthesis	BF_4^- doped PFH[e]	–	–	–	– (blue)	–	–	–	exploited for future LED[b] making		[31]
1990	first polymer LED[b]	IO[f]//PPV[g]//Al	0.05	–	–	– (green-yellow)	–	–	–	for cheap large area. luminescent devices	low EQE[a] due to poor design	[3]
1991	first PF[d] based LED[b]	ITO[c]//PFH[e]//Mg:Ag	–	–	–	470 (blue)	–	–	–		efficiency and brightness not reported	[8]
1991	Shottky diodes from semiconducting polymers		–	–	–	–	–	–	–	simple direct casting without subsequent processing or heat treatment	solubility of polymer enhanced via chemical modification	[5]
1992	copolymer based LED[b]	ITO[c]//PPV[g]-*co*-PDMPV[h]// Al	> 1	–	–	– (green-blue to orange-red)	–	–	–	improved EQE[a] for electroluminescence	–	[4]
1994	MEH[i]-PPV[g] based LED[b]	ITO[c]//MEH[i]-PPV[g]//Ca	0.4	–	330 (6 V)	–	–	–	–	–	–	[32]

(Continued)

Table 6.1 Chronological development and performances of luminescent polymer light-emitting devices and displays. (*Continued*)

Year	Particulars	Device architecture	Properties								Remarks	Ref.
			EQE[a] (%, photons/hole)	Luminous efficiency (lm W^{-1})	Brightness (cd m^{-2}) (operating voltage)	Emission wavelength (λ_{max}) (nm) (color)	Luminance efficiency (cd A^{-1})	Photoluminescence quantum efficiency (%)	Lifetime/halflife (h)	Others		
1997	LECs[j] based on BDOH-PF[k]	ITO[c]//BDOH-PF[k]+lithium triflate//Al	4	12 (3.1 V) 8.6 (3.5 V)	200 (3.1 V) >1000 (3.5 V)	–	–	73	–	–	only emits blue-green	[2]
	LECs[j] based on blends of BDOH-PF[k] and PEO[l]	ITO[c]//BDOH-PF[k]+PEO[l]+lithium triflate//Al	2.4	–	400 (4 V)	– (white)	–	–	–	–	relatively inferior EQE[a] and brightness; exact reason unknown behind color change	
1998	MEH[i]-PPV[g] based PLED[m] containing ultrathin polymer coated/derivatized anode	silyl derivatized ITO[c]//MEH[i]-PPV[g]//Ca:Al	–	0.001 (5 V) 0.002 (7 V)	0.002 (5 V) 0.03 (7 V)	–	–	–	–	–	$–CH_2CH_2CH_2NMe_3^+$ treated ITO[c] as anode; poor performance	[33]
		conducting polymer modified ITO[c]//MEH[i]-PPV[g]//Ca:Al	–	0.200–0.220 (5 V) 0.170–0.210 (7 V)	100–400 (5 V) 1000–6000 (7 V)	–	–	–	–	–	$–CH_2CH_2CH_2NMe_3^+$ treated ITO[c] coated with sulfonated polyaniline and polyhexadiemthine	
		semi-conducting polymer modified ITO[c]//MEH[i]-PPV[g]//Ca:Al	–	0.070–0.260 (5 V) 0.200–0.290 (7 V)	0.5–300 (5 V) 80–6000 (7 V)	–	–	–	–	–	$CH_2CH_2CH_2NMe_3^+$ treated ITO[c] coated with PSS[w] and PPV[g] precursor	
		Insulating polymer modified ITO[c]//MEH[i]-PPV[g]//Ca:Al	–	0.200–0.390 (5 V) 0.280–0.410 (7 V)	0.6–110 (5 V) 60–2500 (7 V)	–	–	–	–	–	$CH_2CH_2CH_2NMe_3^+$ treated ITO[c] coated with PSS[w] and polyallylammonium	

(*Continued*)

Table 6.1 Chronological development and performances of luminescent polymer light-emitting devices and displays. (*Continued*)

Year	Particulars	Device architecture	Properties								Remarks	Ref.
			EQE[a] (%, photons/hole)	Luminous efficiency (lm W^{-1})	Brightness (cd m^{-2}) (operating voltage)	Emission wavelength (λ_{max}) (nm) (color)	Luminance efficiency (cd A^{-1})	Photoluminescence quantum efficiency (%)	Lifetime/ halflife (h)	Others		
1998	PTPD[n] and PFO[o] based double-layered LED[b]	ITO[c]//PTPD[n]+PFO[o]// Ca	0.2	0.04 (20 V)	600 (20 V)	436 (blue)	0.25	55 ± 5	–	–		[34]
1999	single-/double-/triple-layered LED[b] based on crosslinkable polymer	ITO[c]//x-DHF[p]//Ca	0.05	0.001	–	– (blue)	–	–	–	–	poor performances of single layered LED[b]	[35]
		ITO[c]//x-HTPA[q]//Ca	0.02	0.004	–	– (red-orange)	–	–	–	–		
		ITO[c]//x-HTPA[q]+x-DHF[p]// Ca	0.3	0.08	–	– (blue and red-orange)	–	–	–	–	better performances for double layered LED[b] than single layered LED[b]	
		ITO[c]//x-DHF[p]+3-Oxd[r]// Ca	0.3	0.03	–	425 and 450 (blue)	–	–	–	–		
		ITO[c]//x-HTPA[q]+x-DHF[p]+3-Oxd[r]// Ca	1.48	0.4	5.6–1526 (10 V)	425–450 (blue)	–	–	–	–	best performances for triple layered LED[b]; brightness reduces due to enhanced relative thickness of x-DHF[p]	
1999	Polarized light-emitting PF[d] based LED[b]	ITO[c]//(17% ST638[s]/83% PI[t]):PFEH[u]//Ca:Al	–	–	45 (19 V)	477 (blue)	–	–	–	–	ST638[s] and PI[t] as hole conducting filler and inert matrix, respectively	[15]

(*Continued*)

Table 6.1 Chronological development and performances of luminescent polymer light-emitting devices and displays. (*Continued*)

Year	Particulars	Device architecture	Properties								Remarks	Ref.
			EQE[a] (%, photons/ hole)	Luminous efficiency (lm W^{-1})	Brightness (cd m^{-2}) (operating voltage)	Emission wavelength (λ_{max}) (nm) (color)	Luminance efficiency (cd A^{-1})	Photoluminescence quantum efficiency (%)	Lifetime/ halflife (h)	Others		
1999	PF[d] LED[b] (Anode modification effect)	unmodified ITO[c]// PEDOT[v]+PSS[w]// Ca:Al	–	6 (4.4 V)	–	– (green)	6.70	–	120/–	–	PSS[w] as dopant; oxygen-plasma; modified ITO[c] performs better than aquaregia modified ITO[c]	[36]
		oxygen plasma Modified ITO[c]// PEDOT[v]+PSS[w]// Ca:Al	–	5.5 (4.45 V)	–	– (green)	6	–	335/–	–		
		aquaregia modified ITO[c]// PEDOT[v]+PSS[w]// Ca:Al	–	4.6 (> 9.1 V)	–	– (green)	5.40	–	< 0.15/–	–		
2000	LED[b] having blend of homo- and co-polymers of BDOH-PF[k] end-capped with anthracenes	ITO[c]//anthracene-BDOH-PF[k] + anthracene-BDOH-PF[k]-*co*-anthracene+lithium triflate//Al	–	–	100 (4–4.5 V)	425–475 (blue)	–	–	–	–	anthracene end capping/ copolymerization prevents excimer formation	[37]

(*Continued*)

Table 6.1 Chronological development and performances of luminescent polymer light-emitting devices and displays. (*Continued*)

Year	Particulars	Device architecture	Properties								Remarks	Ref.
			EQE[a] (%, photons/hole)	Luminous efficiency (lm W^{-1})	Brightness (cd m^{-2}) (operating voltage)	Emission wavelength (λ_{max}) (nm) (color)	Luminance efficiency (cd A^{-1})	Photoluminescence quantum efficiency (%)	Lifetime/halflife (h)	Others		
2000	PF[d] based LED[b]	ITO[c]//PF[d] based polymer bearing benzothiodiozyl group//alkali metals and alkali metal halides	5	20 (2.6 V) 14.5 (3.2 V)	100 (2.6 V) 1000 (3.2 V)	525–550 (green)	–	–	–	excellent color stability	blue color emitting devices give lower efficiencies than green color emitting device	[6]
		ITO[c]//PF[d] based polymer bearing amine groups//alkali metals and alkali metal halides	2	3.4 (3.6 V) 2.4 (5.3 V)	100 (3.4 V) 1000 (5.3 V)	425–475 (blue)	–	–	–	–		
		ITO[c]//PF[d] based polymer bearing anthracene groups//alkali metals and alkali metal halides	2	1.3 (3.7 V) 1.2 (4.8 V)	100 (3.7 V) 1000 (4.8 V)	425–475 (blue)	–	–	–	–		
2002	effect of PPV[g] as EBL[z] in PF[d] blend based PLED[m]	ITO[c]//PEDOT[v]+PSS[w]/F8BT[x]+PFO[o]//Ca	–	–	0.1 (3.5 V) 2600 (9 V)	–	2.1	–	–	–	PSS[w] doped PEDOT[v] as HIL[y]; no EBL[z]; F8BT[x]+PFO[o] blend as EML[ab]	[38]
		ITO[c]//PEDOT[v]+PSS[w]/PPV[g]/F8BT[x]+PFO[o]//Ca	–	–	3500 (12.5 V)	–	4.1	–	–	–	PPV[g] as EBL[z]	

(*Continued*)

Table 6.1 Chronological development and performances of luminescent polymer light-emitting devices and displays. (*Continued*)

Year	Particulars	Device architecture	Properties								Remarks	Ref.
			EQE[a] (%, photons/ hole)	Luminous efficiency (lm W^{-1})	Brightness (cd m^{-2}) (operating voltage)	Emission wavelength (λ_{max}) (nm) (color)	Luminance efficiency (cd A^{-1})	Photolumine scence quantum efficiency (%)	Lifetime/ halflife (h)	Others		
2003	PLED[m] containing heterojunctions in EML[ab] constituting of polymer blends	oxygen plasma modified ITO[c]// PEDOT[v]+PSS[w]/ TFB[aa]+F8BT[x]//Ca:Al	–	> 19 (2.1 V)	100 (2.1 V)	– (green)	–	–	>> 1000/–	–	TFB[aa]+F8BT[x] or PFB[ac]+F8BT[x] blend as EML[ab]; PSS[w] doped PEDOT[v] as HIL[y]	[39]
		oxygen plasma modified ITO[c]// PEDOT[v]+PSS[w]/ PFB[ac]+F8BT[x]//Ca:Al	–	–	<100 (2.1 V)	–	–	–	–	–		
2003	variable light-emitting domains of EML[ab] in blend of conjugated MEH[i]-PPV[g] with unconjugated PMMA[ad]	ITO[c]//PEDOT[v]+PSS[w]/ MEH[i]-PPV[g]+ PMMA[ad]//Ca:Al	–	–	–	–	–	–	–	highest efficiency	PSS[w] doped PEDOT[v] as the HIL[y]; Blend having 75% MEH[i]-PPV[g] (conjugated emitting) and 25% PMMA[ad] (unconjugated)	[40]

(*Continued*)

Table 6.1 Chronological development and performances of luminescent polymer light-emitting devices and displays. (*Continued*)

Year	Particulars	Device architecture	Properties								Remarks	Ref.
			EQE[a] (%, photons/hole)	Luminous efficiency (lm W^{-1})	Brightness (cd m^{-2}) (operating voltage)	Emission wavelength (λ_{max}) (nm) (color)	Luminance efficiency (cd A^{-1})	Photoluminescence quantum efficiency (%)	Lifetime/halflife (h)	Others		
2005	increasingly efficient PLED[m] by endcapping of EML[ab]	ITO[c]//PEDOT[v]/amorphous-PFO[o]-end-TBP[ae]//CsF/Al	–	–	2500	–	0.74	–	–	–	PEDOT[v] as HIL[y]	[41]
		ITO[c]//PEDOT[v]/β-phase-PFO[o]-end-TBP[ae]//CsF/Al	–	–	8200	–	1.23	–	–	–		
		ITO[c]//PEDOT[v]/PFO[o]-end-Oxd[r]//CsF/Al	–	–	3400	–	0.56	–	–	–		
		ITO[c]//PEDOT[v]/PFO[o]-end-TAZ[af]//CsF/Al	–	–	4000	–	1.67	–	–	–		

(*Continued*)

Table 6.1 Chronological development and performances of luminescent polymer light-emitting devices and displays. (*Continued*)

Year	Particulars	Device architecture	Properties								Remarks	Ref.
			EQE[a] (%, photons/hole)	Luminous efficiency (lm W^{-1})	Brightness (cd m^{-2}) (operating voltage)	Emission wavelength (λ_{max}) (nm) (color)	Luminance efficiency (cd A^{-1})	Photoluminescence quantum efficiency (%)	Lifetime/ halflife (h)	Others		
2005	suppression of fluorenone defects by copolymerization with non-planar copolymers	ITO[c]//PEDOT[v]/PFH[e]//Ca/Al	–	–	–	–	0.51	0.2	–	–	PEDOT[v] as HIL[y]; DHDBO[ag], DHDBT[ah], DHDBTS[ai], DHODBA[aj], and DHHPDBA[ak] copolymers to introduce nonplanarity	[42]
		ITO[c]//PEDOT[v]/PFH[e]-*co*-DHDBO[ag]//Ca/Al	–	–	–	–	0.24	0.5	–	–		
		ITO[c]//PEDOT[v]/PFH[e]-*co*-DHDBT[ah]//Ca/Al	–	–	–	–	0.03	0.05	–	–		
		ITO[c]//PEDOT[v]/PFH[e]-*co*-DHDBTS[ai]//Ca/Al	–	–	–	–	0.09	0.5	–	–		
		ITO[c]//PEDOT[v]/PFH[e]-*co*-DHODBA[aj]//Ca/Al	–	–	–	–	0.01	0.25	–	–		
		ITO[c]//PEDOT[v]/PFH[e]-*co*-DHHPDBA[ak]//Ca/Al	–	–	–	–	0.1	0.09	–	–		
2008	application of HIL[y] in PLED[m] for enhanced brightness	Au//F8BT[x]//ITO[c]	–	–	200 (8 V)	–	–	–	–	–	very low brightness; F8BT[x] as EML[ab]	[43]
		Au//F8BT[x]/TiO_2//ITO[c]	–	–	2500 (6 V)	–	0.1	–	–	–	better brightness; TiO_2 as EIL[al]	
		Au//MoO_3/F8BT[x]/TiO_2//ITO[c]	0.2	–	5700 (8 V)	–	0.6	–	–	–	brightness enhanced further; MoO_3 as HIL[y]	

(*Continued*)

Table 6.1 Chronological development and performances of luminescent polymer light-emitting devices and displays. (*Continued*)

Year	Particulars	Device architecture	Properties								Remarks	Ref.
			EQE[a] (%, photons/hole)	Luminous efficiency (lm W^{-1})	Brightness (cd m^{-2}) (operating voltage)	Emission wavelength (λ_{max}) (nm) (color)	Luminance efficiency (cd A^{-1})	Photoluminescence quantum efficiency (%)	Lifetime/halflife (h)	Others		
2009	air-stable PLED[m] of enhanced lifetime	Au//MoO_3/PPV[g] co-polymer/Cs_2CO_3/ZnO//FTO[am]	–	–	20000 (10 V)	–	8	–	100000/–	–	excellent lifetime; ZnO as EIL[al]; Cs_2CO_3 as HBL[an] and electron injection accelerator; PPV[g] co-polymer as EML[ab]; MoO_3 as HIL[y]	[44]
2010	ETL[ao] containing PLED[m]	ITO[c]//PEDOT[v]:PSS[w]/EML[ab]//Al	0.02–0.13	–	21.6 (16.7 V) 4 (13.9 V)	–	0.1 (16.7 V) 0.2 (13.9 V)	–	–	–	ETL[ao] absent; PSS[w] doped PEDOT[v] as HIL[y]; triplet emitter FIrpic[ay]-doped PVK[aq]+OXD-7[ar] blend as EML[ab]	[45]
		ITO[c]//PEDOT[v]:PSS[w]/EML[ab]/ETL[ao]//Al	4–6.9	–	4020 (16.6 V) 1060 (14 V)	–	7.3 (16.6 V) 12.2 (14 V)	–	–	–	TQB as ETL[ao] is vacuum deposited	
		ITO[c]//PEDOT[v]:PSS[w]/EML[ab]/ETL[ao]//Al	15.5	–	2790 (10 V)	–	28.3 (10 V)	–	–	–	ETL[ao] is solution processed and not vacuum deposited	
2011	ETL[ao] containing solution processed PLED[m] emitting yellow light	ITO[c]//PEDOT[v]:PSS[w]/EML[ab]//CsF/Al	–	–	–	–	–	–	–	all the parameters are lesser than the PLED[m] containing ETL[ao] and HBL[an]	PEDOT[v]:PSS[w] as HTL[ap] and HIL[y]; Blend of PVK[aq], OXD-7[ar], and $Ir(FP)_3^{az}$ as EML[ab]	[46]
		ITO[c]//PEDOT[v]:PSS[w]/EML[ab]/ETL[ao]/HBL[an]//LiF/Al	12.7	12.5	23926	–	41.7	–	–	–	mixture of TmPyPB[bg], TAZ[af], and TPBI[as] as ETL[ao]; Bphen (Figure 6.5a) as HBL[an]	

(*Continued*)

Table 6.1 Chronological development and performances of luminescent polymer light-emitting devices and displays. (*Continued*)

Year	Particulars	Device architecture	Properties								Remarks	Ref.
			EQE[a] (%, photons/ hole)	Luminous efficiency (lm W^{-1})	Brightness (cd m^{-2}) (operating voltage)	Emission wavelength (λ_{max}) (nm) (color)	Luminance efficiency (cd A^{-1})	Photoluminescence quantum efficiency (%)	Lifetime/ halflife (h)	Others		
2011	hybrid bipolar EML[ab] based on aryl F8 structure	ITO[c]//ZnO/Cs_2CO_3/ aryl-F8/MoO_3//Au	0.5	–	1000 (10 V)	–	1 (11 V)	–	–	–	EML[ab] thickness (optimized) = 500 nm	[47]
		ITO[c]//ZnO/Cs_2CO_3/ aryl-F8+TFB[aa]/ MoO_3//Au	2.5–3.1	–	1000 (11–14.8 V)	–	5–5.4 (12–19 V)	–	–	–	TFB[aa] as Hole trapping layer; EQE[a] enhanced in presence of TFB[aa]	
2012	PLED[m] containing GO[at] as HTL[ap]	ITO[c]//PPV[g]//LiF/Al	0.6 (8.4 V)	0.6 (6.6 V)	700 (16 V)	–	1.4 (8.4 V)	–	–	–	HTL[ap] is absent	[48]
		ITO[c]//PEDOT[v]:PSS[w]/ PPV[g]//LiF/Al	3.5 (9.2 V)	3.9 (5.2 V)	33800 (12.6 V)	–	8.7 (9.6 V)	–	–	–	PEDOT[v]:PSS[w] as HTL[ap]	
		ITO[c]//GO[at]/PPV[g]// LiF/Al	3.3 (8.2 V)	4.2 (5 V)	31400 (12.4 V)	–	8.8 (9.4 V)	–	–	–	GO[at] as HTL[ap] (thickness = 2.0 nm)	
			6.7 (6.8 V)	11 (4.4 V)	39000 (10.8 V)	–	19.1 (6.8 V)	–	–	–	GO[at] as HTL[ap] (thickness = 4.3 nm)	
			5 (at 7.4 V)	8.6 (at 4 V)	28500 (11.2 V)	–	13.9 (7.4 V)	–	–	–	GO[at] as HTL[ap] (thickness = 5.2 nm)	

(*Continued*)

Table 6.1 Chronological development and performances of luminescent polymer light-emitting devices and displays. (*Continued*)

Year	Particulars	Device architecture	Properties								Remarks	Ref.
			EQE[a] (%, photons/hole)	Luminous efficiency (lm W^{-1})	Brightness (cd m^{-2}) (operating voltage)	Emission wavelength (λ_{max}) (nm) (color)	Luminance efficiency (cd A^{-1})	Photoluminescence quantum efficiency (%)	Lifetime/halflife (h)	Others		
2014	single layered inverted-structured PLED[m] containing amine modified ZnO-R[au]	Au//MoO_3/F8BT[x]/ZnO-R[au]-2-ME[av]+EA[aw]//FTO[am]	17.8	–	53400 (20.6 V)	–	61.6	–	–	–	inverted-structured PLED[m] of enhanced luminance and efficiencies spontaneously formed ZnO-R[au] as EIL[al]; 2-ME[av] + EA[aw] as polar solvent interlayer; F8BT[x] as EML[ab]; MoO_3 as HIL[y]	[49]
2017	role of annealing and solvents in solution processed PLED[m] of enhanced efficiency	ITOc//PEDOT[v]:PSS[w]:PFI[ax]/PVK[aq]:$Ir(ppy)_3$[ba]/TPBI[as]//LiF/Al	5.5	–	5467	–	17.2	–	–	EML[ab] spin-coated by chlorobenzene; annealed	$Ir(ppy)_3$[ba] doped PVK[aq] as EMLab; HIL[y] = PEDOT[v]:PSS[w]:PFI[ax] (a terpolymer of PEDOT[v], PSS[w], and PFI[ax]); TPBI[as] as ETL[ao]	[50]
2017	flexible and transparent PLED[m]	CP[bb]//ZnO/PEIE[bc]/EML[ab]/MoO_x//CP[bb]	–	–	3200 (12 V)	– (yellow)	2.8	–	–	–	anode and cathode are made of conducting polymers; ZnO = EIL[al]; MoO_x as HIL[y]	[51]
			–	–	1200 (14 V)	– (yellow)	2.7	–	–	–		

(*Continued*)

Table 6.1 Chronological development and performances of luminescent polymer light-emitting devices and displays. (*Continued*)

Year	Particulars	Device architecture	Properties								Remarks	Ref.
			EQE[a] (%, photons/ hole)	Luminous efficiency (lm W^{-1})	Brightness (cd m^{-2}) (operating voltage)	Emission wavelength (λ_{max}) (nm) (color)	Luminance efficiency (cd A^{-1})	Photoluminescence quantum efficiency (%)	Lifetime/ halflife (h)	Others		
2018	perovskite–polymer based hetero-structured LED[b] of huge EQE[a]	ITO[c]//MZO[bd]/PEIE[bc]/ PPBH[be]/TFB[aa]+PFO[o]/ MoO_3//Au	~ 20	–	–	–	–	~ 96	46/–	at 0.1–1 mA cm^{-2}	PPBH[be] as EML[ab]; MZO[bd] as ETL[ao] cum HBL[an]; TFB[aa]+PFO[o] as HTL[ap] cum EBL[z]; MoO_3 as HIL[y]; ITO[c] = cathode; Au = anode	[52]
2018	thermally activated delayed fluorescence emitting polymers	ITO[c]//PEDOT[v]:PSS[w]/ EML[ab]/DPEPO[bf]/ TmPyPB[bg]/Liq[bh]//Al	7.3	–	–	478 (blue)	13.5	–	–	–	i. synthesized by ring-opening metathesis polymerization ii. EML[ab] = copolymer of two different norbornene monomers; PEDOT[v]:PSS[w] as HIL[y]; DPEPO[bf] = exciton blocking layer; TmPyPB[bg] as ETL[ao]; Liq[bh] = EIL[al]; ITO[c] as anode and Al = cathode	[53]
2018	effect of organic dye doped EIL[al] in inverted PLED[m]	Al//MoO_3/EML[ab]/ ZnO:PBI-H//ITO[c]	–	–	21050	–	15.4	–	–	–	PBI-H (Figure 6.5b) is the organic dye	[54]

(*Continued*)

Table 6.1 Chronological development and performances of luminescent polymer light-emitting devices and displays. (*Continued*)

			Properties									
Year	Particulars	Device architecture	EQE[a] (%, photons/hole)	Luminous efficiency (lm W^{-1})	Brightness (cd m^{-2}) (operating voltage)	Emission wavelength (λ_{max}) (nm) (color)	Luminance efficiency (cd A^{-1})	Photoluminescence quantum efficiency (%)	Lifetime/halflife (h)	Others	Remarks	Ref.
2018	white light-emitting PLED[m]	ITO[c] //PEDOT[v]:PSS[w] /PVK[aq] /PBD[bi] /Ir(ppy)$_3$[ba] /Ir(piq)$_2$ (acac)[bj] /B2PPQ[bk] //LIF/	–	–	12000	–	7	–	–	–	PEDOT[v]:PSS[w] as HIL[y]; Ir(piq)$_2$(acac)[bj] as red emitter (λ = 620 nm); Ir(ppy)$_3$[ba] as green emitter (λ = 508 nm); B2PPQ[bk] as blue emitter (λ = 457 nm); PBD[bi] = ETL[ao]; PVK[aq] = EML[ab]	[57]

[a]External quantum efficiency, [b]light-emitting diode, [c]indium tin-oxide, [d]polyfluorenes, [e]poly-(9,9-dihexylfluorene), [f]indium oxide, [g]poly-*p* phenylene vinylene, [h]poly-(2,5 dimethoxy *p* phenylene vinylene), [i]poly(2-methoxy, 5-(2'-ethyl-hexoxy)-1,4-phenylenevinylene, [j]Light-emitting electrochemical cell, [k]poly[9,9-bis(3,6-dioxaheptyl)-fluorene-2,7-diyl], [l]poly-ethylene oxide; [m]polymer light-emitting diode, [n]polymeric triphenyldiamine derivative, [o]poly-(9,9-dioctylfluorene), [p]crosslinked 9,9-di-n-hexylfluorene, [q]poly-[(4-n-hexyltriphenyl)amine], [r]*tris*-oxadiazole, [s]4,4',4''-tris(1-naphthyl)-N-phenyl-amino)-triphenylamine, [t]poly(phenoxyphenylimide), [u]poly(9,9-di(ethylhexyl)fluorene), [v]poly(3,4-ethylene dioxythiophene), [w]poly-(styrene sulfonate), [x]poly 9,9'-dioctylfluorene-alt-benzothiadiazole, [y]hole injection layer, [z]electron blocking layer, [aa]poly 9,9'-dioctylfluorene-*co*-N-(4-butylphenyl)diphenylamine, [ab]emitting layer, [ac]poly 9,9'-dioctylfluorene-*co-bis*-N,N'-(4-butylphenyl) *bis*-N,N'- phenyl-1,4-phenylenediamine, [ad]poly-(methyl methacrylate), [ae]para-tert-butyl phenyl, [af]triazole, [ag]5,7-dihydro-dibenz[*c,e*]oxepin, [ah]5,7-dihydro-dibenz[*c,e*]thiepin, [ai]5,7-dihydro-dibenz[*c,e*]thiepinsulfone, [aj]5,7-dihydro-*N*-octyl-dibenz[*c,e*]azepin, [ak]5,7-dihydro-*N*-(4 hexylphenyl)-dibenz[*c,e*]azepin, [al]electron-injection layer, [am]fluorine-doped tin oxide, [an]hole blocking layer, [ao]electron transport layer, [ap]hole transport layer, [aq]poly(N-vinylcarbazole), [ar]1,3-bis(2-(4- *tert*-butylphenyl)-1,3,4-oxadiazo-5-yl)benzene, [as]1,3,5-tris(1-phenyl-1H-benzimidazol-2-yl)benzene, [at]graphene oxide, [au]ripple-shaped nanostructure of ZnO, [av]2-methoxyethanol, [aw]ethanolamine, [ax]tetrafluoroethylene-perfluoro-3,6-dioxa-4-methyl-7-octene-sulfonic acid, [ay]bis(3,5-difluoro-2-(2-pyridyl)phenyl-(2-carboxypyridyl)iridium, [az]*tris*[2-(9,9-dioctyl-9H-fluoren-2-yl)pyridinato-C3,N]-iridium(III), [ba]1,3,5-tris(4-phenylquinolin-2-yl)benzene, [bb]conducting polymer, [bc]polyethylenimine, [bd]magnesium-alloyed zinc oxide, [be]perovskite–polymer bulk heterostructure, [bf]bis(2-(diphenylphosphino)phenyl)ether oxide, [bg]1,3,5-tri(m-pyrid-3-yl-phenyl)benzene, [bh]8-hydroxyquinolinolatolithium, [bi]2-(4-biphenylyl)-5-(4-tert-butylphenyl)-1,3,4-oxadiazole, [bj]bis(1-phenylisoquinoline)(acetylacetonate)iridium(III), and [bk]oligoquinoline 6,6'-bis(2-p-biphenyl)-4-phenylquinoline).

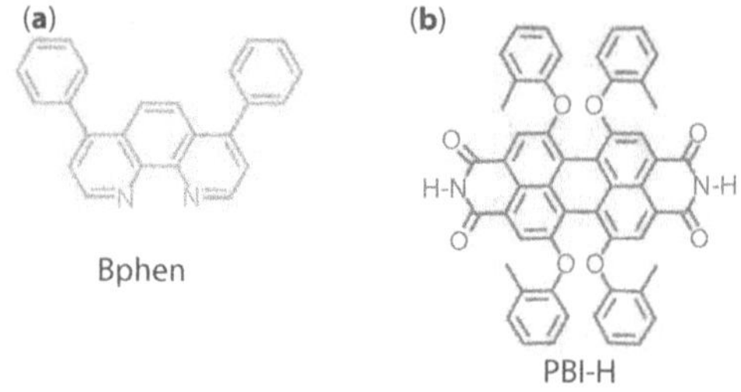

Figure 6.5 Molecular structures of (a) Bphen and (b) PBI-H.

other contemporary conjugated polymers. In addition, excellent alignability of the fully conjugated PF-polymer backbone was exploited to fabricate polarized light emitter at a reasonable brightness [15]. In addition to the enhancement of efficiencies, another objective was to tune the color of the emitted light from these LED materials [16–21]. Over the past two decades, many reviews have been published on the development of organic-/polymeric-LED materials [22–25]. In the last decade, several works have established polymer LEDs suitable as a cost-friendly material for large-area displays including industrial applications in display and lighting [26, 27]. Thereafter, as a result of continuous upgradation, the luminous efficiency and external quantum efficiency of polymer LEDs have reached as high as 61.6 cd A^{-1} and 17.8%, respectively, along with enhancement of lifetime to more than 10^4 h [28–30].

6.3 Basic Principles Behind Luminescence of Polymers

The basic operating principle of a light-emitting device, such as LED, is the spontaneous emission of optical radiation in the form of photons *via* radiative recombination of electron-hole pairs through consumption of electrical energy. Thus, LED devices can be comparable to a kind of transducer, converting electrical energy to light. In fact, the emission spectra obtained from LEDs can be ranging from UV to near IR region, depending on the constitution and construction of the light-emitting device. Both band-based and exciton-based models have been proposed to explain recent measurements relating to the performance of light-emitting diode and device. Sometimes, experimental outcomes have been explained on the basis of band model of the electronic structure of semiconducting luminescent π-conjugated polymers [32]. The light-emitting capability of a conjugated polymer-based device is originated from the effective delocalization of π electron cloud throughout the entire macromolecule [3]. In fact, transduction of electricity to light was explained on the basis of self-localized charge states, which was interconnected to the formation of polaron. Such polaron concept was given by Lev Landau in 1933, relating to formation of a quasiparticle *via* interaction of a moving electron with the atoms, leading to lowering in electron mobility and increase the electron's effective mass. During the interaction of electron and atoms, the charge of electron is screened by the moving atoms or phonon in a crystal. Thus, polarons can be expressed as composite particle made of electron and its virtual phonons or group of atoms. In particular, such polaron should better be represented as electron polaron. In contrast, in a semiconductor, formation of hole polaron is also possible. Thus, a hole polaron can be expressed as a composite made of holes and virtual phonons or group of atoms. In contrast, polaronic exciton or polaron exciton is a bound pair consisting of an electron polaron and hole polaron. Accordingly, polaronic exciton or polaron exciton is uncharged, whereas polarons and bipolarons are singly and doubly charged, respectively. In case of light-emitting conjugated polymer of poly-*p* phenylene vinylene, the photoluminescence was explained by polaronic exciton concept, which is an extension of polaron concept. According to Burroughes *et al.*, the electroluminescence was produced as a result of formation of polaronic exciton or polaron exciton *via* recombination of electron polaron and hole polaron generated from injected electrons and holes at the opposite sides of the structure. In this context, they investigated the charge content of the involved polarons, and eventually they ruled out the generation and role of bipolarons in the formation of polaronic exciton or polaron exciton. In contrast, singly charged polarons played the dominating role in the energy emission from the polymeric diode. In this context, poor quantum efficiency of this newly developed polymeric diode was associated with migration of excited states or excitons to the defect sites located at the polymer/thin metal interface of the light-emitting diode [3]. Therefore, if the migration of the excitons can be receded, these excitons cannot reach easily to the quenching sites. Thus, the variation at the molecular level raised the ability of the material to trap excitons [4], and thus, the migration of those excitons was hindered because of the presence of mixed structures carrying either conjugated or unconjugated or both conjugated and non-conjugated moieties with variable $\pi \rightarrow \pi^*$ gaps. Structural varieties were also introduced while synthesizing semiconducting polymer, i.e., poly(2-methoxy, 5-(2′-ethylhexoxy)-1,4-phenylenevinylene [5]. Herein, the basic strategy was to produce an intra-chain exciton-type luminescence center comprising of neutral bipolaron made of a positive and a negative polaron. In fact, Braun and Heege analyzed two mechanisms behind the formation of neutral bipolaron. Between those two mechanisms, tunneling-injection mechanism dominated over the other mechanism related to thermal excitation enabled injection of electrons from the metal into the conduction band of the polymer [5]. In case of tunneling-injection mechanism, the electrons tunnels through the barrier to reach the upper polaron level (Figure 6.6). Thereafter, a radiative decay occurs though structural relaxation to the excited state of the neutral bipolaron. In contrast, the second mechanism is associated with formation of negative polaron by self-localization of electron and phonon, followed by recombination of this negative polaron with a positive polaron to form the excited state of the neutral bipolaron exciton. Thereafter, non-radiative decay starts, when excited state of the neutral bipolaron exciton starts loosing energy to produce ground-state of the neutral bipolaron exciton.

While investigating the reasons behind the electroluminescence shown by polyfluorines, it has been inferred that upon excitation by the UV light, the blue electroluminescence emanating from the solution of polyfluorine consists of three major peaks at 420, 448, and 472 nm, assigned to the 0-0, 0-1, and 0-2 intrachain singlet transitions, of which the 0-0 transition is the most intense. However, at the time of excitation by UV, the

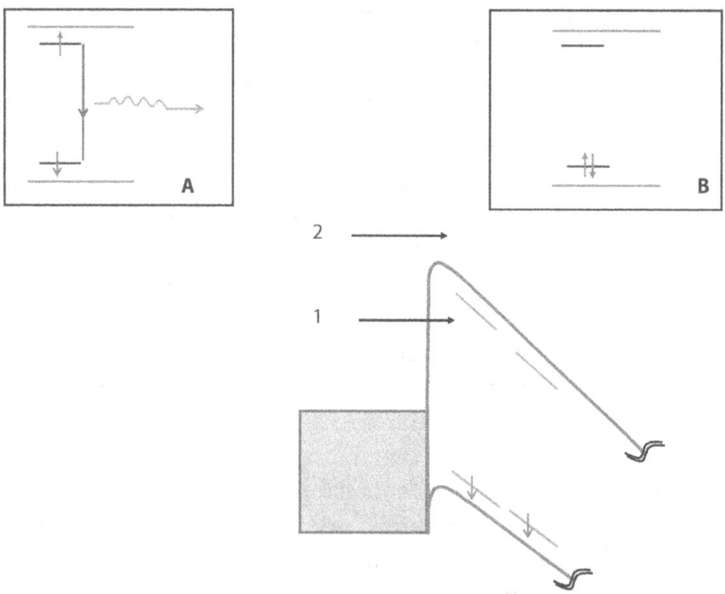

Figure 6.6 Two different mechanisms of luminescence.

electrons showed a π→π* transition. Since the energy of the π→π* transition depends upon the conjugation length, a distribution of conjugation lengths is responsible for producing a broadened absorption spectrum constituting of a distribution of energies [7].

6.4 Classification of Polymer Light-Emitting Diode

6.4.1 Classification Based on the Type of Components

In case of hybrid PLED, the major components are a member of hybrid structure consisting of both polymer and nonorganic layers sandwiched between two metal electrodes. Otherwise, for pure PLED, the entire structure including the emissive and transporting layers is made of organic components. According to Lu *et al.*, hybrid LED structures endow almost five times higher efficiency compared to apure LED device [47].

6.4.2 Classification Based on the Device Architecture

This classification is based on the relative position of the emissive and transporting layers with respect to the entire device. Accordingly, a PLED can be of following types: a) bottom-emitting conventional, b) bottom-emitting inverted, c) top-emitting conventional, and d) top-emitting inverted. In case of either conventional or inverted bottom emitting PLEDs, the device emits through a transparent or semitransparent bottom electrode (anode), whereas a top-emitting PLED giving off light through a transparent or semitransparent top electrode as cathode (Figures 6.7 and 6.8).

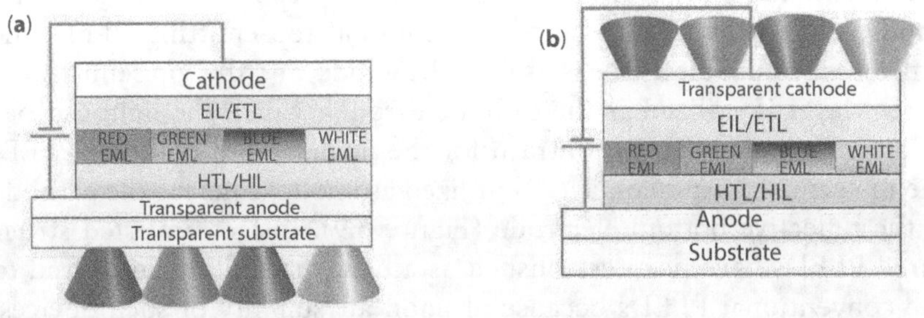

Figure 6.7 Architecture of conventional (a) bottom-emitting and (b) top-emitting PLEDs containing EML, EIL, ETL, HIL, and HTL sandwiched within anode and cathode [56].

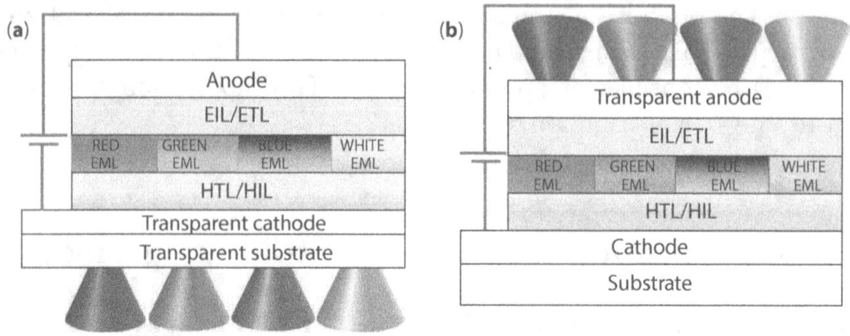

Figure 6.8 Architecture of inverted (a) bottom-emitting and (b) top-emitting PLEDs containing EML, EIL, ETL, HIL, and HTL sandwiched within anode and cathode [56].

Notably, either a bottom-emitting or top-emitting conventional PLED consists of different layers, such as EIL/ETL, EML, HTL/HIL, sandwiched within cathode and anode. Herein, either HIL or HTL is positioned just adjacent to the anode, and the functions of HIL and HTL are to improve the number of injected holes from the anode. Similarly, both EIL and ETL are located exactly adjacent to the cathode, and the roles of EIL and ETL are to increase the number of injected electrons from the cathode. In this context, EML, the meeting point for the recombination of hole and electron carriers, is positioned in between HIL/HTL and EIL/ETL to produce colors of different band gaps (Figures 6.7 and 6.8). Additionally, to improve further the efficiencies of the devices, often HBL and EBL are added to the device (not shown in the figure), and these layers are used to confine most of the carriers to the EML, leading to enhanced luminous efficiency of the LED [56]. In addition to the conventional bottom- and top-emitting PLEDs, inverted varieties of both of these PLEDs are also available. The key difference between a conventional and an inverted PLED is the locations of cathode and anode. For instance, in case of conventional top-emitting PLED, the reflective bottom electrode is used as the anode, and the top semi-transparent layer is utilized as the cathode through which the light escapes (Figures 6.7 and 6.8). In contrast, for the inverted top-emitting PLED, the top semi-transparent layer is utilized as anode, whereas the cathode is the reflective bottom electrode (Figure 6.9). In fact, inverted structured PLEDs have been established as a better substitute compared to the conventional PLEDs because of good air stability of such devices. Herein, high work function metal is used as anode, whereas air stable metal oxide layers are employed as EIL and HIL [49].

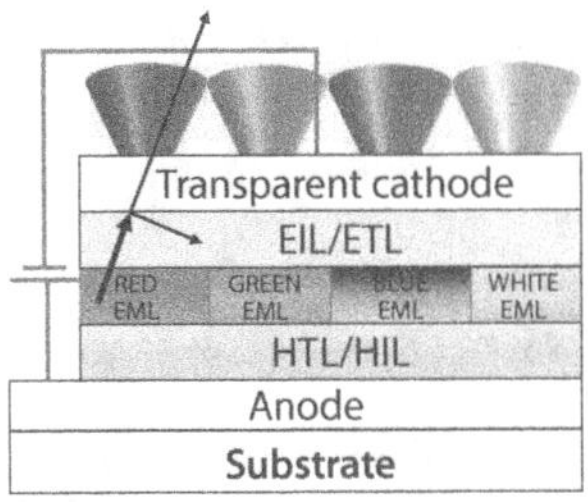

Figure 6.9 Loss of out-coupling efficiency due to diversion of EML at the EIL made metal oxides of high refractive index.

6.4.3 Classification Based on the Charge Carriers

Depending on the work function of the added electrode, PLED devices can be either single carrier or bipolar types.

6.4.3.1 Single Carrier Device

A single carrier device is usually prepared by incorporating electrodes of almost similar work functions [47]. Again, a single carrier device can be sub-classified into hole only device and electron only device, depending on the value of the work functions of the added electrodes as well as the nature of the intermediate layers. For instance, a device having structures of ITO//PEDOT:PSS/aryl-F8:TFB/MoO_3//Au support only hole injection from either of the contacts [47]. This is because both PEDOT:PSS and MoO_3 can function as hole injecting layers. The hole mobility in such devices can be retarded adding hole traps in the form of TFB [47]. Accordingly, it has been observed that the hole current has been decreased significantly with increasing amount of TFB additive. If hole trap like TFB has been added in a small amount (0.1%) to aryl-F8, the current reduced remarkably by almost three orders of magnitude. However, such pronounced decrease has been noted to retard on further addition of TFB.

In contrast, electron-only devices are usually fabricated by sandwiching the semiconducting layer between two thermally evaporated low work function metals. In this case, the bottom layer makes a thin oxide layer for reducing the extraction of true electron current. For instance, ITO//ZnO/Cs_2CO_3/aryl-F8:TFB//Ca/Al can operate as an electron-only device [47]. In fact, a hole trapping agent, such as TFB, can reduce the mobility of electrons in an electron-only device, as TFB itself has low electron mobility because of the presence of unconjugated segment within the molecule.

Thus, the electron current also dropped four- to five-fold on introduction of TFB in aryl-F8. In addition, the HBL like Cs_2CO_3 is capable to block the movement of holes within the device [44].

6.4.3.2 *Bipolar Devices*

Both hybrid and pure-type PLEDs can be made to function as bipolar devices. For instance, ITO//PEDOT:PSS/aryl-F8:TFB/Ca//Al is a pure type bipolar device consisting of PSS doped PEDOT as hole injecting layer and Ca//Al component is the characteristics of electron only device. Similarly, ITO//ZnO/Cs_2CO_3/aryl-F8:TFB/MoO_3//Au is a hybrid bipolar LED containing ZnO as EIL and MoO_3 as HIL.

6.4.4 Classification Based on the Color of Emission

Since the discovery of PLEDS, color tuning and accordingly achieving LEDs emitting variegated color remain as an interesting field of research for the last few decades. Among the various families of light-emitting polymer families, the emission color of polyfluorenes can be tuned over the entire visible range by introducing narrow-band gap comonomer into the polyfluorene backbone [57]. The basic principle involved in determining the characteristic color emission from a particular PLED is the energy of the quanta which is inversely related to the λ_{max} of the emitted light as a result of hole-electron recombination at the emitting layer. Thus, the characteristic λ_{max} for blue, green, and red color are noted to lie within 425–475, 525–550, and 600–650 nm. Since, the λ_{max} of the emitted light is dependent on the HOMO-LUMO gap, and such gap can be reduced or enhanced mostly by the structural alteration of polymeric emissive layer by encouraging or interrupting the π–π conjugation, respectively.

6.4.4.1 *Green and Blue Color Emitting PLEDs*

It may be recalled that among all the LEDs, the green color emitting LEDs were discovered and reported for the first time by Tang and VanSlyke in 1987 [1]. Thereafter, PLEDs invented in 1990 by Burroughes *et al.* were also noted to emit green-yellow light [3]. Followed by these works, different workers have reported various green color emitting PLEDs [4, 6, 36, 39]; however, such documentations are comparatively less frequent compared to the widely reported blue color emitting PLED [15, 34, 35, 37, 53]. In fact, the emitted color from the EML is component specific. For instance, PFO and F8BT emit blue and green fluorescence, respectively.

Notably, sometimes, green color emission from PLEDs was treated as a drawback for the fluorene-based EMLs arising from unwanted keto defects that eventually hamper long-time storage stability of the PLEDs. However, no such limitations have been reported in case of blue color emitting PLEDs.

6.4.4.2 Red Color Emitting PLED

Compared to green light-emitting PLEDs, the device efficiency of red emitters is typically low. Accordingly, acquiring high-efficiency red emissions from a PLED device still remains a great challenge to date. Among the polyfluorene family, F8TBT or poly((9,9-dioctylfluorene)-2.7-diyl-alt-[4,7-bis(3-hexylthien-f-yl)-2,1,3-benzothiadiazole]-2′,2″-diyl)) is an alternating copolymer, which emits saturated red color. In fact, this copolymer is derived from fluorene (F8) and TBT units (Figure 6.10).

Herein, the major strategy behind achieving a saturated red color is the incorporation of a narrow band-gap monomeric segment, such as TBT, in between two wide band-gap fluorene segments, so that the narrow-bandgap segment can therefore function as an exciton trap, to facilitate efficient intramolecular energy transfer from the fluorene segment to the TBT. In fact, such narrow band gap characteristics of TBT is originated from the coexistences of thiophene as aromatic-donor unit and 10π- or 14π-electron-heterocycles as o-quinoid-acceptor within the TBT, leading to higher HOMO and lower LUMO (Figure 6.11), which ultimately gives rise to a saturated low energy red emission of higher λ_{max} [58].

Thus, the increased incorporation of the TBT component has led to greater performance of PLED devices. Moreover, it has been realized that the alkyl substitution plays a significant role in increasing the device efficiency. Moreover, the spatial orientation of the alkyl chains can influence the intra- and intermolecular interactions, and thus tune the optical and electrical properties. In fact, the emission maxima of F8TBT-in (Figure 6.5b) and F8TBT-out (Figure 6.5b) are recorded at 636 and 655 nm, respectively [57]. Such blue-shift in the emission maxima of F8TBT-in is directly related to

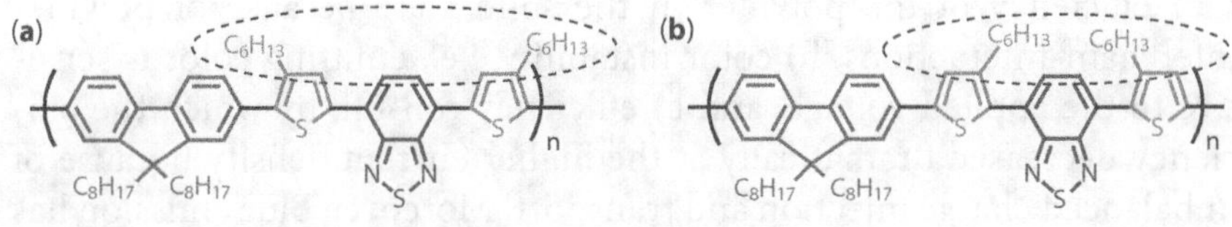

Figure 6.10 Copolymers of F8 and TBT (a) F8TBT-out and (b) F8TBT-in having different side chains [57].

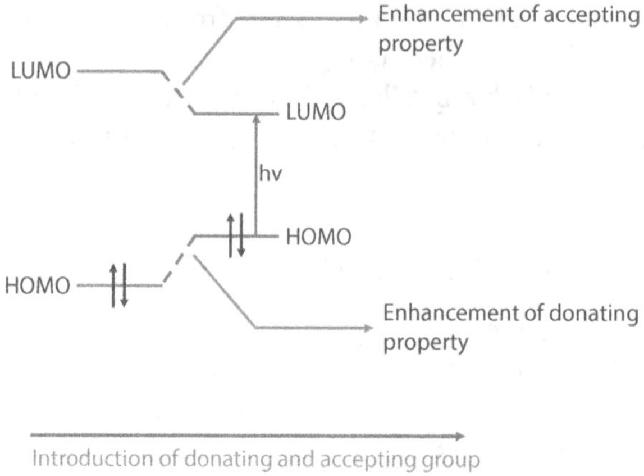

Figure 6.11 Red colored emission by reduction of HOMO-LUMO gap.

the interruption of the π–π conjugation coupled with reduced interchain interaction. In this respect, in the solid state, the F8TBT-in copolymer shows appreciably enhanced PLQE over F8TBT-out. Since alkyl side chains in F8TBT-in are closer to each other, close packing of polymer chains should be prevented, leading to reduction in the emission quenching.

6.4.4.3 *White Color Emitting PLED*

Nowadays, white organic light-emitting diodes are drawing attractive attention in solid-state lighting and other applications. Such devices can be broadly categorized into small-molecule and polymer-based devices. The small-molecule devices are often fabricated by thermal evaporation, by which multiple layers are deposited simultaneously, whereas the polymer-based devices can be processed either by spin-casting, screen printing, or inkjet printing techniques. In fact, all of these techniques are advantageous in terms of cost and large-area displays. However, white light-emitting polymer devices suffer from the following drawbacks: a) low overall efficiency because of poor photoluminescence quantum efficiency of the polymer in the solid state because of polymer interchain-interactions, b) color instability, i.e., emitting color is sensitive to the applied voltage, and c) efficiency roll-off, by which the efficiency decreased dramatically at the higher current density because of unbalanced charge injection and transport. Moreover, blue emission has been remained as the bottleneck for the performance of polymer-based white LEDs [55].

6.5 Dependence of Various Performance Parameters on Structural Factors

6.5.1 Brightness

Brightness of a source is interpreted as the visual perception resulted from radiating or reflecting light emanating from that source. Though the term is related to luminance and can be expressed by the same unit (e.g., cd m^{-2}), brightness and luminance are not the same. Luminance of a radiating or reflecting light source is the luminous energy per unit time per unit solid angle per unit projected source area. In contrast, brightness is the visual perception by a human being as a result of luminance of the radiating or reflecting light source.

6.5.2 Efficiencies

There are different kinds of efficiencies, such as luminous efficiency, luminance efficiency, internal quantum efficiency, external quantum efficiency, electroluminescence efficiency, photoluminescent quantum efficiency, and out-coupling efficiency, representing the overall performance potential of a particular PLED device. For instance, luminous efficiency of a light source is the energy efficiency of that particular sources, which can be interpreted as the amount of light produced for unit of consumed electricity. Usually, luminous efficiency is measured in lumens per watt (lm W^{-1}). In contrast, the efficiency of PLEDs is often reported in the form of their external quantum efficiency (EQE) or in candelas per amp (cd A^{-1}). In fact, external quantum efficiency of a PLED depends on two major factors, i.e., the extent of internal quantum efficiency and out-coupling efficiency. In this regard, photoluminescent quantum efficiency, singlet formation *via* spin statics, and a high fraction of recombination of electrons and holes in the emissive layer are the deciding factors behind the internal quantum efficiency. **Importantly, efficiencies of a PLED depend on following factors.**

6.5.2.1 Characteristics of EML

6.5.2.1.1 Thickness

Thickness of EML can be an important parameter to decide the stability and operational voltage of a bipolar PLED. For instance, Lu *et al.* have varied the thickness within 300–600 nm of the EML in a bipolar hybrid PLED device, such as ITO//ZnO/Cs_2CO_3/aryl-F8:TFB/MoO_3//Au [47]. It has been observed that the luminance efficiency and stability of EML have been deteriorated, if thickness of the EML is lower than 500 nm. In fact, the thinner device has been noted to break-down quickly compared to the thicker devices. Such instability of a thinner device is because of the high current density arising from the doped emissive polymer region at either end of the emissive layer, along with inherent roughness of ITO//ZnO/Cs_2CO_3 [47]. In contrast, if the thickness is too high, i.e., 600 nm, higher operating voltages are required. Therefore, thickness of a particular type of EML should be optimized for a particular hybrid-type PLED device to ensure substantial stability and lower operating voltage of the device.

6.5.2.1.2 Blending

In contrast to the single-emitting layers, devices fabricated with a blend of emitting polymers show strong phase segregation because of the low entropy of mixing of various polymer components. Often, blending is carried out involving a high-electron transport material with high-hole transport material for achieving an acceptable charge-carrier balance [59]. Moreover, these blends can be classified into two categories depending upon the stoichiometric ratio of the components: the first type of blend contains almost equal quantities of individual components, whereas the second type constitutes of a small amount of one component as dopant in a large amount of the other. Earlier, workers have opted for blending two different polymers to derive individual advantages of both in the blend [38]. For instance, PFO possesses an inherent high hole mobility, whereas F8BT has got high electron mobility in combination with high electron affinity. Therefore, attempt has been made to receive the advantages of both in the form of a blend of PFO and F8BT envisaging a good balance between hole and electron mobility. According to Chappell *et al.*, the best results are obtained when the stoichiometric ratio of 19:1 is maintained in favor of PFO [60]. In fact, blue and green fluorescence are emitted by PFO and F8BT, respectively, with quantum efficiencies around 50%. It has been observed that the photoluminescence spectrum of PFO:F8BT blend is almost similar to the individual photoluminescence spectrum of F8BT (Figure 6.12).

Such phenomenon has been ascribed to the Förster transfer or dipole-dipole coupling. By such process, the photogenerated excitons on PFO molecules can readily be transferred to F8BT molecules over a distance of ~5 nm. Moreover, blending of PFO with F8BT has resulted almost similar electroluminescent spectra. In fact, blending of PFO with F8BT has resulted greater electroluminescence efficiency compared to individual F8BT [61]. However, such enhancement in efficiency of PLED device

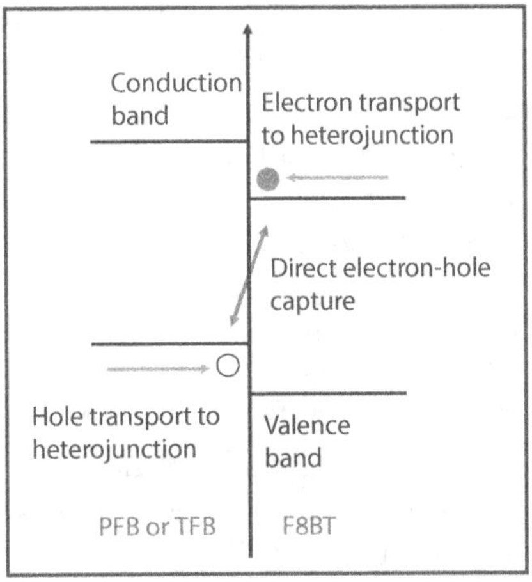

Figure 6.12 Electrons and holes are transported through the suitable phases and meet at the heterojunction connecting F8BT with either PFB or TFB.

is dependent on the individual components of the blend present as EML [45]. For instance, the excellent performance has been obtained when a blend of TFB and F8BT has been used as EML [39]. In contrast, such performance deteriorates when a combination of PFB and F8BT is used as EML (Table 6.1). In this regard, heterojunctions have been formed in both the blended EMLs [39, 60]. Since F8BT has got high electron mobility [38], majority of the electron transport occurs through F8BT phase, whereas hole transport mostly takes place through either TFB or PFB phase. Finally, both the transported holes and electrons reach and meet at the heterojunctions or the interfacial zone (Figure 6.12). Moreover, the TFB/F8BT device shows yellow-green emission at a significantly low threshold voltage (i.e., <2.1 V). This is because of the two reasons: a) from the charge-separated state, the neutral state can easily be formed by electron-hole capture at the heterojunction without any prior injection of electrons and holes in the TFB and F8BT phases, respectively; and b) another highly stable interfacial state or exciplex has been formed *via* conformational relaxation of the charge-separated state [39]. However, when a small amount of TFB has been mixed with aryl modified F8, long-lived red-shifted exciplex emission is absent, suggesting absence of exciplex formation in arylE8/TFB blend [47]. Accordingly, this observation has contradicted the earlier results [39]. However, this observation confirms that the emission is coming from bulk aryl-F8 in polymer blends, and excitons are not trapped at heterojunctions [47].

Another group of researchers conducted a study on the influence of phase domain and overall microstructure of polymer blends on the efficiency and performance of LEDs [40]. Depending on the relative proportion of light-emitting conjugated MEH-PPV component in a polymer blend of MEH-PPV and unconjugated PMMA as EML, alteration of morphological characteristics including phase inversion, bicontinuous morphology formation, and discretization of either MEH-PPV or PMMA phase becomes evident. For instance, MEH-PPV/PMMA blend comprising of 75% MEH-PPV and 25% PMMA envisaged continuous MEH-PPV phase surrounding the discrete PMMA phase. Accordingly, the highest efficiency has been reached among all the blends. In contrast, completely different morphology is observed when the proportion of MEH-PPV is less than 50%. Herein, discrete MEH-PPV phases emit luminescence and appear collectively as nano- or micro-LEDs envisioned as separate small light sources (Figure 6.13) [40].

Recently, blending driven change in the energy landscape of the PLED architecture has been reported by Hassan *et al.* [62]. As a result of blending, the overall charge transport properties have been influenced significantly, leading to elevated performance level of the LED. It has been realized that deep hole traps have been generated in the region constituting of poly(para-phenylene vinylene) derivative because of substantial difference in the HOMO levels of poly(9,9-dioctylfluorene) and poly(para-phenylene vinylene). These hole traps eventually have retarded the movement of holes within hole dominated device. By such energy landscape analyses, it is possible to achieve the optimum balance between the electron and hole

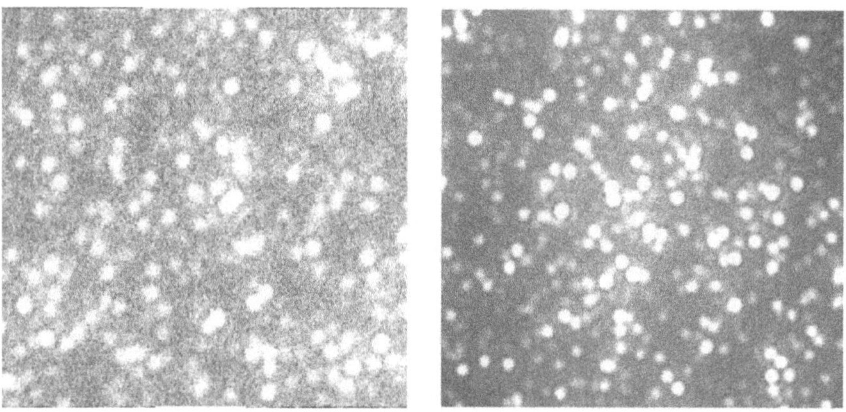

Figure 6.13 In the MEH-PPV/PMMA blends carrying lesser MEH-PPV envisage discrete MEH-PPV phases as nano- or micro-LEDs envisioned as separate small light sources [40].

currents, leading to efficient formation of excitons within the emission zone. In fact, such emission zone is spatially shifted away from the cathode surface minimizing the detrimental quenching effects.

6.5.2.1.3 Chemical Treatment

6.5.2.1.3.1 Solvent Treatment

Sometimes, efficiencies of an LED have been improved treating the nanostructure of PLED by amine-based polar solvent [49]. In fact, the added solvent reduces the contact barrier between the poly(9,9′-dioctyl-fluorene-co-benzo-thiadiazole) or F8BT and the ZnO layer. In fact, the contact barrier at the polymer-ZnO interface arises from the interfacial dipole effect at the junction. However, in presence of the added solvent, the already existed interfacial dipole effect at that junction gets reduced after the absorption of amine (–NH_2) and hydroxyl (O–H) functionalities of ethanolamine on rippled structured ZnO (Figure 6.14).

6.5.2.1.3.2 Endcapping

Moreover, the modification of polymeric EML by endcapping is an established technique to modify the emission characteristics of polyfluorenes [37, 41]. The main objective of such endcapping is to prevent aggregate and excimer formation, as excimers or aggregates deteriorate the luminance efficiency. Earlier, anthracene [37] and triarylamines [63] have been successfully employed as endcappers to prevent excimer formation in enhancing the device efficiency. In fact, among the various endcapped EMLs (Table 6.1), device bearing PFO-end-TAZ as EML demonstrates the best device performance because of the larger size and nonplanar structure

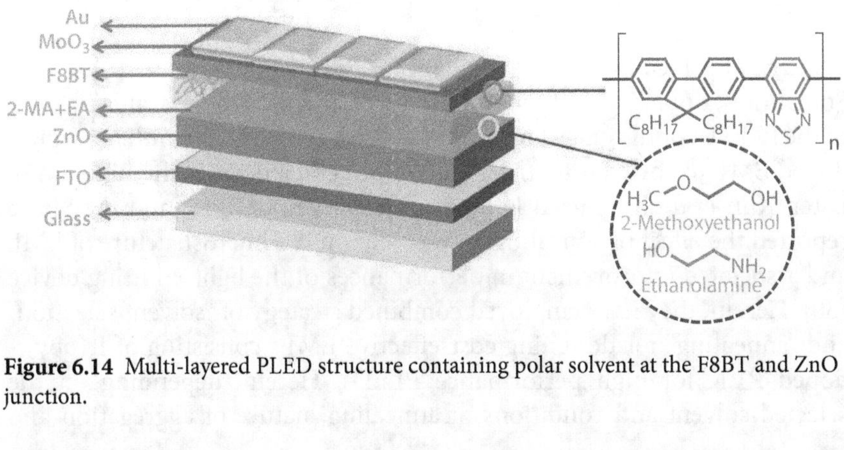

Figure 6.14 Multi-layered PLED structure containing polar solvent at the F8BT and ZnO junction.

of triazole preventing the formation of aggregates, along with prevalent electron-transporting and hole-blocking abilities of triazole [41]. In fact, such twisted structure of aryl units attached to the polyfluorene prevents the π–π stacking necessary for excimer or aggregate formation [47]. Moreover, in case of device having PFO-end-TAZ as EML, increased deep blue emission is originated from the incomplete energy transfer from the amorphous matrix to the β-phase.

6.5.2.1.3.3 Copolymerization

Addition of nonplanar moieties in the EML structure can cause suppression of unwanted green emission from polyfluorenes emitting blue lights (Table 6.1). In fact, such emission from polyfluorenes was originated from fluorenone defects in the microstructure of polyfluorenes [42]. These fluorenone defects are comparable to the ketone defects observed during device operation, and these result in color impurity, as ketones are electron-withdrawing groups and have a low-lying LUMO [47]. By the incorporation of non-planar copolymers comprising of the twisted biphenyl moieties, intermolecular separations have been increased, leading to restrained movements of the generated excitons to defective site of polyfluorenes. In fact, by incorporating twisted biphenyl moieties carrying variable functionalities at the C9 position, inter-ring torsion angles are increased to a different extent, and such inter-ring torsion angles can be well correlated with the suppression of green emission as well as enhancement of the blue emission [42]. In fact, biphenyl torsion angle is noted to control the degree of intrachain conjugation and affects the intermolecular packing in the solid state. Accordingly, large biphenyl torsion angle ensures poor interchain or supramolecular interaction, leading to high-energy optical transitions in the isolated molecules.

6.5.2.1.3.4 Doping

Sometimes, doping technique is followed to enhance the overall efficiency of the PLED. For instance, Earmme *et al.* have used triplet emitter (FIrpic)-doped EML to enrich brightness and output efficiencies of the PLED [45]. Later, Kim *et al.* have used $Ir(ppy)_3$ doped PVK as EML, and they have reported the effect of annealing and solvent on the microstructure of EML and associated improvement in performances of the light-emitting device [50]. Herein, they have employed combined strategy of "solvent selection" and "annealing" for designing cost-effective EMLs consisting of $Ir(ppy)_3$-doped PVK for high performance PLEDs. Herein, depending on the selected solvent and conditions of annealing, nature of aggregation and shape of aggregates can be altered (Figure 6.15), which ultimately influence overall performances and efficiencies of PLED.

In general, device efficiency can be improved *via* reducing the surface roughness of a device by virtue of annealing (Figure 6.16). Moreover, shape of the aggregate (i.e., needle or spherical) depends on the nature of applied

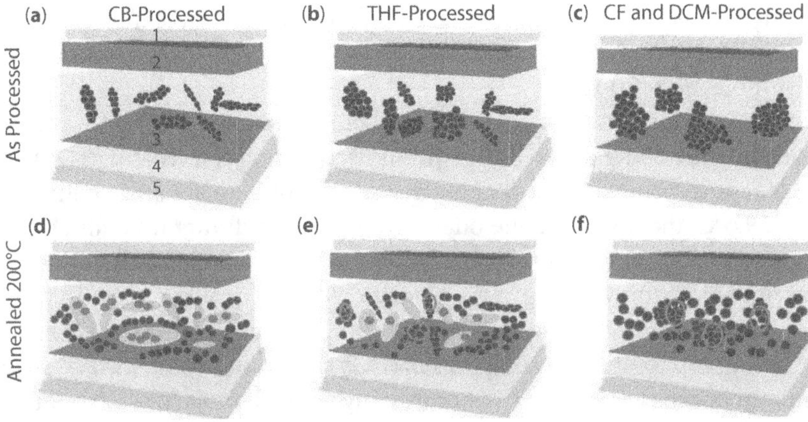

Figure 6.15 Morphological changes, shape of the aggregates, i.e., (a) nodular, (b) nodular + spherical, and (c) spherical, and brightness of the doped EMLs depending on the applied solvent, i.e., (a) CB, (b) THF, and (c) CF & DCM, and annealing, i.e., (d), (e), and (f) process [50].

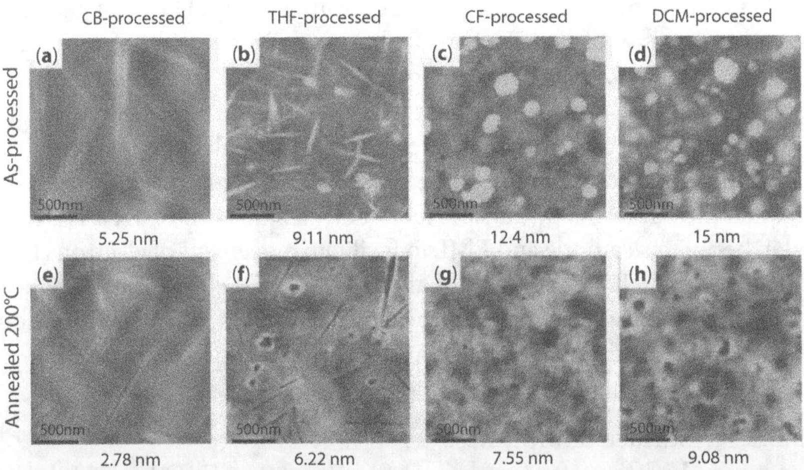

Figure 6.16 AFM images of doped EML coats envisaging changes in shape of aggregates, i.e., (a) nodular, (b) nodular + spherical, and (c, d) spherical, and surface roughness of Ir(ppy)3 doped EMLs spin coated by applying different solvents, i.e., (a) CB , (b) THF, (c) CF, and (d) DCM, followed by annealing, i.e., (e–h) [50].

solvent used at the time of solution coating. In this regard, surface roughness of the EML has been significantly reduced as a result of easier diffusion of needle-like Ir(ppy)$_3$ aggregates in EML (Figure 6.15), leading to higher efficiency of PLEDs. In contrast, spherical shaped Ir(ppy)$_3$ aggregates have been noted face constraints in diffusing through EML (Figure 6.15), leading to higher surface roughness and deteriorated efficiency of the concerned PLEDs (Figure 6.16). In fact, higher surface-to-volume ratio for needle shaped aggregates facilitates reduction in the surface energy *via* easier molecular diffusion.

In fact, the formation of either needle- or spherical-shaped aggregates is guided by the boiling point of the solvent used during spin coating of EML. In this regard, being a faster evaporating solvent of lower boiling point, i.e., b.p. = 39.6°C, dichloromethane offers the minimum drying time for aggregation of Ir(ppy)$_3$ through π-π stacking interactions among phenylpyridine rings in EML coats, resulting in the formation of spherical aggregates having the minimum surface energy. In contrast, sufficient aggregation time is available during spin coating by applying relatively slow evaporating solvent, such as chlorobenzene (b.p. = 131°C). Accordingly, Ir(ppy)$_3$ molecules get enough time to order in the form of needle in a dried coating layer of EML.

6.5.2.2 *Characteristics of EIL/ETL*

The basic function of EIL is to facilitate the injection of electrons from the conduction band of the metal oxide cathodes to LUMO of the EML [43]. The necessity of the EIL is related to the minimization of mis-match between conduction band of metal oxide and LUMO levels in the organic electron transport moieties, so that the operation of the device remains uninterrupted. In contrast, the function of ETL molecules [45], such as 1,3,5-tris(4-phenylquinolin-2-yl)benzene or TQB, is almost similar except the difference in terms of transporting ability. Either EIL or ETL is deposited in between a cathode and EML to facilitate movement of electrons from cathode to EML. It can be noted that brightness and efficiencies have been significantly increased by the incorporation of ETL in between Al cathode and the EML constituted of triplet emitter (FIrpic)-doped PVK blend.

6.5.2.2.1 Surface Smoothness

Sometimes, higher surface roughness of EIL, such as ZnO layer [49], may lead to optical quenching and trapping of photons, leading to reduced out-coupling efficiency, and thus, poor external quantum efficiency. In contrast, higher roughness of the ETL coated cathode is advantageous to achieve better performance of the PLED. In this regard, the performance of ETL solution-deposited Al cathode is superior compared to vacuum-deposited Al cathode by ETL [45]. Such enhanced performance of solution deposited cathode is ascribed to the higher roughness resulted from solution mediated coat having oriented self-aggregation of molecules. The solution-processed high quality TQB/Al interface contributes appreciably in efficient injection of electrons into devices. In fact, the nature of TQB/Al interface is likely similar to polyquinoline/Al interface constituted by chemical bonding of Al to the N heteroatoms of the TQB based ETL layer.

6.5.2.2.2 Refractive Index

Poor out-coupling efficiency of a PLED can be related to the presence of metal oxide components of high refractive index as EIL. These metal oxide layers are used to reflect, trap, and divert the majority of the light generated from the polymeric EML, leading to escape low fraction of photons from the device [49]. Such problem is a sort of waveguiding problem, by which the waves are diverted from their desired direction by some impeding elements.

6.5.2.2.3 Phase Separation

The extent of phase separation and incompatibility of components in a blend can be a deciding factor behind performances of PLEDs. For instance, the brightness of the emitted light has been noted to enhance compared to PLED having mere components, when 1:1 combination of TPBI/TmPyPB and TPBI/TAZ have been used as ETL [46], reducing operational voltage. However, such parameters become unfavorable when 1:1 combination of TPBI/Biphen has been used as ETL. In this case, the tendency of phase separation in a blend is the contributing factor behind the deteriorated performance of TPBI/Biphen, which does not happen either in the case of TPBI/TmPyPB or TPBI/TAZ. Interestingly, performance potentials have been enhanced further when a blend of TPBI:TmPyPB:TAZ in 1.3:0.7:1 ratio has been used as ETL (Table 6.1).

6.5.2.2.4 Shape of the Nanostructure

The efficiency can be significantly improved by changing the shape of the EIL nanostructure of PLED in a favorable form. In this regard, Lee *et al.* have introduced a spontaneously formed ripple-shaped nanostructure of ZnO-based EIL in PLEDs, imparting encouraged results in terms of external quantum efficiency as well as luminance efficiency [49]. In fact, the rippling structure in this hybrid LED has equipped with some periodicity. However, since the structure is randomly oriented, a wide distribution of period is obtained. Such random orientation and associated wide distribution in periods help function as out-coupler for the wave guided light to escape as a broad range of emission wavelengths, regardless of direction, leading to enhanced out-coupling efficiency.

6.5.2.2.5 Chemical Treatment

Sometimes, in order to achieve higher performance potential in PLEDs, the surface of the EIL, such as ZnO, is modified with polyethylenimine (PEI) to reduce the work function of ZnO from 4.1 to 3.4 eV [51]. Recently, ZnO layer has been doped with organic dye, which has resulted dramatic enhancement in device performance. Herein, the electron injection ability has been promoted through elevated electron injection by the added dye molecules, leading to increased electron-hole recombination efficiency, when derived PPV has been used as light-emitting material [54].

6.5.2.3 Characteristics of HIL/HTL

The application of transition metal oxides as HILs can greatly reduce the injection barrier for hole injection [43]. In presence of HIL, device can be turned on at low voltage, suggesting significant contribution of the flow of holes facilitated by HIL. In fact, by the incorporation of transition metal oxide, i.e., MoO_3, enhanced brightness has been observed in the PLED, as it has been assumed that a charge transfer complex might be formed with organic materials, leading to increased local carrier density and associated formation of an Ohmic contact. In the absence of MoO_3, a significant gap exists between the work function of gold and the HOMO of the F8BT. However, once MoO_3 has been added as HIL [47], the additional charge-injection layer has decreased the pre-existing barrier, and accordingly, Ohmic contact is generated. If the work function of gold is kept closer to the HOMO energy of polymeric EML, no efficient hole injection has been observed in the absence of a thin layer of MoO_3 within 5–20 nm [44]. Moreover, the hole-injecting MoO_3 layer can also function as additional barrier against moisture, as the EML is sandwiched and protected between two metal oxide layers constituting of MoO_3 (HIL) and ZnO (EIL).

Earlier, in some devices, HILs have been doped with polystyrene sulfonic acid [36, 64]. In this regard, PEDOT:PSS is usually used as HTL for reducing the contact barrier between ITO and active semiconducting polymers of EML [38–40, 46]. This is because the fact that PEPOT:PSS is a good HTL envisaging high electric conductivity. At the same time, such layer reduces the roughness of ITO upon coating [48]. However, PEDOT:PSS as HTL possesses several drawbacks including corrosion of ITO electrode during coating of the electrode by the highly acidic aqueous solution of PEDOT:PSS, leading to deterioration of performance and long-term stability of the device. Moreover, significant quenching of radiative excitons usually occurs in between PEDOT:PSS and emissive polymer interface. In the past, several attempts have been made either *via* replacement of the PEDOT:PSS layers or by incorporation of a thin buffer layer in between PEDOT:PSS and EML interface [65–69]. Though some problems have been overcome by such treatment, the processing of such device remained complicated. Therefore, a solution processible GO as HTL has been introduced by Lee *et al.* [48]. If applied at an optimum thickness, the GO interlayer is capable to prevent significantly the quenching of the radiative excitons in between polymer EML and GO layer. Accordingly, a better balance between electron and hole injection can be achieved by blocking the electrons from EML to ITO, leading to enhanced performance and efficiency of PLEDs when GO layer of 4.3 nm has been applied as HTL (Table 6.1). If the thickness of the applied GO layer is less than 4.3 nm, both performance and efficiency of the PLED is reduced because of the poor electron-blocking In contrast, if the thickness is greater than 4.3 nm, degraded device performance and efficiency are in consequence to increased Ohmic contact resistance, resulting in receded hole injection/transfer. Recently, a thermally crosslinkable hole-transporting poly(indenofluorene-*co*-triphenylamine) copolymer containing vinyl-functionalized triphenylamine moiety has been used as HTL in multi-layered PLEDs [70], resulting in the elevated performance properties.

6.5.2.4 *Characteristics of HBL and EBL*

These layers are strategically incorporated in the LED to confine most of the carriers to the EML, leading to enhanced luminous efficiency of the LED [56]. The major function of HBL is to block the holes with utmost efficiency, leading to lowering the loss of holes at the EML-metal oxide interface. Such hole blocking can be achieved by placing a low molecular weight hole-blocking molecule, such as 1,3,5-tris(2-*N*-phenylbenzimidazolyl) benzene, between metal oxide and polymeric EML [44]. However, 1,3,5-tris(2-*N*-phenylbenzimidazolyl) benzene is soluble in solvents used for processing light-emitting polymer of the EML. Thus, there is every possibility for these 1,3,5-tris(2-*N*-phenylbenzimidazolyl) benzene molecules to get dissolved while depositing or coating the light-emitting polymer during device making. As a better alternative, a thin layer of Cs_2CO_3 is often deposited in between EIL and EML, resulting in a significant increase in efficiency attributed to an increase in electron injection combined with efficient blocking of holes [44]. As per the work of Bolink, the assembled multilayered hybrid device shows operational lifetimes up to 100,000 h, and thus, this is an ideal representative of stable LEDs. In addition, excellent brightness and luminance efficiency prove the potential of Cs_2CO_3 as a hole-blocking/electron-injecting layer (Table 6.1). Moreover, in presence of Cs_2CO_3, the required turn-on voltage has been increased compared to the device devoid of such HBL component, which can be ascribed to the presence of additional resistance endowed by the insulating Cs_2CO_3 layer [44]. Since Cs^{1+} belongs to the alkali metal group, it has an inherent tendency to diffuse into the polymer, introducing interfacial traps for holes, which in turn lower the mobility of holes [47]. It is tantamount to note that the thickness of Cs_2CO_3 layer should be sufficient to ensure full film coverage. Therefore, the thickness of the Cs_2CO_3 layer should be maintained higher than the atomic radius of Cs (i.e., 0.26 nm). However, full film coverage is not the dominant factor in determining the efficiencies and other parameters of the PLED. Therefore, it has been anticipated that there is a possible reaction between Cs_2CO_3 and the surrounding polymeric and ZnO layers, leading to formation of new material having higher electron injection capability.

Earlier, Morgado *et al.* reported that the inclusion of PPV layer as EIL resulted doubling of the electroluminescence efficiency, along with a reduction of the light-onset voltage [38]. For instance, in the absence of PPV layer, the electroluminescence efficiency was 2.1 cd A^{-1}, which has been increased to 4.1 cd A^{-1} when PPV as EIL was incorporated (Table 6.1). Moreover, with inclusion of PPV layer, the threshold current for light emission has been reduced significantly from about 7 to 0.04 mA cm^{-2}, indicating attainment of better hole/electron balance within the EML. Recently, LED device based on quasi-two-dimensional and three-dimensional (2D/3D) perovskites and an insulating polymer has been synthesized demonstrating excellent EOE (~20%) [52]. Herein, a dual functioning ETL cum HBL layer has been incorporated, constituting of MZO, a tunable wide-bandgap n-type metal oxide. In the same work, a blend of TFB and PFO has been used as HTL cum EBL layer.

6.5.2.5 *Characteristics of Cathode*

Notably, the transmittivity of the cathode can influence the efficiency of the PLEDs. In fact, transparency of cathode is a valuable criterion for conventional top emitting and inverted bottom emitting PLED types. For instance, metal oxide-based cathodes possess the higher transparency, and thus, better transmittivity compared to the metal-based cathodes. The cathodes made from metal oxides, such as ITO, IZO, and IO, are transparent in the visible wavelength region, possessing the transmittivity of more than 80%, whereas the metal-based cathodes, constituted of metals, such as Mg:Ag, Ag, Sm, and Al, are semi-transparent envisaging the transmittivity values often <70%. Thus, the greater transmittivity of the cathode should assist in better escape of the emitting light from the EML of PLED. Recently, a group of researchers have invented highly flexible see-through PLEDs consisting of conducting nanofibrillar polymeric transparent cathodes [51]. However, they have not disclosed the exact constitution of the conducting polymer cathode. Earlier, though GO has been used as HTL, Kwon *et al.* have utilized N doped graphene as a transparent cathode of low work function [71]. Herein, the added amine dopant has significantly enhanced the device efficiency, i.e., 13.8 cd A^{-1}, compared to undoped graphene, i.e., 2.74 cd A^{-1}.

6.5.2.6 *Characteristics of Anode*

A few workers have got satisfactorily improved efficiencies and lifetime when they have used oxygen plasma modified ITO replacing untreated ITO as anode [36, 38]. However, when the same ITO anode is modified by aquaregia, lifetime and efficiency of the device have been deteriorated compared to the untreated device [36]. Moreover, encouraging results have been obtained when ITO has been derivatized by silyl coupling layer and then coated with multiple polyelectrolyte layer(s) (Table 6.1) [33]. In this regard, both luminous efficiency and luminance deteriorate when an ITO anode is derivatized by silyl coupling layer. However, when such silylated ITO anode is coated with polyelectrolyte layers, performances of the devices improve. Irrespective of the thickness of the polymeric layer, conducting polymer modified ITO anode is noted to enhance both carrier injection and electroluminescence efficiency, leading to higher LED brightness. In contrast, improvements of brightness and electroluminescence efficiency are heavily dependent on the thickness of the polymer coatings on either semi-conductive or insulating polymer coated ITOs. Moreover, compared to insulating polymer layer coat, semiconducting polymer layers pose as a much weaker barrier against the charge transmission, since the oligo-*p*-phenylenevinylene segments of semiconducting polymer layers is able to participate in the charge transport. Accordingly, the performance of the semiconducting polymer coated ITO containing devices bearing tends is the intermediate between the conducting polymer and insulating polymer coated devices.

Transparency of anode is a valuable criterion for inverted top emitting and conventional bottom emitting PLED types. However, recently a group of researchers invented highly flexible see-through PLEDs, in which they have used nanofibrillar polymeric anodes [51].

6.6 Life Time and Stability

Stability and life time of a PLED depends on several factors, including the air stability of the metal oxide layers incorporated in EIL and HIL [49]. Usually, these devices are constructed of air-sensitive metals or charge-injection layers that require rigorous encapsulation to prevent the degradation [43, 44]. Because of their encouraging combined properties, such as high transparency, low resistance, and air stability, metal oxides are promising candidates as good charge injecting components. If the selected metal oxide-based cathodes are unreactive, there is no need to encapsulate the device. Accordingly, the cost of such device should be lower than the PLED containing encapsulated cathodes. In this context, TiO_2 and ZnO have been employed as air-stable cathodes, as an alternative to low work function metals in PLED [43]. Sometimes, PLED containing fluorene-based EMLs suffer from keto defects leading to unwanted green emission, which can hamper long-time storage stability. Moreover, unsaturated polymeric EML-based PLEDs can undergo degradation mechanism by loss of conjugation in the presence of singlet oxygen radicals, leading to blue shift in the emission maxima and unwanted quenching. Moreover, quantum efficiency roll-offs is another kind of usual problem encountered in the presence of high current densities. In fact, such EQE roll-off at high current densities is attributed to imbalance between the numbers of holes and electrons in the emitting layer and the nonradiative exciton quenching.

6.7 Recent Developments, Challenges, and Constraints

Very recently, perovskite-polymer–based hetero-structured LEDs constituting of quasi-two-dimensional and three-dimensional (2D/3D) perovskites and an insulating polymer have been fabricated, which demonstrated very high EQEs of up to 20.1% within 0.1–1 mA cm^{-2} current densities [52]. In fact, some major limitations of purely perovskite-based LEDs, such as strong leakage current and deteriorated performances because of rough or discontinuous nature of the perovskite surface, can be maneuvered by combining the insulating polymer with the perovskites [72]. Such an improved EQE is attributed to the effective suppression of bulk and interfacial non-radiative relaxation processes. However, the device lacks desirable brightness level and stability (T_{50} = 46 h in air at the maximum EQE) and thus requires further improvement. The major strategy behind inclusion of quasi-2D/3D nanostructures is to confine charge carriers. Thus, the efficiencies of perovskite-based LEDs are comparable to the available high-performance OLEDs and QD LEDs. However, encouraging results have been obtained when a new architecture, i.e., "insulator-perovskite-insulator" has been introduced while replacing the earlier HIL/HTL composed of PEDOT:PSS or blend of PEDOT:PSS of a conventionally used perovskite-based LEDs [72]. Another group of workers have developed a kind of highly efficient, i.e., the maximum EQE = 7.3%, blue thermally activated delayed fluorescence polymers by ring-opening metathesis polymerization of norbornene [53]. Basic principle behind such thermally activated delayed fluorescence is the small energy gap (ΔE_{ST}) between triplet (T_1) state and the lowest excited singlet state (S_1), facilitating thermally aided reverse intersystem crossing of excitons from triplet (T_1) state to the radiative lowest excited singlet state (S_1). Herein, the major strategy was to elevate the level of the T_1 state to minimize the ΔE_{ST}, along with confinement of the triplet excitons in TADF units to suppress the back transfer of the triplet energy from emissive units to the backbone. In fact, small ΔE_{ST} is favorable for transformation of triplet excitons to radiative singlet excitons by reverse intersystem crossing process. Herein, an exciton blocking layer has been incorporated that is constituted of DPEPO having very high T_1 state (Table 6.1). Another group of scientists have prepared a series of transparent highly deformable PLEDs utilizing conducting polymer electrodes [51]. In doing so, they have used a highly conductive nanofibrillated CP as deformable transparent electrodes. Thus, this type of device may emit light in all directions and thus cannot be categorized under either top or bottom emitting PLEDs. Moreover, people are engaged in developing LED emitting near infrared of enhanced intensity being accomplished by aggregation induced emission [73]. In this regard, as near infrared emitters, PbS-based quantum dot LEDs have been proven to be more useful and efficient compared to the near infrared emitting PLEDs [74]. Therefore, in near future, quantum dot-based PLEDs may bring about interesting outcomes compared to the traditional PLEDs. In addition, increasing attention is being paid in developing white light-emitting PLEDs [55]. While developing the white light-emitting PLEDs, one of the major strategies is to admix different layers emitting three different primary colors, so that the finally emitted color becomes white. For instance, according to Xin *et al.* [55], the device structure consists of a blue-emitting oligoquinoline as ETL and a poly(*N*-vinyl carbazole) layer doped with phosphorescent green and red iridium emitters, such as $Ir(ppy)_3$ and $Ir(piq)_2(acac)$, respectively (Table 6.1). In addition to the substantial luminous efficiency and brightness, these devices demonstrate excellent color stability as well as less rolling off. However, such device still has large room for further improvement in brightness, sustainability, and efficiencies through fabrication in an inert atmosphere and/or by employing more efficient phosphorescent dyes. Modified forms of quinoline-based iridium complexes are also employed for PLED emitting near infrared within 690 to 780 nm [75]. In this regard, superior EQE max and almost negligible efficiency roll-off for the device containing TmPyPB as ETL are associated with facilitated electron-transport by TmPyPB layer. Recently, Xuan *et al.* have fabricated quantum dot containing polymer composite as a component of white light-emitting LED [76]. The main objective behind such pervoksite-based polymer composite preparation is to protect the hydrolysis-prone pervoksite component engulfed within highly hydrophobic polymer network, so that fabricated polymer composite can acquire sufficient stability against hydrolysis, and thus, the quantum dot components may retain the desired level of quantum efficiencies for a longer period of time. In fact, the composite is capable of emitting stable, tunable, bright green light, and when red-emitting $K_2SiF_6:Mn^{4+}$ is admixed with the green emitting composite, along with blue light-emitting chip, white light is produced from this device with high luminous efficiency of 50 lm W^{-1}. In contrast to Xuan *et al.* [76], a different approach has been adopted by Wu *et al.* [77], as they have improved the efficiency, stability, and durability of white light-emitting diode further by applying a thick coat on the pervoksite particles. By this method, they have achieved the higher stability and purity of the polymeric coating around the discrete pervoksite phases. Such green light-emitting polymer coated pervoksite has been combined with red-emitting phosphor on blue InGaN chip. Very recently, to avoid the addition of inorganic component, Shao *et al.* have fabricated a series of tunable PLEDs capable to emit thermally activated delayed fluorescence at a reasonably higher EQE of 15.8%, 17.1%, and 10.1% for blue, red, and green emissions, respectively [78]. The structure of EML has been constituted of bipolar poly(arylene phosphine oxide) hosts based on electron transporting arylphosphine oxide units coupled with hole transporting carbazole units. The fabricated EML can be modulated to achieve the desired triplet energy levels and good carrier transport capability by structural alterations of the added spacers with different twist angles or variable linkage patterns or different heteroatoms or conjugation lengths. In this regard, enhanced triplet energy can be achieved by affecting the conjugation either *via* reducing the conjugation length or by inserting the heteroatom within the conjugated spacer. Moreover, conjugation can be reduced by replacing ortho/para linkage pattern with meta linkages or by enhancing the twist angles. PLEDs based on perovskite-polymer–based hetero-structures are increasingly utilized for developing low-cost and high-performance photon sources for lighting, display, and communication technologies. Moreover, thermally activated delayed fluorescence strategy may be adopted in large scale to produce highly efficient polymers for the solution-processed blue light-emitting devices. On the other hand, near infrared and short-wave infrared light-emitting diodes have the capability to serve a broad range of applications, including surveillance, night vision, bio-imaging, remote sensing, and spectroscopy.

6.8 Conclusions

Though discovered three decades ago, rapid technological advancement of PLEDs has been observed in the 21st century, leading to scaled up and diversified applications of PLEDs in manufacturing cost-effective large display devices. The efficiencies, including EQE, luminance efficiency, and external photoluminescent quantum efficiencies, and performance parameters, such as luminance or brightness and lifetime, have been improved remarkably over the past decade and are still in progress. The device architecture has been engineered to introduce new layer(s) or modification of layer(s) to achieve the optimum balance between hole and electron mobility and so as to attain balanced charge injection and transport. It is possible to modulate the nature and wavelength of the emitted light from the EML by structural alteration of the polymer chains *via* blending, doping, and chemical treatment to incorporate spacer and/or narrow/wide band gap moieties. Otherwise, the hole/electron transporting/blocking nature the adjacent layers and/or constitution of electrodes can suitably be altered to achieve PLEDs of enhanced efficiency, longevity, and diversity in terms of color emission. Very recently, work is in progress in the field of quantum dot/pervoksite containing white light-emitting PLED and PLEDs emitting thermally activated delayed fluorescence or near infrared light. However, red or near infrared or white light-emitting devices are still suffering from lower efficiencies compared to blue or green colored light-emitting PLEDs, and there is enough ground to improve the efficiency level and longevity further for the most recently developed PLEDs through architectural changes *via* diversified single/multi-layer insertion/modification(s). In this regard, insertion of intrinsically fluorescent multipolymers as a component the emitting layer may bring about auto-fluorescent PLED device of enhanced efficiency and output.

References

1. Tang, C.A. and Slyke, V., Organic Electroluminescent Diodes. *Appl. Phys. Lett.*, 51, 913–915, 1987.
2. Yang, Y. and Pei, Q., Efficient Blue-Green and White Light-emitting Electrochemical Cells Based on Poly[9,9-bis(3,6-dioxaheptyl)-fluorene-2,7-diyl]. *J. Appl. Phys.*, 81, 3294–3298, 1997.
3. Burroughes, J.H., Bradley, D.D.C., Brown, A.R., Marks, R.N., Mackay, K., Friend, R.H., Burns, P.L., Holmes, A.B., Light-emitting Diodes Based on Conjugated Polymers. *Nature*, 374, 539–541, 1990.
4. Burn, P.L., Holmes, A.B., Kraft, A., Bradley, D.D.C., Brown, A.R., Friend, R.H., Gymer, R.W., Chemical Tuning of Electroluminescent Copolymers to Improve Emission Efficiency and Allow Patterning. *Nature*, 356, 47–49, 1992.
5. Braun, D. and Heeger, A.J., Visible Light Emission from Semiconducting Polymer Diodes. *Appl. Phys. Lett.*, 58, 1982, 1991.
6. Millard, I.S., High-efficiency Polyfluorene Polymers Suitable for RGB Applications. *Synth. Met.*, 111-112, 119–123, 2000.
7. Neher, D., Polyfluorene Homopolymers: Conjugated Liquid-crystalline Polymers for Bright Blue Emission and Polarized Electroluminescence. *Macromol. Rapid Commun.*, 22, 1365–1395, 2001.
8. Ohmori, Y., Uchida, M., Muro, K., Yoshino, K., Blue Electroluminescent Diodes Utilizing Poly (Alkylfluorene). *Jpn. J. Appl. Phys.*, 30, 1941–1943, 1991.
9. Pei, Q.B. and Yang, Y., Efficient Photoluminescence and Electroluminescence from a Soluble Polyfluorene. *J. Am. Chem. Soc.*, 118, 7416–7417, 1996.
10. Miteva, T., Meisel, A., Knoll, W., Nothofer, H.G., Scherf, U., Müller, D.C., Meerholz, K., Yasuda, A., Neher, D., Improving the Performance of Polyfluorene-Based Organic Light-Emitting Diodes via End-Capping. *Adv. Mater.*, 13, 565–570, 2001.
11. Bernius, M., Inbasekaran, M., Woo, E., Wu, W.S., Wujkowski, L., Fluorene-based polymers-Preparation and Applications. *J. Mater. Sci. Mater. Electron.*, 11, 111, 2000.
12. Bernius, M.T., Inbasekaran, M., O'Brien, J., Wu, W.S., Progress with Light-emitting Polymers. *Adv. Mater.*, 12, 1737, 2000.
13. Inbasekaran, M., Woo, E., Wu, W.S., Bernius, M., Wujkowski, L., Fluorene Homopolymers and Copolymers. *Synth. Met.*, 111, 397, 2000.

14. Bradley, D.D.C., Grell, M., Grice, A., Tajbakhsh, A.R., O'Brien, D.F., Bleyer, A., Polymer Light emission: Control of Properties through Chemical Structure and Morphology. *Opt. Mater.*, 9, 1, 1998.
15. Grell, M., Knoll, W., Lupo, D., Meisel, A., Miteva, T., Neher, D., Nothofer, H.G., Scherf, U., Yasuda, A., Blue Polarized Electroluminescence from a Liquid Crystalline Polyfluorene. *Adv. Mater.*, 11, 671–675, 1999.
16. Palilis, L.C., Lidzey, D.G., Redecker, M., Bradley, D.D.C., Inbasekaran, M., Woo, E.P., Wu, W.W., Bright and Efficient Blue and Green Light-emitting Diodes Based on Conjugated Polymer Blends. *Synth. Met.*, 111, 159–163, 2000.
17. Ahn, T., Song, S.Y., Shim, H.K., Highly Photoluminescent and Blue-Green Electroluminescent Polymers: New Silyl- and Alkoxy-substituted Poly(*p*-phenylenevinylene) Related Copolymers Containing Carbazole or Fluorene Groups. *Macromolecules*, 33, 6764–6771, 2000.
18. He, Y. and Kanicki, J., High-efficiency Organic Polymer Light-emitting Heterostructure Devices on Flexible Plastic Substrates. *Appl. Phys. Lett.*, 76, 661, 2000.
19. Stephan, O., Collomb, V., Vial, J.C., Armand, M., Blue-green Light-emitting Diodes and Electrochemical Cells Based on a Copolymer Derived from Fluorine. *Synth. Met.*, 113, 257–262, 2000.
20. Zhu, F.R., Zhang, K.R., Guenther, E., Jin, C.S., Optimized Indium Tin Oxide Contact for Organic Light Emitting Diode Applications. *Thin Solid Films*, 363, 314–317, 2000.
21. Jiang, X.Z., Liu, S., Ma, H., Jen, A.K.Y., High-performance Blue Light-emitting Diode Based on a Binaphthyl-containing Polyfluorene. *Appl. Phys. Lett.*, 76, 1813, 2000.
22. Segura, J.L., The Chemistry of Electroluminescent Organic Materials. *Acta Polym.*, 49, 319–344, 1998.
23. Kraft, A., Grimsdale, A.C., Holmes, A.B., Electroluminescent Conjugated Polymers-seeing Polymers in a New Light. *Angew. Chem. Int. Ed.*, 37, 402–428, 1998.
24. Mitschke, U. and Bäuerle, P., The Electroluminescence of Organic Materials. *J. Mater. Chem.*, 10, 1471–1507, 2000.
25. Grimsdale, A.C., Chan, K.L., Martin, R.E., Jokisz, P.G., Holmes, A.B., Synthesis of Light-emitting Conjugated Polymers for Applications in Electroluminescent Devices. *Chem. Rev.*, 109, 897–1091, 2009.
26. Minshall, T., Seldon, S., Probert, D., Commercializing a Disruptive Technology Based Upon University IP through Open Innovation: A Case Study of Cambridge Display Technology. *Int. J. Innov. Technol. Manag.*, 4, 225–239, 2007.
27. Holmes, A., Electronics: Polymers Light the Way. *Nature*, 421, 800–801, 2003.
28. Kabra, D., Lu, L.P., Song, M.H., Snaith, H.J., Friend, R.H., Efficient Single-Layer Polymer Light-Emitting Diodes. *Adv. Mater.*, 22, 3194–3198, 2010.

29. Sekine, C., Tsubata, Y., Yamada, T., Kitano, M., Doi, S., Recent Progress of High Performance Polymer OLED and OPV Materials for Organic Printed Electronics. *Sci. Technol. Adv. Mater.*, 15, 034203, 2014.
30. Lee, B.R., Jung, E.U., Park, J.S., Nam, Y.S., Min, S.H., Kim, B.-S., Lee, K.-M., Jeong, J.-R., Friend, R.H., Kim, J.-S., Kim, S.O., Song, M.H., Highly Efficient Inverted Polymer Light-emitting Diodes Using Surface Modifications of Zno Layer. *Nat. Commun.*, 5, 4840, 2014.
31. Fukuda, M., Sawada, K., Yoshino, K., Fusible Conducting Poly(9-alkylfluorene) and Poly(9,9-dialkylfluorene) and Their Characteristics. *Jpn. J. Appl. Phys.*, 28, 1433–1435, 1989.
32. Parker, I.D., Carrier Tunneling and Device Characteristics in Polymer Light-emitting Diodes. *J. Appl. Phys.*, 75, 1656, 1994.
33. Ho, P.K.H., Granström, M., Friend, R.H., Greenham, N.C., Ultrathin Self-assembled Layers at the ITO Interface to Control Charge Injection and Electroluminescence Efficiency in Polymer Light-Emitting Diodes. *Adv. Mater.*, 10, 769–774, 1998.
34. Grice, A.W., Bradley, D.D.C., Bernius, M.T., Inbasekaran, M., Wu, W.W., Woo, E.P., High Brightness and Efficiency Blue Light-Emitting Polymer Diodes. *Appl. Phys. Lett.*, 73, 629, 1998.
35. Chen, J.P., Klaerner, G., Lee, J.-I., Markiewicz, D., Lee, V.Y., Miller, R.D., Scott, J.C., Efficient, Blue Light-emitting Diodes using Cross-linked Layers of Polymeric Arylamine and Fluorine. *Synth. Met.*, 107, 129–135, 1999.
36. Kim, J.S., Friend, R.H., Cacialli, F., Improved Operational Stability of Polyfluorene-based Organic Light-emitting Diodes with Plasma-treated Indium–Tin–Oxide Anodes. *Appl. Phys. Lett.*, 74, 3084, 1999.
37. Lee, J.I., Hwang, D.H., Park, H., Do, L.M., Chu, H.Y., Zyung, T.R., Miller, D., Light-emitting Electrochemical Cells Based on Poly(9,9-bis(3,6-dioxaheptyl/-fluorene-2,7-diyl). *Synth. Met.*, 111, 195–197, 2000.
38. Morgado, J., Friend, R.H., Cacialli, F., Improved Efficiency of Light-emitting Diodes Based on Polyfluorene Blends Upon Insertion of a poly(p-phenylenevinylene) electron confinement layer. *Appl. Phys. Lett.*, 80, 2436–2438, 2002.
39. Morteani, A.C., Dhoot, A.S., Kim, J.-S., Silva, C., Greenham, N.C., Murphy, C., Moons, E., Ciná, S., Burroughes, J.H., Friend., R.H., Barrier-free Electron-hole Capture in Polymer Blend Heterojunction Light-emitting Diodes. *Adv. Mater.*, 15, 1708–1712, 2003.
40. Iyengar, N.A., Harrison, B., Duran, R.S., Schanze, K.S., Reynolds, J.R., Morphology Evolution in Nanoscale Light-emitting Domains in MEH-PPV/PMMA blends. *Macromolecules*, 36, 8978–8985, 2003.
41. Hung, M.-C., Liao, J.-L., Chen, S.-A., Chen, S.-H., Su, A.-C., Fine Tuning the Purity of Blue Emission from polydioctylfluorene by End-capping with Electron-deficient Moieties. *J. Am. Chem. Soc.*, 127, 14576–14577, 2005.
42. Lim, S.-F., Friend, R. .H., Rees, I. .D., Li, J., Ma, Y., Robinson, K., Holmes, A.B., Hennebicq, E., Beljonne, D., Cacialli, F., Suppression of Green Emission in

New Class of Blue-emitting Polyfluorene Copolymer with Twisted Biphenyl Moieties. *Adv. Funct. Mater.*, 15, 981–988, 2005.

43. Bolink, H.J., Coronado, E., Repetto, D., Sessolo, M., Barea, E.M., Bisquert, J., Garcia-Belmonte, G., Prochazka, J., Kavan, L., Inverted Solution Processable OLEDs Using a Metal Oxide as an Electron Injection Contact. *Adv. Funct. Mater.*, 18, 145–150, 2008.
44. Bolink, H.J., Coronado, E., Orozco, J., Sessolo, M., Efficient Polymer Light-emitting Diode Using Air-stable Metal Oxides as Electrodes. *Adv. Mater.*, 21, 79–82, 2009.
45. Earmme, T., Earmme, T., Ahmed, E., Jenekhe, S.A., Solution-processed Highly Efficient Blue Phosphorescent Polymer Light-emitting Diodes Enabled by a New Electron Transport Material. *Adv. Mater.*, 22, 4744–4748, 2010.
46. Ye, T., Shao, S., Chen, J., Wang, L., Ma, D., Efficient Phosphorescent Polymer Yellow-light-emitting Diodes Based on Solution-processed Small Molecular Electron Transporting Layer. *ACS Appl. Mater. Interfaces*, 3, 410–416, 2011.
47. Lu, L.P., Kabra, D., Jhonson, K., Friend, R.H., Charge-carrier Balance and Color Purity in Polyfluorene Polymer Blends for Blue Light-emitting Diodes. *Adv. Mater.*, 22, 144–150, 2012.
48. Lee, B.R., Kim, J.-W., Kang, D., Lee, D.W., Ko, S.-J., Lee, H.J., Lee, C.-L., Kim, J.Y., Shin, H.S., Song, M.H., Highly Efficient Polymer Light-emitting Diodes Using Graphene Oxide as a Hole Transport Layer. *ACS Nano*, 6, 2984–2991, 2012.
49. Zhao, J., Zhou, G., Yan, K., Xie, J., Li, Y., Liao, L., Jin, Y., Liu, K., Hsu, P., Wang, J., Cheng, H., Cui, Y., Air-stable and Freestanding Lithium Alloy/Graphene Foil as an Alternative to Lithium Metal Anodes. *Nat. Nanotechnol.*, 12, 993–999, 2017.
50. Kim, Y.-T., Seol, J.B., Kim, Y.-H., Ahn, H.-J., Park, C.-G., Correlation of controllable aggregation with light-emitting property in polymer blend optoelectronic devices. *Small*, 13, 1–10, 2017.
51. Kee, S., Kim, N., Park, B., Kim, B.S., Hong, S., Lee, J.-H., Jeong, S., Kim, A., Jang, S.-Y., Lee, K., Highly Deformable and See-through Polymer Light-emitting Diodes with All-conducting-polymer Electrodes. *Adv. Mater.*, 30, 1703437, 2017.
52. Zhao, B., Bai, S., Kim, V., Lamboll, R., Shivanna, R., Auras, F., Richter, J.M., Yang, L., Dai, L., Alsari, M., She, X.-J., Liang, L., Zhang, J., Lilliu, S., Gao, P., Snaith, H.J., Wang, J., Greenham, N.C., Friend, R.H., Di, D., High-efficiency Perovskite–polymer Bulk Heterostructure Light-emitting Diodes. *Nat. Photonics*, 12, 783–789, 2018.
53. Zeng, X., Luo, J., Zhou, T., Chen, T., Zhou, X., Wu, K., Zou, Y., Xie, G., Gong, S., Yang, C., Using Ring-opening Metathesis Polymerization of Norbornene to Construct Thermally Activated Delayed Fluorescence Polymers: High-efficiency Blue Polymer Light-emitting Diodes. *Macromolecules*, 51, 1598–1604, 2018.

54. Luo, Y., Yu, T., Nian, L., Liu, L., Huang, F., Xie, Z., Ma, Y., Photoconductive Cathode Interlayer for Enhanced Electron Injection in Inverted Polymer Light-emitting Diodes. *ACS Appl. Mater. Interfaces*, 10, 11377–11381, 2018.
55. Xin, H., Yan, W., Jenekhe, S.A., Color-Stable White Organic Light-emitting Diodes Utilizing a Blue-emitting Electron-transport Layer. *ACS Omega*, 3, 12549–12553, 2018.
56. Chen, S., Deng, L., Xie, J., Peng, L., Xie, L., Fan, Q., Huang, W., Recent Developments in Top-emitting Organic Light-emitting Diodes. *Adv. Mater.*, 22, 5227–5239, 2010.
57. Lu, L.P., Finlayson, C.E., Kabra, D., Albert-Seifried, S., Song, M.H., Havenith, R.W.A., Wilhelm, G.T., Huck, T.S., Friend, R.H., Influence of Side-chain Position on the Optoelectronic Properties of a Red-emitting Conjugated Polymer. *Macromol. Chem. Phys.*, 214, 967–974, 2013.
58. Kitamura, C., Tanaka, S., Yamashita, Y., Design of Narrow-bandgap Polymers: Syntheses and Properties of Monomers and Polymers Containing Aromatic-Donor and o-quinoid-acceptor Units. *Chem. Mater.*, 8, 570–578, 1996.
59. Nimith, K.M., Satyanarayan, M.N., Umesh, G., Enhancement in Fluorescence Quantum Yield of MEH-PPV: BT Blends for Polymer Light Emitting Diode Applications. *Opt. Mater.*, 80, 143–148, 2018.
60. Chappell, J., Lidzey, D.G., Jukes, P.C., Higgins, A.M., Thompson, R.L., O'Connor, S., Grizzi, I., Fletcher, R., O'Brien, J., Geoghegan, M., Jones, R.A.L., Correlating Structure with Fluorescence Emission in Phase-separated Conjugated Polymer Blends. *Nat. Mater.*, 2, 616–621, 2003.
61. Morgado, J., Moons, E., Friend, R.H., Calcialli, F., De-mixing of Polyfluorene-based Blends by Contact with Acetone: Electro- and Photo-luminescence Probes. *Adv. Mater.*, 13, 810–814, 2001.
62. Hassan, M.U., Liu, Y.-C., Yetisen, A.K., Butt, H., Friend, R.H., Energy Landscape of Vertically Anisotropic Polymer Blend Films toward Highly Efficient Polymer Light-emitting Diodes (PLEDs). *Adv. Funct. Mater.*, 28, 1705903, 2018.
63. Mieva, T., Meisel, A., Knoll, W., Nothofer, H., Scherf, U., Müller, D.C., Meerholz, K., Yasuda, A., Neher, D., Improving the Performance of Polyfluorene-based Organic Light-emitting Diodes via End-capping. *Adv. Mater.*, 13, 565–570, 2001.
64. Cao, Y., Yu, G., Zhang, C., Menon, R., Heeger, A.J., Polymer Light-emitting Diodes with Polyethylene Dioxythiophene-polystyrene Sulfonate as the Transparent Anode. *Synth. Met.*, 87, 171–174, 1997.
65. Yan, H., Lee, P., Armstrong, N.R., Graham, A., Evmenenko, G.A., Dutta, P., Marks, T.J., High-performance Hole-transport Layers for Polymer Light-emitting Diodes: Implementation of Organosiloxane Cross-linking Chemistry in Polymeric Electroluminescent Devices. *J. Am. Chem. Soc.*, 127, 3172–3183, 2005.

66. Yang, X.H., Muller, D.C., Neher, D., Meerholz, K., Highly Efficient Polymeric Electrophosphorescent Diodes. *Adv. Mater.*, 18, 948–954, 2006.
67. Niu, Y.H., Liu, M.S., Ka, J.W., Bardeker, J., Zin, M.T., Schofield, R., Chi, Y., Jen, A.K.Y., Crosslinkable Hole-transport Layer on Conducting Polymer for High-efficiency White Polymer Light-emitting Diodes. *Adv. Mater.*, 19, 300–304, 2007.
68. Lim, Y., Park, Y.S., Kang, Y., Jang, D.Y., Kim, J.H., Kim, J.J., Sellinger, A., Yoon, D.Y., Hole Injection/Transport Materials Derived from Heck and Sol-gel Chemistry for Application in Solution-processed Organic Electronic Devices. *J. Am. Chem. Soc.*, 133, 1375–1382, 2011.
69. Han, T.H., Choi, M.R., Woo, S.H., Min, S.Y., Lee, C.L., Lee, T.W., Molecularly Controlled Interfacial Layer Strategy toward Highly Efficient Simple-structured Organic Light-emitting Diodes. *Adv. Mater.*, 24, 1487–1493, 2012.
70. Zhong, Z., Wang, X., Guo, T., Cui, J., Ying, L., Peng, J., Cao, Y., Crosslinkable Triphenylamine-based Hole-transporting Polymers for Solution-processed Polymer Light Emitting Diodes. *Org. Electron.*, 53, 35–42, 2018.
71. Kwon, S.-J., Han, T.-H., Kim, Y.-H., Ahmed, T., Seo, H.-K., Kim, H., Kim, D.J., Xu, W., Hong, B.H., Zhu, J.-X., Lee, T.-W., Solution-Processed n-Type Graphene Doping for Cathode in Inverted Polymer Light-emitting Diodes. *ACS Appl. Mater. Interfaces*, 10, 4874–4881, 2018.
72. Shi, Y., Wu, W., Dong, H., Li, G., Xi, K., Divitini, G., Ran, C., Yuan, F., Zhang, M., Jiao, B., Hou, X., Wu, Z., A Strategy for Architecture Design of Crystalline Perovskite Light-emitting Diodes with High Performance. *Adv. Mater.*, 30, 1800251, 2018.
73. Baysec, S., Minotto, A., Klein, P., Poddi, S., Zampetti, A., Allard, S., Cacialli, F., Scherf, U., Tetraphenylethylene-BODIPY Aggregation-induced Emission Luminogens for Near-infrared Polymer Light-emitting Diodes. *Sci. China Chem.*, 61, 932–939, 2018.
74. Pradhan, S., Stasio, F.D., Bi, Y., Gupta, S., Christodoulou, S., Stavrinadis, A., Konstantatos, G., High-Efficiency Colloidal Quantum Dot Infrared Light-Emitting Diodes via Engineering at the Supra Nanocrystalline Level. *Nat. Nanotechnol.*, 14, 72–79, 2019.
75. Fu, G., Zheng, H., He, Y., Li, W., Lüa, X., He, H., Efficient Near-infrared (NIR) Polymer Light-emitting Diodes (PLEDs) Based on Heteroleptic Iridium(III) Complexes with Postmodification Effect of Intramolecular Hydrogen Bond or BF_2-chelation. *J. Mater. Chem. C*, 6, 10589–10596, 2018.
76. Xuan, T., Huang, J., Liu, H., Lou, S., Cao, L., Gan, W., Liu, R.-S., Wang, J., Super-Hydrophobic Cesium Lead Halide Perovskite Quantum Dots-polymer Composites with High Stability and Luminescent Efficiency for Wide Color Gamut White Light-emitting Diodes. *Chem. Mater.*, 31, 1042–1047, 2019.
77. Wu, H., Wang, S., Cao, F., Zhou, J., Wu, Q., Wang, H., Li, X., Yin, L., Yang, X., Ultrastable Inorganic Perovskite Nanocrystals Coated with a Thick

Long-chain Polymer for Efficient White Light-emitting Diodes. *Chem. Mater.*, 31, 1936–1940, 2019.

78. Shao, S., Wang, S., Xu, X., Yang, Y., Lv, J., Ding, J., Wang, L., Jing, X., Wang, F., Bipolar Poly(arylene phosphine oxide) Hosts with Widely Tunable Triplet Energy Levels for High-efficiency Blue, Green, and Red Thermally Activated Delayed Fluorescence Polymer Light-emitting Diodes. *Macromolecules*, 52, 3394–3403, 2019.

7

Polymer Liquid Crystal Devices and Displays

Nimra Shakeel, Mohd Imran Ahamed and Naushad Anwar*

Department of Chemistry, Aligarh Muslim University, Aligarh, India

Abstract

Liquid crystals (LCs) are considered as an intermediate state between the solid and liquid phase and are referred as to mesophase. Polymer liquid crystals (PLCs) offer a unique combination of properties including excellent solder resistance, barrier and mechanical properties, low viscosity, thermal expansion, flammability and water absorption properties, fast cycle times, good solvent resistance, very low mould shrinkage, moulding to tight tolerances, and high continuous use temperature. These are the excellent properties that are common to PLCs. The difference arises in their processing conditions including performance at high temperature, mechanical properties, and, most importantly, the cost effect. Potential applications of PLCs arise when a unique combination monomer molecules directly to form polymer molecule that show an excellent properties such as complex extrusion profiles, precision moulded small components, high strength/modulus fibers, film exhibiting excellent barrier properties, and novel composites. LCs act as prototype self-organize supramolecular soft material and tunable solvents that is why it has been grown a wide variety of applications in various fields such as LC display technology, bio-sciences, in materials science, etc. Research community has described the substantial developments in LCs nano-science embracing contemporary nano-materials including nano-rods, nanoparticles, nano-platelets, nano-tubes, etc. In view of these, LCs is considered as "fourth state of matter", and recently, in relation with current science, LCs plays an important role in nanoscience and nanotechnology in the near future. In this chapter, we have reported the general idea about PLCs, their brief synthesis, and applications in various fields in the form of LCs devices and displays.

**Corresponding author*: naushadanwar85@gmail.com

Inamuddin, Rajender Boddula, Mohd Imran Ahamed and Abdullah M. Asiri (eds.) Polymers for Light-Emitting Devices and Displays, (177–196) © 2020 Scrivener Publishing LLC

Keywords: Liquid crystals, polymer lquid crystals, LC devices and display, applications

7.1 Introduction

Liquid crystals (LC) are the class of soft materials representing distinct and promising functions that have come in area of research from more than hundred years ago and incessantly growing in the forefront fields of both in basic research and also in applied science [1–3]. LCs have potential applications in various technologies including watches, calculators, computers displays, LCD televisions, controllable lenses and lasing, band gap structure, drug delivery applications, holographic data recording, controllable photonic implications, sensors, etc. [4–6]. The structure and morphology of LCs shows an impressive modification with extrinsic factors. Figure 7.1 shows the different classifications of LCs systems [7]. The change in temperature and pressure, and electric and magnetic field may alter the molecular anisotropy (i.e., the orientation, dispersion, and double refraction) of LCs affect the optic and optoelectronic modification [8]. Due to the susceptible characteristics of LCs, they show nonlinear optical properties that it can easy to be activated by low optical fields. Furthermore, investigation of optical properties including dielectric constant, extinction

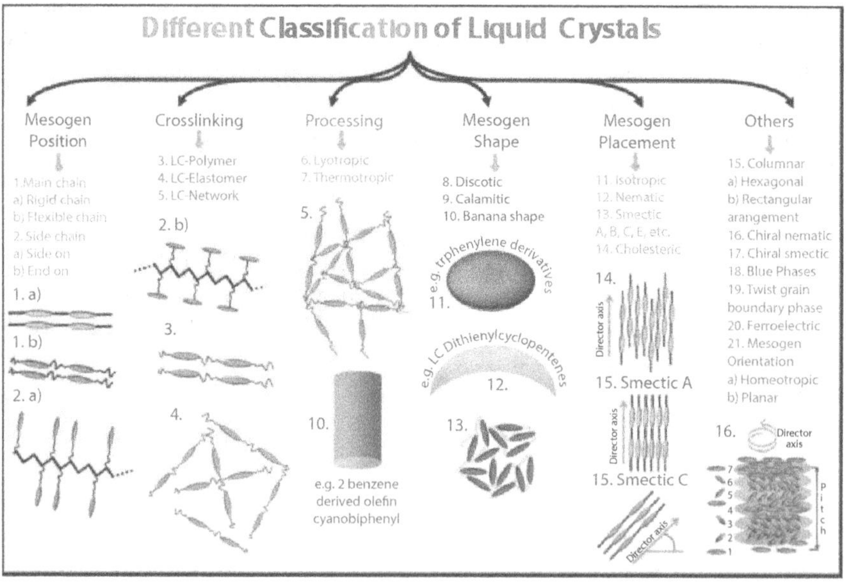

Figure 7.1 Different classifications of liquid crystals [6].

coefficient, refractive index, etc., provides valuable information which may be used to improve LCs-based optic and optoelectronic technologies. Such type of optical characteristics may change the electronic structure, and structure and efficiency of the ingratitude technological inferences. LCs have capacitated a lot of electro-optical devices such as LCDs [2], tunable Fresnel lenses [9], universal optical phase modulators [3], nanosecond light modulators [10], broad-spectrum tunable color reflectors, and beam steering devices [11]. These devices involve to a transparent conductors to apply voltage and let light through. The mostly used transparent conductor, as indium tin oxide (ITO), is costly and a brittle inorganic material and involves in vacuum for deposition. Tremendous attempt is to be made in the development of such type of transparent conductors that are usually using as a soft matrices to host conducting materials like as graphene [12–14], silver nano-wires [15–17], conducting polymers, [18–21] carbon nano-tubes [22–24] and nano-particulate metal oxides [23].

LCs [22–27] are an anisotropic fluids that attributed spontaneous positioning the subsequent order of rod-like molecular structure, known as nematic phase, and the other having sometimes position of low-dimensional order, known as the smectic phases (A and C) as shown in Figure 7.2 [28]. These are only the scratches surface of LCs, besides these, it has been reviewed that there are about 30 types of thermodynamically distinct LCs are in account and have been reported by the researchers nowadays [29]. Along with rod like, it can also be represented a non–rod-like shape-anisotropic nature such as bent-core or disks [30] types of molecules [31]. From

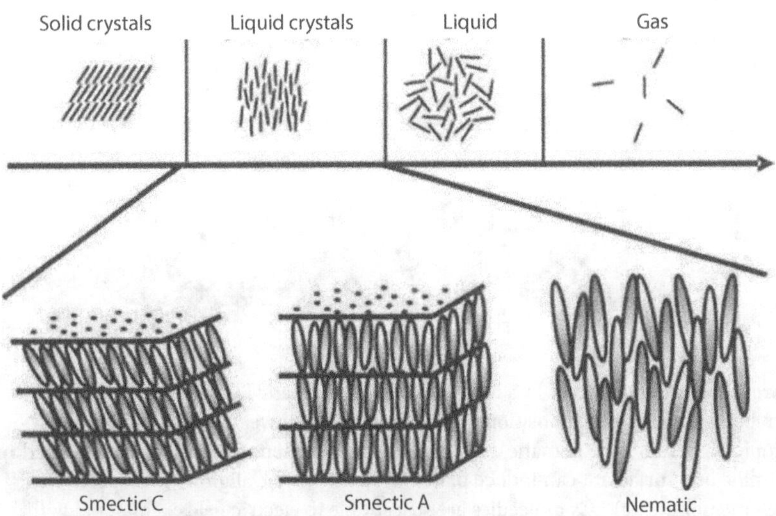

Figure 7.2 Various phases of liquid crystals [28].

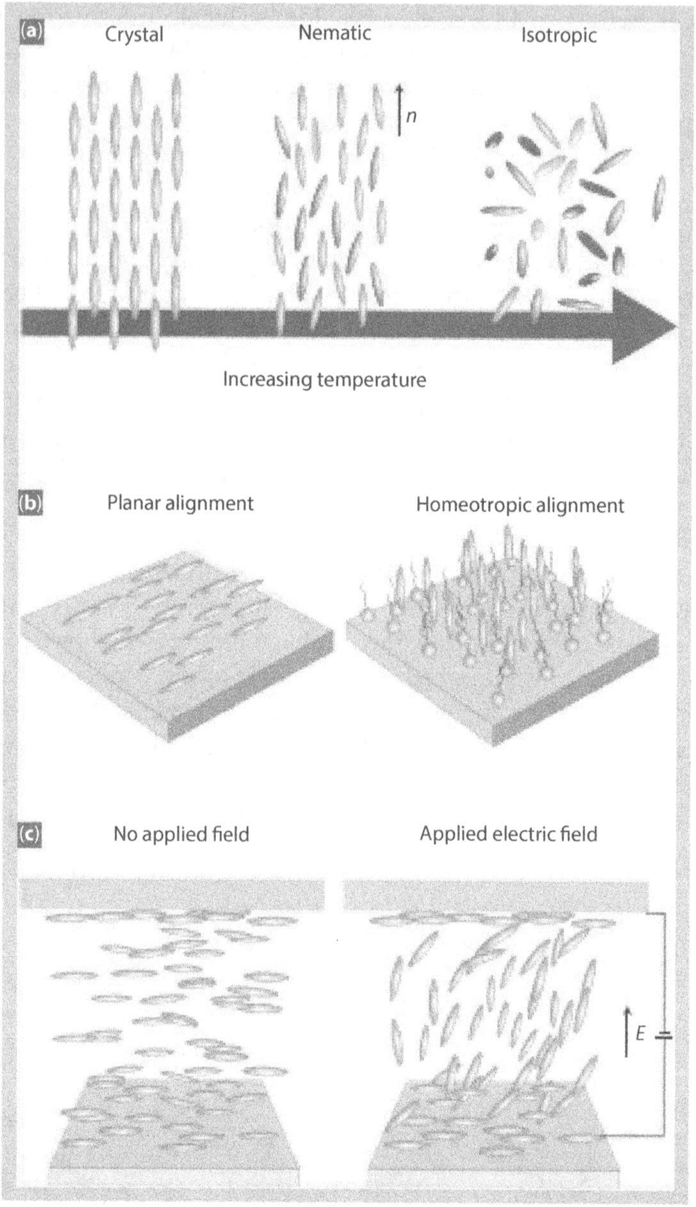

Figure 7.3 Fundamentals of LCs materials. (a) The nematic LCs phase has a positional order, but lacks long-range positional order, and exists between the crystalline and isotropic phases. n is the nematic director. (b) The treatment of a surface with rubbed polyimide or a surfactant can induce planar or homeotropic alignment, respectively, in LCs membrane. (c) LCs molecules are susceptible to electric fields; a material with a positive dielectric anisotropy will align along the direction of an applied field [34].

the last three decades, study on the LCs systems, it could be seen the fact about LCs materials that possess tendency to show chirality, i.e., innovating a non-superimposed molecular symmetry that can be seen in the structure and properties of the LCs phase (smectic A and C, Figure 7.1) [32].

Thermotropic LCs are of either rod-like also known as calamitic (TCLCs) or disk-like as discotic (DTLCs). The optical anisotropy of LCs molecules gives it a birefringence that is employed to approximate in all optical devices. Commonly used phase of LCs is the "nematic" phase, where the bulk systems have low positional order. The directions along where the axes of LCs molecule align are as the nematic director, n. Figure 7.3a represents the comparative phases of nematic with crystalline and isotropic phases. Cholesteric-type LCs contain an intrinsic helicity. Figure 7.3b describes the LCs material that is well aligned where as Figure 7.3c has been drawn with the consideration of applied electric field that reveals LCs molecules those are of negative dielectric anisotropy are perpendicular, whereas those with a positive dielectric anisotropy are directed towards the applied electric field [33, 34].

PLCs directed one into the area of inter-disciplinary science incorporating the physics and chemistry of LCs, along with in material science, the electronic industries, and polymer technology and also a subject area of interests for the researchers [35–37]. Study on LCs and PLCs systems are an interesting topic in various aspects like experimental investigation, theoretical modeling, and the practical applications. Polymers with high-molecular mass usually contain a long flexible chain, while LCs are small rigid molecule with cognition property. Under reliable circumstances, these molecular candidates are chemically combined to form a new side- or main-chain polymer known as polymer liquid crystals (PLCs). One of the other class of composite materials of LCs known as polymer dispersed liquid crystals (PDLCs) as composite materials that contain LCs droplets in a solid polymer matrix [38]. Normally, the formation of LC droplets *via* a phase separation process initiated by a UV or thermal polymerization [28] and the droplets are inconstantly dispersed inside the polymer matrix. Because of the electric switch ability property of LCs, the index mismatch between LCs droplets and the polymer matrix surrounding can be pitched by an applied external field. Over the past decades, a lot of research works have been performed on transparent display technology. Various transparent display technologies have been used polymer network liquid crystal (PNLC), cholesteric liquid crystals (ChLCs), organic light-emitting diodes (OLEDs), plasma display, electro-wetting display, liquid crystal display (LCD), and polymer dispersed liquids crystals (PDLCs) [39–43]. In all of these display technologies, only the PDLCs technology possibly to acquire transparent display devices. The PDLCs are used in sharp windows that are furnishing with glassy device able to electrically change either their transmittance or scattering [44, 45].

Recently, the composite materials of LCs are perceptible due to their quality to transferring long range positional order on the isotropic/anisotropic substances including carbon nano-tubes (graphene and different NPs) [1, 5]. The LCs in nematic phase (NLCs) are favorable for photonic and optical applications [46]. Doping of LCs with nano-particles substantially affects the electro-optic demonstration appliances in color range, increased birefringence [47], alteration the optical properties of photonic crystal fibers [48], the laser speckle pattern reduction, and the angle for viewing [49], improved optical non-linearity [50, 51], enhanced photorefractive responses [52], and by the strong development of two-beam coupling [47, 50].

The scattering of CNTs in NLCs displays an excellent class of soft matters with different electronic and optical property that utilizes in memory, sensor, and optoelectronic devices that upgrade the electro-optical capacity of NLCs. Moreover, a lot of properties have been investigated by the research community in doping of CNTs in NLCs system such as faster response, improvement of electrical conductivity, and dielectric anisotropy [5, 53–57], lower threshold voltage and low effective concentration of ions [46], improvement of the positional order in the LCs [58], time reduction of optoelectronic response [59], reduced controlling voltage, decrease the voltage of dc driving, and improve the switching behavior, also improvement of electrical as well as dielectric properties [60–64]. Some other excellent properties include power consumption, low operating voltage, and slight thickness; LCDs have been widely used in various appliances. Beside these, the slow response of LCs molecules is in account as the reason to failure or major disadvantages of LCDs in dynamic display applications. In the patterned vertical alignment (PVA) display modes, various precocious display modes, multi-domain vertical alignment (MVA), and in-plane switching (IPS) modes are the prevalent technologies in the current LCDs industry because they are provided an excellent full viewing angle [65–71]. In this book chapter, we have discussed the general idea about PLCs, their brief synthesis and applications in various fields in the form of LCs devices and displays.

7.2 History and Progress

In 1888, F. Reinitzer, a botanist, investigated in his experiments while he was working to extract cholesterol from carrots in order to find out the correct chemical formula of extracted compounds and determine whether that was the same material that was found in the cells of various animals. To attain this purpose, he has isolated it from human gallstones that are as the two compounds now known as cholesteryl acetate (CA) and cholesteryl benzoate (CB). Both are the organic compounds in solid crystals form at room temperature and he has been measured the melting point of observed compounds. While he has heated the compounds, he observed that the color reflections from the compounds during melting. Moreover, he also demonstrated two distinguishable melting points of organic compounds, *viz.*, cholesteryl benzoate and cholesteryl acetate. Melting of cholesteryl benzoate observed at 145.5°C becomes viscous and cloudy, but while increase in temperature above upto 178.5°C, nature of CB was changed and became clear and isotropic, as like as liquid state. The phenomena were reversible. The CA melted at 94.8°C and showed clarity at 114.3°C, and however, when it was cooled, the color was appeared has been observed. Discovery by Reinitzer on LCs developed from a specified peculiarity to a high interdisciplinary area of research with LCDs, having the broad applications to date. Moreover, likewise to use of thermotropic-type LCs for display technologies have also enormous applications that will probably be in business now and upcoming future, *viz.*, photo-voltaic devices, organic LEDs, gas sensors, organic field effect transistors, etc. [72, 73].

Based on Reinitzer's concept, O. Lehmann has worked on the determination of a new class of material, termed as LCs phase. Thereafter, Lehmann restrained this different or unusual materials as drop-like [74] that would become the specific materials in today's technologies known as PDLCs, which have been discovered, approximately, one century later [75, 76]. Research work on LCs system has grown quickly in the years of 1960s and 1970s, during which it declared for displays and devices highly drove the basic and applied science and technology. Nowadays, the LCDs field has spread dramatically, starting with monitor desktop, laptops, and now discriminating the broad area of television market.

7.3 Polymer Liquid Crystal: An Overview

PLCs are one of the areas promoted by demand to their own right to elucidate the nature and excellent features of macromolecular substances in LCs state. Beside this, the interest also towards PLCs is accounted for expectation to creating a polymeric system, along with an excellent property of low molecular mass LCs and macromolecular substance, making it spontaneous to produce fibers, membranes, and coating with extraordinary features. The thermotropic LCs possessing low molecular masses has been well known to require special hermetic protective shells (microcapsules, electro-optical cells, etc.) that maintain the shape of LCs and also protect the LCs materials from external abstraction. A lot of studies represented the synthesis of PLCs, and few are also described by us. The main attention of the researchers was focused to grow the various synthetic routes for comprising only the calamitic mesogenic groups that are responsible for the formation of LCs phase into main as well as the side branches of the polymer chains. These are the two principle path that could be employed to obtain thermotropic polymers with the propertied of LCs:

(i) With the polymer backbone, mesogenic groups are attached that lead to a side chain polymers of liquid crystals, and
(ii) Addition of the mesogenic groups into a polymer backbone leads to main polymer chain of liquid crystals [77–79].

Stupp *et al.* [80] have been reported the chemically synthesized liquid crystal polymer using various monomer units and chemically characterized of the regular mesogen were as follows:

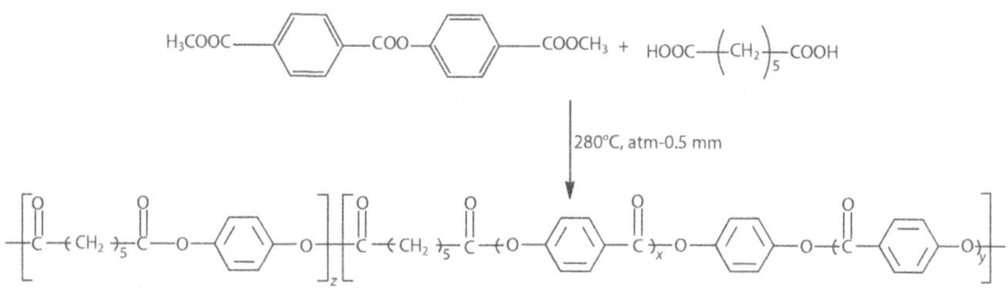

Scheme 7.1 Synthesis of the chemically disordered liquid crystal polymer [80].

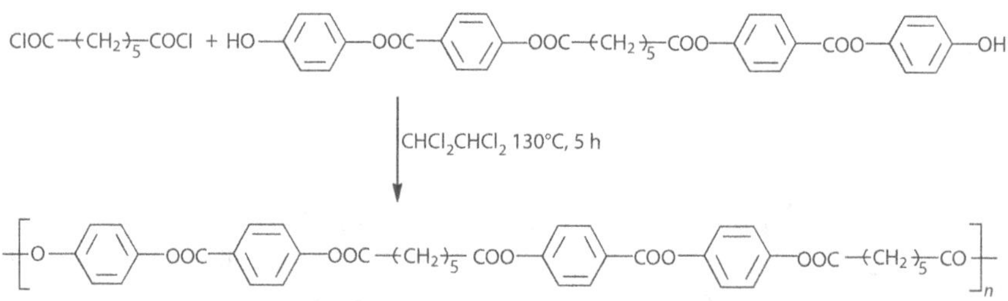

Scheme 7.2 Synthesis of the chemically ordered liquid crystal polymer [80].

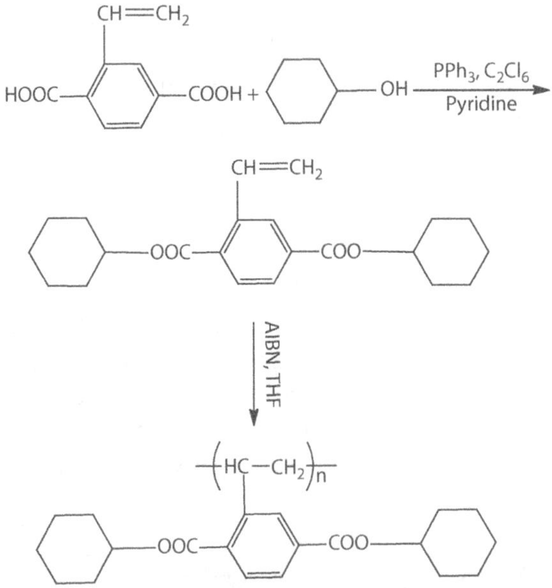

Scheme 7.3 Synthetic route of polymer from monomer unit [81].

The other synthetic route for the synthesis of PLCs has been reported by Zhang *et al.* [81] The group has been taken vinyl terephthalic acid and cyclohexanol monomers and after polymerization by the following pathway to synthesized poly[di(cyclohexyl) vinylterephthalate] in which both of the flexible spacers and conventional rigid mesogenic units are not incorporated.

7.4 Applications of PLCs

7.4.1 PLCs as Laser Sources

The various phases and repeated structures of LCs materials subsequent with the spectroscopic and optoelectronic applications are investigated as a laser source. These repeated structures and phases including ferroelectric LCs, holographic PDLCs, and blue and cholesteric phases of LCs can form photonic crystals with clear band-gap. Because of their tunability properties, the LCs lasers are mostly used in the area of biology and medicine. Sensitivity towards thermo-electrical areas, these compounds may be easy tuned over a wide range of wavelengths [82–85]. Under laser protective eyewears such as goggle glasses are used for protecting medical, industrial, and other personnel from laser radiation. Laser sources have been used in medical and industries to create a growing demand for windows to view galleries or in tower observations that provide both laser protection and "good" viewing. These potencies for LCs lasers for medical and biological photographing and therapeutic has been a growing area for young researchers.

7.4.2 PLCs as Dynamic Lenses

The use of PLCs materials as dynamic lenses is to procreate the effects of lens systems that are naturally occurred. The theory of bio-inspired optics creates an optical system that was altered by the nature for an indefinite quantity of applications. Figure 7.4 describes a single LCs droplet in nematic phase that may acs as a simple lens in which the focal length of the lens is adjusted by an electric field. Moreover, advanced designation of electrode plans make it potential to fabricate and tunable LCs as negative lenses [86–89].

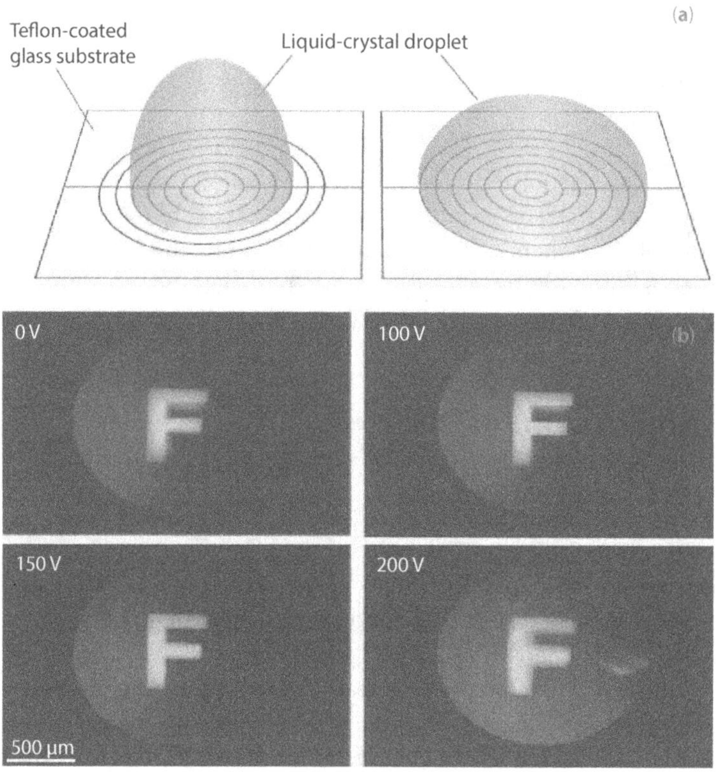

Figure 7.4 LCs lenses. As shown schematically (a) and experimentally, looking at an object "F" at a constant length from the lens (b), a LCs droplet can serve as a tunable lens through the use of a well-designed electrode pattern [88].

7.4.3 PLCs as Biosensors

Scientific communities have been performed a large set of research works to developed advanced biosensor devices using these LCs materials for bedside diagnostics, industrials, and lab applications. LCs materials having birefringence qualities, immoderate sensitivity, and interactions towards surface are put to excellent role play in biosensors. LCs materials used in optical sensors could obviate the mode for tags or markers, as the LCs molecule acts to affecs the optical abstraction of signals of the structure or biological methods investigation. It is beneficial that numerous biological systems such as cholesterols, phospholipids, cell membranes, DNA, and so on exist in LCs phases. Brake *et al.* have been observed when phospholipids molecules are introduced to a LCs/aqueous interface, a change is occurred in the alignment of LCs materials [90–93]. Figure 7.5 represents

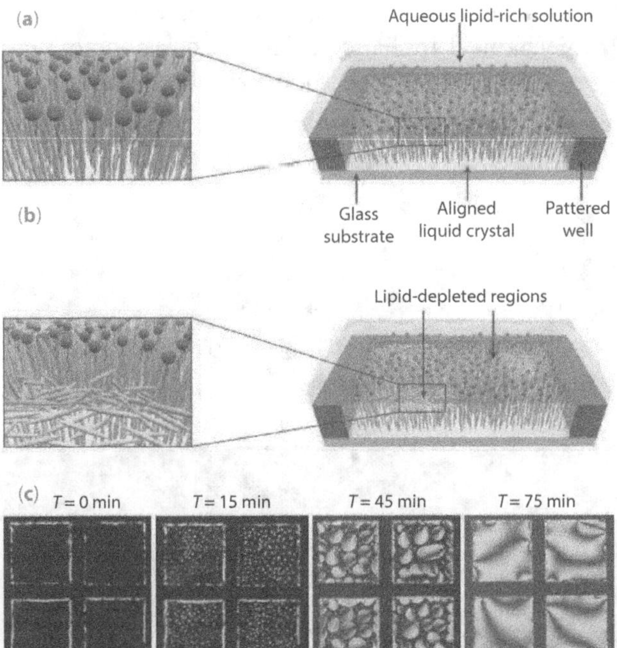

Figure 7.5 Liquid-crystal biosensors. (a, b) A well-aligned liquid-crystal optical biosensor (a) undergoes a change in optical appearance after an enzymatic process removes the aligning lipid material (b). (c) Optical microscopy images show the changes in the optical texture of such a biosensor undergoing an enzymatic reaction; dark regions represent homeotropic alignment, and bright regions represent a tilted or planar alignment. Each of the wells is ~283 μm wide with a depth of ~20 μm. Images in (c) courtesy of Nicholas L. Abbott, University of Wisconsin.

the perception of an enzymatic activity by which a biosensor carrying an enzyme laden LCs layers detect the activity of enzyme when a lipid-containing aqueous solution is introduced. Figure 7.5a depicts the homeotropic alignment of LCs molecules before the enzymatic reaction at the aqueous interface, whereas Figure 7.5b shows the orientational change in planer alignment of LCs molecules when lipid-containing aqueous solution is introduced. During these transitions, when the alignments are viewed between cross polarizers, the optical visualization of biosensor will change as shown in Figure 7.4c [54]. LCs materials have also been used in the immobilization of peptides, image protein-binding events, and determination of enzymatic activity at a LCs/aqueous interface [94–97].

7.4.4 PLCs as Actuator Devices

LCs materials such as nematic gels, elastomers, and reactive mesogens are available as a new prospect for microelectromechanical and bio-microelectromechanical appliances. Nematic gels contain fluid behaviur as it possesses a network of weak polymer; the nature of elastomers is

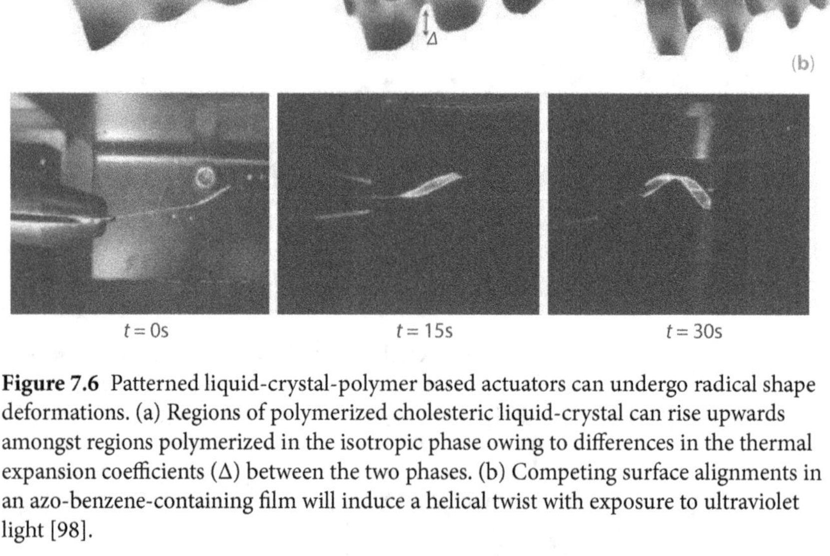

Figure 7.6 Patterned liquid-crystal-polymer based actuators can undergo radical shape deformations. (a) Regions of polymerized cholesteric liquid-crystal can rise upwards amongst regions polymerized in the isotropic phase owing to differences in the thermal expansion coefficients (Δ) between the two phases. (b) Competing surface alignments in an azo-benzene-containing film will induce a helical twist with exposure to ultraviolet light [98].

rubber-like that makes it deformable easily and the structures of reactive mesogens are most rigid and having the largest cross-linking densities saturated with LC solvent [99–102]. Any of these substances may change their simplest conformation when they are contact with order-disorder transition temperature into their isotropic state. Figure 7.6a represents a photo aligned structures of reactive mesogen films in cholesteric materials. When reactive mesogen is exposed to ultraviolet radiation at different time intervals, i's conformations become bend helical as shown in Figure 7.6b. The deformation in the induced shape of films can be removed either when the film is heated to the isotropic phase or allowed it to relax at their state of equilibrium or influence to light of different wavelength. Progressive applications may be deduced from these effects;an LCs elastomers film has been investigated by Lopez *et al.* having the property to 'floats' forth from a source of light, and its movements correspond tg those of flatfish. Actuation in electric field can also be stimulated in PLCs and in presence of strong electric fields; Ferroelectric PLCs can show large lateral electrostriction. Some other applications for PLCs actuators have been demonstrated such as mixers, micro-pumps and valves [103–107].

7.5 Conclusions

The present chapter describes the general idea about PLCs, their brief synthesis, and applications in various fields in the form of LCs devices and displays. The research on LCs materials is a rapidly growing field and has an important and not yet fully realized potential with significant impact on other disciplines of science. A whole network of new, advanced and complex structure of mesophase has become acquirable by an appropriate designed of comparatively simple LCs molecules. PLCs networks have been found in various applications such as optical, optoelectronic, in medicals as biosensors, etc., as well as described in this chapter. In particular, PLCs have also been used in monitor and television displays, mobile display, etc., which have been widely introduced.

References

1. Tie, W., Bhattacharyya, S.S., Zhang, Y., Zheng, Z., Lee, T.H., Lee, S.W., Kim, T.H., Lee, Y.H., Lee, S.H., Field-induced stretching and dynamic reorientation of functionalized multiwalled carbon nanotube aggregates in nematic liquid crystals. *Carbon*, 96, 548, 2016.

2. Wu, S.T. and Yang, D.K., *Fundamentals of liquid crystal devices*, p. 1, John Wiley amd Sons, Chichester, England.
3. Kawatasuki, N. and Ono, H., Synthesis of poly(benzylmethacrylate)-co-perfluorooctylethylmethacrylate) and electrooptical properties of the resultant polymer/liquid crystal composite films. *J. Appl. Polym. Sci.*, 55, 911, 1995.
4. McManamon, P.F., Shi, J., Bos, P.J., Broadband optical phased-array beam steering. *Opt. Eng.*, 44, 128004–1, 2005.
5. Tripathi, P.K., Misra, A.K., Manohar, S., Gupta, S.K., Manohar, R., Improved dielectric and electro-optical parameters of ZnO nanoparticle (8% Cu^{2+}) doped nematic liquid crystal. *J. Mol. Struct.*, 1035, 371, 2013.
6. Khoo, I., *Liquid Crystals*, 2nd Edn., John Wiley and Sons, Inc., Hoboken, New Jersey.
7. Rastogi, P., Njuguna, J., Kandasubramanian, B., Exploration of elastomeric and polymeric liquid crystals with photothermal actuation: A review. *Eur. Polym. J.*, 121, 109287, 2019.
8. Vertogen, G. and de Jeu, W.H., *Thermotropic liquid crystals; Fundamentals*, p. 45, Springer Sci. Business Media, Springer, Berlin, Heidelberg.
9. Pishnyak, O., Sato, S., Lavrentovich, O.D., Electrically tunable lens based on a dual-frequency nematic liquid crystal. *Appl. Opt.*, 45, 4576, 2006.
10. McManamon, P.F., Shi, J., Bos, P.J., Demonstration of a slow-light laser radar. *Opt. Eng.*, 44, 128004, 2005.
11. Lu, S.Y. and Chien, L.C., Electrically switched color with polymer-stabilized blue-phase liquid crystals. *Opt. Lett.*, 35, 562, 2010.
12. Kim, K.S., Zhao, Y., Jang, H., Lee, S.Y., Kim, J.M., Kim, K.S., Ahn, J.H., Kim, P., Choi, J.Y., Hong, B.H., Large-scale pattern growth of graphene films for stretchable transparent electrodes. *Nature*, 457, 706, 2009.
13. Eda, G. and Chhowalla, M., Blue photoluminescence from chemically derived graphene oxide. *Adv. Mater.*, 22, 505, 2010.
14. Wobkenberg, P.H., Eda, G., Leem, D.S., de Mello, J.C., Bradley, D.D.C., Chhowalla, M., Anthopoulos, T.D., Reduced graphene oxide electrodes for-large area organic electronics. *Adv. Mater.*, 23, 1558, 2011.
15. Blake, P., Brimicombe, P.D., Nair, R.R., Booth, T.J., Jiang, D., Schedin, F., Ponomarenko, L.A., Morozov, S.V., Gleeson, H.F., Hill, E.W., Geim, A.K., Novoselov, K.S., Graphene-based liquid crystal device. *Nano Lett.*, 8, 1704, 2008.
16. Liang, J., Li, L., Niu, X., Yu, Z., Pei, Q., Elastomeric polymer light-emitting devices and displays. *Nat. Photonics*, 7, 817, 2013.
17. Shian, S. and Clarke, D.R., Electrically tunable window device. *Opt. Lett.*, 41, 1289, 2016.
18. Levermore, P.A., Jin, R., Wang, X., Chen, L., Bradley, D.D.C., de Mello, J.C., Vapor phase oxidative synthesis of conjugated polymers and applications. *J. Mater. Chem.*, 18, 4414, 2008.

19. Huang, J., Wang, X., deMello, A.J., de Mello, J.C., Bradley, D.D.C., Materials and applications for large area electronics: Solution-based approaches. *J. Mater. Chem.*, 17, 1043, 2007.
20. Na, S.I., Kim, S.S., Jo, J., Kim, D.Y., Efficient and flexible ITO-free organic solar cells using highly conductive polymer anodes. *Adv. Mater.*, 20, 4061, 2008.
21. Yodyingyong, S., Zhang, Q., Park, K., Dandeneau, C.S., Zhou, X., Triampo, D., Cao, G., ZnO nanoparticles and nanowire array hybrid photoanodes for dye-sensitized solar cells. *Appl. Phys. Lett.*, 96, 073115, 2010.
22. Chen, Z., To, J.W.F., Wang, C., Lu, Z., Liu, N., Chortos, A., Pan, L., Wei, F., Cui, Y., Bao, Z., A Three-Dimensionally Interconnected Carbon Nanotube–Conducting Polymer Hydrogel Network for High-Performance Flexible Battery Electrodes. *Adv. Energy Mater.*, 4, 1400207, 2014.
23. Chen, D., Liang, J., Pei, Q., Flexible and stretchable electrodes for next generation polymer electronics: A review. *Sci. China Chem.*, 59, 659, 2016.
24. McCoul, D., Hu, W., Cao, M., Mehta, V., Pei, Q., Recent advances in stretchable and transparent electronic materials. *Adv. Electron. Mater.*, 2, 1500407, 2016.
25. Collings, P.J. and Hird, M., *Introduction to liquid crystals: Chemistry and physics*, 1st Edn., Taylor Francis, London, 1997.
26. Chandrasekhar, S., *Liquid Crystals*, 2nd edn, Cambridge University Press, Cambridge, 1992.
27. de Gennes, P.G. and Prost, J., *The Physics of liquid crystals*, 2nd edn, Clarendon Press, Oxford, 1993.
28. Dierking, I., Recent developments in polymer stabilised liquid crystals. *Polym. Chem.*, 1, 1153, 2010.
29. Dierking, I., *Textures of liquid crystals*, Wiley-VCH, Weinheim, 2003.
30. Chandrasekhar, S. and Ranganath, G.S., Discotic liquid crystals. *Rep. Prog. Phys.*, 53, 57, 1990.
31. Pelzl, G., Diele, S., Weissflog, W., Banana-shaped compounds-A new field of liquid crystals. *Adv. Mater.*, 11, 707, 1999.
32. Kitzerow, H.S. and Bahr, C., *Chirality in liquid crystals*, Springer, Berlin, 2001.
33. Smalyukh, I.I., Kuzmin, A.N., Kachynski, A.V., Prasad, P.N., Lavrentovich, O.D., Optical trapping of colloidal particles and measurement of the defect line tension and colloidal forces in a thermotropic nematic liquid crystal. *Appl. Phys. Lett.*, 86, 021913, 2005.
34. Woltman, S.J., Jay, G.G.D., Crawford, G.G.P., Liquid-crystal materials find a new order in biomedical applications. *Nat. Mater.*, 6, 929, 2007.
35. Shibaev, V.P., Liquid-crystalline polymer systems: From the past to the present. *Polym. Sci. Ser. A*, 56, 727, 2014.
36. Goodby, J., Kato, T., Tschierske, C., Gleeson, H., Raynes, P., *Handbook of liquid crystals*, 2nd edn., p. 8, Wiley-VCH Verlag GmbH, Weinheim, 2014.

37. Li, Q., *Nanoscience with liquid crystals-From self-organized nanostructures to applications*, p. 135, Springer, New York, 2014.
38. de Gennes, P.G. and Prost, J., *The physics of liquid crystals*, 2nd Edn., Oxford University Press, Oxford, U.K.
39. Riedl, T., Gorrn, P., Kowalsky, W., Transparent electronics for see-through amoled displays. *J. Disp. Technol.*, 5, 501, 2009.
40. Lo, K.L., Tsai, Y.H., Cheng, W.Y., Shiu, J.W., Chen, J.L., Recent development of transparent electrowetting display, in: *Proc. Soc. Inf. Display*, p. 123, 2013.
41. Lee, S.M., Oh, S.H., Choi, K.C., Highly transparent SU-8 photoresist barrier rib for a transparent AC plasma display panel. *J. Disp. Technol.*, 7, 40, 2011.
42. Lin, C.H., Lo, W.B., Liu, K.H., Liu, C.Y., Lu, J.K., Sugiura, N., Novel transparent LCD with tunable transparency, in: *Proc. Soc. Inf. Display*, p. 1159, 2012.
43. Satoh, E., Asaoka, Y., Deguchi, K., Ihara, I., Minoura, K., Fujiwara, S., Miyata, A., Itoh, Y., Iyama, Y., Shibasaki, M., Kikuchi, K., Kubo, M., 60-inch highly transparent see-through active matrix display without polarizers, in: *Proc. Soc. Inf. Display*, p. 1192, 2010.
44. Su, C.W. and Chen, M.Y., Polymer-dispersed liquid crystal applied in active-matrix transparent display. *J. Disp. Technol.*, 10, 683, 2014.
45. Cupelli, D., Nicoletta, F.P., Manfredi, S., Vivacqua, M., Formoso, P., Filpo, G.D., Chidichimo, G., Self-adjusting smart windows based on polymer-dispersed liquid crystals. *Solar Energy Mater. Solar Cells*, 93, 2008, 2009.
46. Tikhonov, E.A. and Ilchishin, I.P., Resonance nonlinear optical properties of dye-doped liquid crystals under pulse excitation: Insight into early experiments. *J. Mol. Liq.*, 267, 73, 2018.
47. Buchnev, O., Dyadyusha, A., Kaczmarek, M., Reshetnyak, V., Reznikov, Y., Enhanced two-beam coupling in colloids of ferroelectric nanoparticles in liquid crystals. *J. Opt. Soc. Am. B*, 24, 1512, 2007.
48. Budaszewski, D., Siarkowska, A., Chychłowski, M., Jankiewicz, B., Bartosewicz, B., Dąbrowski, R., WoliNski, T.R., Nanoparticles-enhanced photonic liquid crystal fibers. *J. Mol. Liq.*, 267, 271, 2018.
49. Kobayashi, S., Shiraishi, Y., Toshima, N., Furue, H., Takeishi, K., Takatsu, H., Chang, K.H., Chien, L.C., Further study of optical homogeneous effects in nanoparticle embedded liquid-crystal devices. *J. Mol. Liq.*, 267, 303, 2018.
50. Podoliak, N., Bartczak, D., Buchnev, O., Kanaras, A.G., Kaczmarek, M., High optical non-linearity of nematic liquid crystals doped with gold nanoparticles. *J. Phys. Chem. C*, 116, 12934, 2012.
51. Dalir, N., Javadian, S., Dehghani, Z., High optical nonlinearity of nematic liquid crystal doped with graphen oxide. *J. Mol. Liq.*, 244, 103, 2017.
52. Khoo, I.C., Williams, Y.Z., Lewis, B., Mallouk, T., Photorefractive Cd-Se and gold nanowire-doped liquid crystals and polymer-dispersed-liquid-crystal photonic crystals. *Mol. Cryst. Liq. Cryst.*, 446, 446233, 2006.
53. Lisetski, L.N., Minenko, S.S., Samoilov, A.S., Lebovka, N.I., Optical density and microstructure-related properties of photoactive nematic and cholesteric liquid crystal colloids with carbon nanotubes. *J. Mol. Liq.*, 235, 90, 2017.

54. Singh, D., Singh, U.B., Pandey, M.B., Dabrowski, R., Dhar, R., Improvement of orientational order and display parameters of liquid crystalline material dispersed with single-wall carbon nanotubes. *Mater. Lett.*, 216, 5, 2018.
55. Chen, H.Y. and Lee, W., Electro-optical characteristics of a twisted nematic liquid-crystal cell doped with carbon nanotubes in a DC electric field. *Opt. Rev.*, 12, 223, 2005.
56. Vimal, T., Pandey, S., Gupta, S.K., Singh, D.P., Manohar, R., Enhanced negative dielectric anisotropy and high electrical conductivity of the SWCNT doped nematic liquid crystalline material. *J. Mol. Liq.*, 204, 21, 2015.
57. Cirtoaje, C., Petrescu, E., Moţoc, C., Electricfield effects in nematic liquid crystals doped with carbon nanotubes. *Physica E*, 54, 242, 2013.
58. Sigdel, K.P. and Iannacchione, G.S., Effect of carbon nanotubes on the isotropic to nematic and the nematic to smectic-A phase transitions in liquid crystal and carbon nano-tubes composites. *Eur. Phys. J. E*, 34, 11034, 2011.
59. Jber, N.R., Rashad, A.A., Shihab, M.S., Effects of carbon nanotubes on the physical properties of a nematic liquid crystal N-(4′-methoxybenzylidene)-4-butylaniline. *J. Mol. Struct.*, 1043, 28, 2013.
60. Ion, F.M., Berezovski, C., Berezovski, R., Heimann, G., Moisescu, M., Effects of carbon nanotubes on the electrooptical properties of nematic liquid-crystal cells. *Rom. Rep. Phys.*, 64, 1011, 2012.
61. Lee, W., Wang, C.Y., Shih, Y.C., Effects of carbon nanosolids on the electro-optical properties of a twisted nematic liquid-crystal host. *Appl. Phys. Lett.*, 85, 513, 2004.
62. Kovalchuk, A., Dolgov, L., Yaroshchk, O., Dielectric studies of dispersions of carbon nanotubes in liquid crystals 5CB. *Semicond. Phys. Quantum Electron. Optoelectron.*, 11, 337, 2008, URI: http://dspace.nbuv.gov.ua/handle/123456789/119076.
63. Lisetski, L.N., Minenko, S.S., Ponevchinsky, V.V., Soskin, M.S., Goncharuk, A.I., Lebovka, N.I., Microstructure and incubation processes in composite liquid crystalline material (5CB) filled with multi walled carbon nanotubes. *Mater. Sci. Eng. Technol.*, 42, 5, 2011.
64. Yadav, S.P. and Singh, S., Carbon nanotube dispersion in nematic liquid crystals: An overview. *Prog. Mater. Sci.*, 80, 38, 2016.
65. Lee, C.Y., Tseng, M.C., Yeung, F.S.Y., Ho, T.K., Kwok, H.S., Stereoscopic 3D display by fast response liquid crystal polarization rotator. *SID Symp. Dig. Tech. Pap.*, 43, 121, 2012.
66. Shin, H.K., Lee, J.H., Kim, J.W., Yoon, T.H., Kim, J.C., Fast polarization switching panel with high brightness and contrast ratio for three-dimensional display. *Appl. Phys. Lett.*, 98, 063505–1, 2011.
67. Oh-e, M. and Kondo, K., Electro-optical characteristics and switching behavior of the in-plane switching mode. *Appl. Phys. Lett.*, 67, 3895, 1995.
68. Ohe, M. and Kondo, K., Response mechanism of nematic liquid crystals using the in-plane switching mode. *Appl. Phys. Lett.*, 69, 623, 1996.

69. Takeda, T., Kataoka, S., Chida, H., Tsuda, H., Ohmuro, K., Sasabayshi, T., Koike, Y., Okamoto, K., A super-high image quality multi-domain vertical alignment LCD by new rubbing-less technology. *SID Symp. Dig. Tech. Pap.*, 29, 1077, 1998.
70. Kim, K.H., Lee, K., Park, S.B., Song, J.K., Kim, S.N., Souk, J.H., Domain divided vertical alignment mode with optimized fringe field effect, in: *Proc. 18th Int. Display Research Conf.*, p. 383, 1998.
71. Guan, R.H., Sun, Y.B., Kang, W.X., Rubbing angle effect on in-plane switching liquid crystal displays. *Liq. Cryst.*, 33, 829, 2006.
72. Reinitzer, F., Beiträge zur Kenntniss des Cholesterins. *Monatshefte for chemistry and related parts of other sciences*, 9, 421, 1888.
73. Sluckin, T.J., Dunmur, D.A., Stegemeyer, H., *Crystals that flow: Classic papers from the history of liquid crystals*, 1st Edn., CRC Press, Taylor & Francis Group.
74. Lehmann, O., Liquid crystals: As well as the plasticity of crystals in general, molecular rearrangements and changes of state of matter. *Nature*, 70, 622, 1904.
75. Ren, H.W., Fan, Y.H., Gauza, S., Wu, S.T., Tunable microlens arrays using polymer network liquid crystal. *Opt. Commun.*, 230, 267, 2004.
76. Liu, Y.J., Sun, X.W., Shum, P., Yin, X.J., Tunable fly's-eye lens made of patterned polymer-dispersed liquid crystal. *Opt. Express*, 14, 5634, 2006.
77. Shibaev, V.P., Some new physico-chemical aspects of side chain liquid crystal polymers. *Mol. Cryst. Liq. Cryst.*, 243, 201, 1994.
78. Shibaev, V.P., Comb-shaped liquid-crystal polymers, in: *Liquid crystalline polymers*, N. Plate (Ed.), Plenum Press, New York-London, 1993.
79. Ringsdorf, H. and Schneller, A., Synthesis, structure and properties of liquid crystalline polymers. *Br. Polym. J.*, 13, 43, 1981.
80. Moore, J.S. and Stupp, S.I., Synthesis of a chemically ordered liquid crystal polymer. *Macromolecules*, 21, 1217, 1988.
81. Zhang, D., Liu, Y., Wan, X., Zhou, Q.F., Synthesis of a new side-chain type liquid crystal polymer Poly[dicyclohexyl vinylterephthalate]. *Macromolecules*, 32, 4494, 1999.
82. Kopp, V.I., Zhang, Z.Q., Genack, A.Z., Lasing in chiral photonic structures. *Prog. Quantum Electron.*, 27, 369, 2003.
83. Ford, A.D., Morris, S.M., Coles, H.J., Photonics and lasing in liquid crystals. *Mater. Today*, 9, 36, 2006.
84. Moreira, M.F. and Carvalho, I.C.S., Cholesteric liquid-crystal laser as an optic fiber-based temperature sensor. *Appl. Phys. Lett.*, 85, 2691, 2004.
85. Lin, T.H., Chen, Y.J., Wu, C.H., Fuha, A.Y.G., Cholesteric liquid crystal laser with wide tuning capability. *Appl. Phys. Lett.*, 86, 161120, 2005.
86. Zuccarello, G., Scribner, D., Sands, R., Buckley, L.J., Materials for bio-inspired optics. *Adv. Mater.*, 14, 1261, 2002.
87. Aizenberg, J. and Hendler, G., Designing efficient microlens arrays: Lessons from Nature. *J. Mater. Chem.*, 14, 2066, 2004.

88. Cheng, C.C., Chang, C.A., Yeh, J.A., Variable focus dielectric liquid droplet lens. *Opt. Express*, 14, 4101, 2006.
89. Wang, B., Ye, M.O., Sato, S., Liquid crystal lens with focal length variable from negative to positive values. *IEEE Photon. Technol. Lett.*, 18, 79, 2006.
90. Stewart, G.T., Liquid crystals in biology; I. Historical, biological and medical aspects. *Liq. Cryst.*, 30, 541, 2003.
91. Stewart, G.T., Liquid crystals in biology; II. Origins and processes of life. *Liq. Cryst.*, 31, 443, 2004.
92. Brake, J.M. and Abbott, N.L., An experimental system for imaging the reversible adsorption of amphiphiles at aqueous–liquid crystal interfaces. *Langmuir*, 18, 6101, 2002.
93. Brake, J.M., Daschner, M.K., Luk, Y.Y., Abbott, N.L., Biomolecular interactions at phospholipid-decorated surfaces of liquid crystals. *Science*, 302, 2094, 2003.
94. Clare, B.H. and Abbott, N.L., Orientations of nematic liquid crystals on surfaces presenting controlled densities of peptides: Amplification of protein-peptide binding events. *Langmuir*, 21, 6451–6461, 2005.
95. Luk, Y.Y., Tingey, M.L., Dickson, K.A., Raines, R.T., Abbott, N.L., Imaging the binding ability of proteins immobilized on surfaces with different orientations by using liquid crystals. *J. Am. Chem. Soc.*, 126, 9024, 2004.
96. Tingey, M.L., Wilyana, S., Snodgrass, E.J., Abbott, N.L., Imaging of affinity microcontact printed proteins by using liquid crystals. *Langmuir*, 20, 6818, 2004.
97. Hoogboom, J., Velonia, K., Rasing, T., Rowan, A.E., Nolte, R.J.M., LCD-based detection of enzymatic action. *Chem. Commun.*, 4, 434, 2006.
98. Harris, K.D., Cuypers, R., Scheibe, P., van Oosten, C.L., Bastiaansen, C.W.M., Lub, J., Broera, D.J., Large amplitude light-induced motion in high elastic modulus polymer actuators. *J. Mater. Chem.*, 15, 5043, 2005.
99. Broer, D.J., *Liquid Crystals in complex geometries: Formed by polymer and porous networks*, 1st Edn., G.P. Crawford, and S. Zumer (Eds), p. 239, Taylor and Francis, London.
100. Warner, M., Terentjev, E.M., *Liquid Crystal Elastomers*, (Oxford Univ. Press, New York.
101. Yu, Y., Ikeda, T., Soft actuators based on liquid-crystalline elastomers. *Angew. Chem., Intern. Edn.*, 45, 5416, 2006.
102. Xie, P. and Zhang, R.B., Liquid crystal elastomers, networks and gels: Advanced smart materials. *J. Mater. Chem.*, 15, 2529, 2005.
103. Sousa, M.E., Broer, D.J., Bastiaansen, C.W.M., Freund, L.B., Crawford, G.P., Isotropic islands in a cholesteric sea-patterned thermal expansion for responsive surface topologies. *Adv. Mater.*, 19, 1842, 2006.
104. Elias, A.L., Harris, K.D., Bastiaansen, C.W.M., Broer, D.J., Brett, M.J., Photo patterned liquid crystalline polymers for microactuators. *J. Mater. Chem.*, 16, 2903, 2006.

105. Camacho-Lopez, M., Finkelmann, H., Palffy-Muhoray, P., Shelley, M., Fast liquid-crystal elastomer swims into the dark. *Nat. Mater.*, 3, 307, 2004.
106. Lehmann, W., Skupin, H., Tolksdorf, C., Gebhard, E., Zentel, R., Kruger, P., Lösche, M., F. Kremer, F., Giant lateral electrostriction in ferroelectric liquid-crystalline elastomers. *Nature*, 410, 447, 2001.
107. Buguin, A., Li, M.H., Silberzan, P., Ladoux, B., Keller, P., Micro-actuators: When artificial muscles made of nematic liquid crystal elastomers meet soft lithography. *J. Am. Chem. Soc.*, 128, 1088, 2006.

8

Hybrid Inorganic-Organic White Light Emitting Diodes

Mauro Mosca, Roberto Macaluso and Isodiana Crupi*

Department of Engineering, University of Palermo, Viale delle Scienze, Palermo, Italy

Abstract

This chapter reviews the state of the art of materials, technologies, characterizations, process and challenges concerning hybrid white light-emitting diodes (LEDs). Here, for a "hybrid LED" we mean a device based on a layer of organic phosphors (or a mix of inorganic and organic ones) pumped by a high-energy inorganic LED. Light is emitted by a frequency down-conversion (sometimes simply named color-conversion) process. Benefits and weak spots of this technology are investigated with a special attention for the materials involved into the process of frequency down-conversion, in order to envisage the future impact of the hybrid lighting technology among the well-established inorganic ones.

Keywords: Hybrid inorganic-organic white light emitting diodes (HWLEDs), frequency-down conversion, luminescent polymers and molecular dyes, biomaterials and biomolecules, metal-organic frameworks, carbon dots, color tuning and rendering of HWLEDs, stability of HWLEDs

8.1 Introduction

Compared to traditional light source, light-emitting diodes (LEDs) first of all offer a higher efficiency; then, LEDs allow one to have available lighting of higher quality. High quality means not only more care to the impact of lighting on human health but also color (and color temperature) tunability, high color rendering, dimmable light source, low energy consumption,

**Corresponding author*: mauro.mosca@unipa.it

Inamuddin, Rajender Boddula, Mohd Imran Ahamed and Abdullah M. Asiri (eds.) Polymers for Light-Emitting Devices and Displays, (197–262) © 2020 Scrivener Publishing LLC

and sustainable impact on the environment [1]. For these reasons, white LEDs have become the lighting dominant technology and the main source of light in the global market.

The luminous efficiency of the white LEDs, that is the ratio between the luminous flux (lm) and the optical power (W), continues to advance toward the practical limit of 255 lm W^{-1} for phosphor-based color-converting LEDs [2].

The rise of LEDs in the global market is basically due to the falling prices of LED lamps, and several advantageous government initiatives, which boosts the use of LEDs for private and public lighting. A big bump has been to ban inefficient lighting (incandescent lamps) from the European Union, the United States, and Canada (although these decisions will play a decreasing role in the next years).

Lighting today is responsible for about 15% of global electricity consumption. According several analysts (Evergreen Lighting), the global LED lighting market is expected to be valued at US$ 92.40 billion by 2022 and to witness a shipment of 14.01 billion units by 2022 at a Compounded Average Growth Rate of 13.66% and 21.23%, respectively, between 2016 and 2022 [1].

White light in commercial LED lighting is basically generated in two ways: as additive mixing of light produced by two or more monochromatic LEDs or by pumping a phosphor (or a combination of different phosphors) layer emitting in the visible spectrum region with the primary source coming from a blue, violet, or near-UV LED. In the latter case, the phosphor is designed to partially absorb the light of the primary source, convert it to a

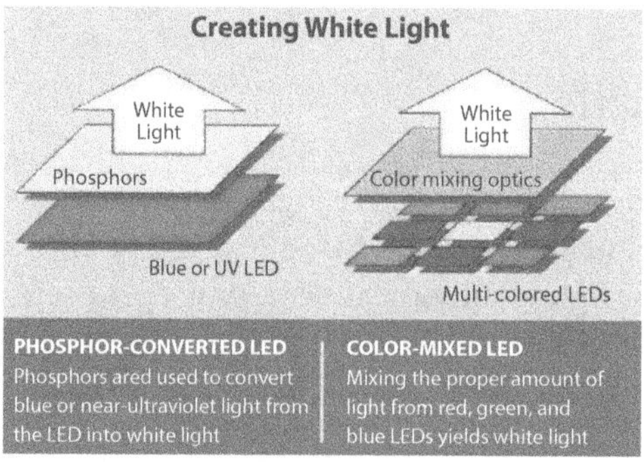

Figure 8.1 Creating white light with LEDs. Reprinted from LED Light Spectrum Enhancement with Transparent Pigmented Glazes, LED Professional Review (Jul 06, 2016); with permission of Luger Research e.U.

longer wavelength light (in general yellow), and re-emit it (cf. Figure 8.1). White LEDs based on multichromatic sources exhibit high values of the chromatic rendering index (CRI), although they have a considerable dependence on the temperature; compared to the first ones, white LEDs based on color-converting phosphors have lower efficiencies, due to the wavelength-conversion losses, but they are indifferent to temperature changes (up to a point) and easy to fabricate [3].

Due to their simplicity and low cost, frequency down-converter white LEDs exploiting the luminescence of inorganic phosphors such as Ce:YAG have stood out in the world market, especially for general purpose lighting applications. However, several factors limit the performance of the phosphor-based LEDs:

1. The wavelength-conversion efficiency of Ce:YAG is around 75%; therefore, losses are consistent [3].
2. Other phosphors, especially those emitting in the red region, have even lower efficiencies due to the poor eye sensitivity at those wavelengths [4].
3. Self-absorption and quenching effects lower the phosphor performances [4].

These elements make the luminous efficiency of current phosphor-converted LEDs be well below the potential maximum value. Moreover:

4. Current phosphors are based on rare-earth metals which are not simply recycled. Moreover, production of rare-earth requires high costs and a detrimental environmental impact for mining and purification processes [5].
5. Inorganic phosphors are often toxic compounds; therefore, their medical use is forbidden [5].
6. Supply of rare earths is becoming more and more precarious [6].
7. The typical grain size of rare earth-based phosphors (tens of micrometers) is not compatible with pixel sizes of the micro-displays [4].

These supplementary reasons have encouraged the scientist to find some alternatives to the inorganic phosphors, especially those based on rare earth elements. Given the constant rise of the organic LEDs (OLEDs), organic luminescent materials have been intensively studied during last 20 years; therefore, it is no surprise that these materials have been also taken into consideration as possible candidates to substitute inorganic phosphors in frequency down-converting white LEDs.

The advantages in using organic luminescent materials to convert pumping light wavelength in white LEDs are numerous. Just to list some of them:

a. Luminescence quantum yield of organic material is often higher than in inorganics [4].
b. Huge choice of different wavelength-converters (which, in principle, allows reaching high values of the CRI) [4].
c. Cost of organic dyes is generally less than cost of inorganic phosphors [7].
d. Usually, organic dyes are not toxic, harmless to human health, and easy to dispose of due to their environmentally friendly nature [7].
e. Luminescent polymers can be easy dissolved in appropriate solvents and directly deposited on the LED chip by low-cost processing methods [4].

Besides these unquestionable benefits, there are two open problems which weigh on the commercial perspectives of substituting traditional phosphors with organic materials in white LEDs. The first is the so-called aggregation-induced quenching of the photoluminescence which occurs in most of organic luminescent materials when in solid form. While these materials in solution exhibit an intense photoluminescence, they do not emit as a solid layer or powder. The second problem is the mediocre thermal and photo-stability of organic dyes, which strongly limits the lifetime of these devices when they are used as down-converter materials. The most evident effect of the material degradation is the bleaching of the converter layer containing the organic dye. In the following sections, it will be shown several different approaches to face these problems.

In this chapter, a frequency-down converter white LED obtained by coupling the (blue or near UV) light irradiated from an InGaN- or GaN-based LED to a single layer or a multilayer containing organic dyes dispersed in a host matrix will be referred as a "hybrid inorganic-organic white light-emitting diode" or, more simply, *Hybrid White LED* (HWLED).

8.2 Hybrid Devices and Other Ambiguities

"Hybrid" is one of the most nonspecific terms, often used to refer to very specific features of a system or device. For instance, in transport vehicles, the noun "hybrid" is used to identify a vehicle with two distinct types of power (e.g., electric and combustion) which work in synergy between them (a combustion engine drives and recharges an electric motor); therefore, sometimes, "hybrid vehicles" also designate bi-fuel vehicles that use two different engines for propulsion (petrol and natural gas, for example). The same type of ambiguity is sometimes found in electronics where a "hybrid" circuit should consist of simple devices or integrated circuits bonded to a printed circuit board or directly to a substrate, while frequently, especially in spoken language, indicates various assemblage of different electronic components: digital and analogic, simple and integrated, etc. Similarly, when "hybrid" is coupled to LEDs, an ambiguity can rise about the actual meaning of this compound word. Before the decisive onset of the research activities on blue OLEDs (around 2004-05 [8]), many scientists investigated on possibility to obtain efficient blue light from inorganic-organic heterojunctions (and then other colors). Thus, the term "hybrid LEDs" was also used to designate these heterojunctions formed by an organic emitting layer and an inorganic either hole transport layer (HTL) or electron transport layer (ETL). In other words, they consisted more properly of OLEDs where either the HTL or the ETL were substituted by an inorganic layer having the same properties of the equivalent organics [9–14]. The reason of the (modest) interest for these kinds of structures was justified by the difficulties in chemical design to obtain a very efficient all-organic blue emitter, due to the high rate of non-radiative recombination [15–17]. The interest rapidly dropped away with the improvements in the OLED technology.

From now on, to distinguish an LED based on organic phosphors (or a mix of inorganic and organic ones) pumped by a high-energy inorganic LED, from a radiative heterojunction formed by an organic emitting layer and an inorganic transport layer, the first will be referred to as an "hybrid LED" (HWLED, for white light emission), while the second as an "hybrid OLED" (cf. Figure 8.2). In this chapter, only hybrid LEDs are discussed.

The list of ambiguities in this subject matter, unfortunately, does not end with the equivocation of the word "hybrid". Other factors that generate confusion are: i) in photometry and in colorimetry, sometimes, the same physical quantity is called in different ways; furthermore, ii) the way in which the physical quantities are defined leaves ample room for misunderstanding or vagueness of meaning. For the first case, the quantity "luminous efficacy" is sometimes confused with the "luminous efficiency". Schubert, in one of the first and most completed textbook on LEDs (2003, 1st edition; 2006, 2nd edition) [3], defines the luminous efficacy of an optical radiation as the conversion efficiency from optical

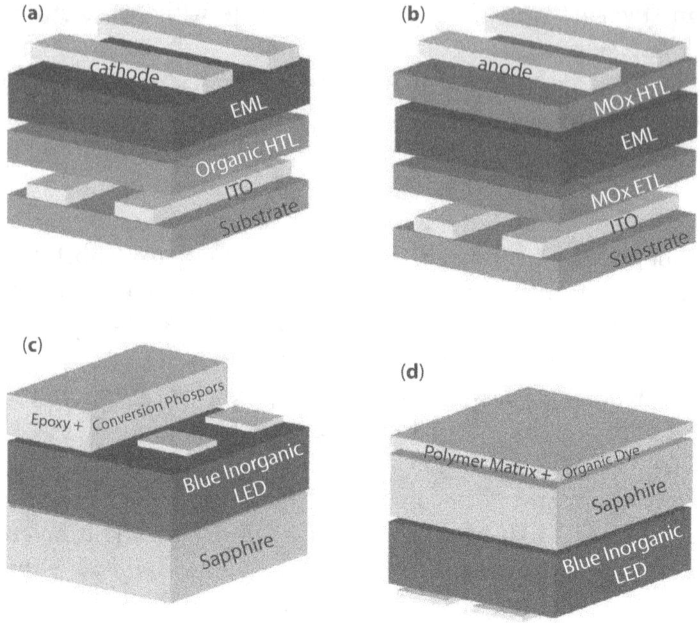

Figure 8.2 Structures of (a) OLEDs, (b) Hybrid OLEDs, (c) Inorganic phosphor-converted LEDs, and (d) Hybrid LEDs. Meaning of the acronyms: EML = organic EMissive Layer, HTL = Hole Transport Layer, ETL = Electron Transport Layer, MOx = Metal Oxide layer. Figures 8.2a and b adapted from M. Sessolo and H. J. Bolink, Hybrid Organic–Inorganic Light-Emitting Diodes, *Adv. Mater.*, 23 (2011) 1829–1845; with permission of John Wiley and Sons.

power to luminous flux, while the luminous efficiency of a light source as the luminous flux of the light source divided by the electrical input power. Khanna, in his more recent (2014) "Fundamentals of Solid-State Lighting", reaffirms the same definition [18], while in "LED lighting" edited in 2015 by Khanh *et al.*, the authors define the luminous efficacy of a source the ratio of luminous flux to electric power [19]. This confusion is also fed by the fact that "efficiency" and "efficacy" are quasi-synonyms and they are measured with the same units (lm W^{-1}). Hence, in specific literature, it could happen that values reported as luminous efficacies could be really referred to the luminous efficiency; unfortunately, not always the scientists give many details about the measurement set-up of the devices. When not clear, in this chapter, it was adopted the same terminology used by the authors of the cited papers.

Concerning the second class of ambiguities, the Color Rendering Index (CRI) is a measure of the ability of a source of light (illuminant) to accurately render the colors of physical objects illuminated by the source when

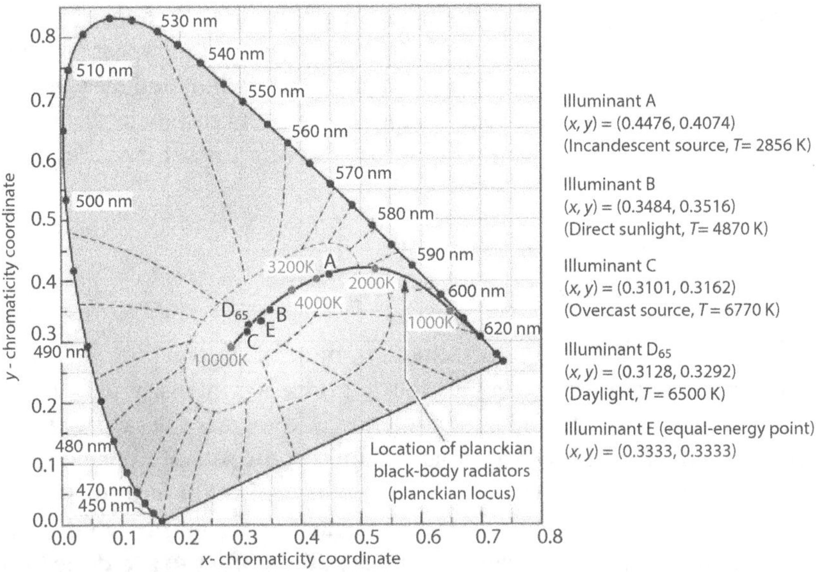

Figure 8.3 Position of the standard white illuminants A, B, C, D_{65}, and E, and their color temperatures in the CIE chromaticity diagram. Reprinted from E. F. Schubert, Light Emitting Diodes, 2nd edition, (2006) Cambridge University Press, New York, US; with permission of Cambridge University Press.

compared to a reference light. This means that the CRI of a source can significantly change if one changes the reference source used to calculate the CRI itself. In many papers, a source having chromatic coordinates (x, y) localized at (0.33, 0.33) is considered a sort of a pure white, thus a "perfect" reference. Actually, this source is located at the so-called equal-energy point (because $x = y = z = 0.33$) and is named *standard illuminant E* (cf. Figure 8.3). This illuminant does not fall in the Planckian locus because is not a black-body radiator, so it is not possible to evaluate its color temperature; its correlated color temperature (CCT) is 5,455 K (approximately, the same of the standard illuminant D_{55}). Its high CCT makes the illuminant E a cool white source that is not necessarily the best reference to evaluate the ability of a source to render the colors. Although, it was demonstrated that only CRI does not well quantify color rendering but it is necessary to couple with another simple metric, such as gamut area index (GAI), for example [20], for cultural and personal reasons people often prefer the color rendering of warm white sources [21, 22]. However, most of authors cited in this chapter report the standard illuminant E as reference for their HWLEDs (in other cases, instead, the reference is not specified making the CRI calculation ambiguous). On the other hand, a typical cool white spectrum contains the most intense components in the yellowish region, while a portion of the typical warm white spectrum falls in the red and far-red region where the eye sensitivity is reduced; hence, the luminous efficiency of a cool white LED results to be always higher than the efficiency of a warm white source having the same value of the power spectral density.

8.3 Necessity of a Host Matrix

As mentioned before in the Introduction, most of organic luminophores (polymers or oligomers) show a high fluorescence intensity in solution but not in the solid state. This phenomenon is known as the "aggregation-induced quenching". This means that the fluorescence quantum yield of the dyes is very low or even zero in solid form (also as a thin film) due to the π-π aggregation, or intermolecular π-π interactions established in the aromatic ring plane, like in perylene diimides [23]. In order to avoid the fluorescence quenching of the fluorophores in the solid state, they should be in a disaggregation state as well as when in solution. However, using luminophores in a liquid form as down-converters for HWLEDs is not possible, or it should be necessary strict requirements for device encapsulation in order to keep a liquid film in contact with the emitting surface of the source and avoid evaporation. Another way to disaggregate the dyes is to embed them into a solid matrix (playing the same role of the solvent in a solution). The choice of a proper host matrix allows the fluorophores to preserve their characteristic luminescent properties. From a chemical point of view, this corresponds to alter the intermolecular distance between the above-mentioned ring planes (notice that a similar effect can be also gained by doping the dyes). In other words, this expedient prevents the formation of π-π stackings between aromatic rings [23–25].

It is worthwhile to point out that the aggregation-induced quenching worsens even more by increasing the concentration of the luminophores, as it is intuitive (there are more interactions between the luminophores, which give rise to the formation of dimers and excimers [24]). Therefore, it is always important to evaluate the optimal value of dye concentration in order to maximize the photoluminescence [23, 26]. A consequent detrimental effect in HWLEDs is that by increasing the deposited quantity of the fluorescent converter on the pumping blue LED, the quantity of blue light absorbed is higher (less blue light is perceived by an external observer), but the fluorescence emission of the converter does not increase (or even decreases) although more blue light is absorbed; thus, the conversion efficiency of the HWLED decreases [27].

As material for host matrix, silica is one of the most advantageous since is simple to prepare; it is non-toxic and biocompatible and can improve the thermal stability of the embedded dye [25]. Polymers—like poly(methyl methacrylate) (PMMA)—are also a good choice since dye molecules can be disaggregated into single molecules if mixed in a polymer solution [28]. The matrices embedding the down-converting dyes are deposited on the primary source with various methods, mainly by spin- and dip-coating.

8.4 Materials for Hybrid LEDs

The year 1997 is a milestone for HWLEDs and for white LEDs in general. In that year, in fact, Schlotter *et al.* [29] proposed luminescent organic dyes as down-converters. They used a 430-nm-LED chip embedded in an epoxy resin matrix containing perylene-based BASF Lumogen® F. Unfortunately, no many details are provided in this pioneering work. Similar studies was led in the same period by Heeger's group [30, 31], who used for the first time a luminescent conjugated polymer to convert the blue emission of an InGaN LED in a white light. Curiously, the work of Schlotter is also one of the first paper (together with the well-known communication by the Nobel-Prize winner Shuji Nakamura at Photonic West '97 [32]) that reports Ce:YAG phosphors as color-converters for white LEDs, which will become the standard commercial solution for down-converting white LEDs.

From these first studies on, the researchers started to give more information about the fabrication (including the arrangement of the dyes in relation to the primary source) and the characterization (emission spectra, chromatic coordinates, luminous flux and efficiency, stability). They are also reported in an excellent review on luminescent materials for LED applications [33]. In addition, the studies on the chemistry of luminescent organic material made considerable advances, stimulated by the promising results in the field of the OLEDs and polymer LEDs (PLEDs). The latter, in particular, gave an impetus to chemistry progress in conjugated polymers luminescence.

8.4.1 Luminescent Polymers

Nowadays, the most promising luminescent polymers suitable for down conversion of UV/blue LED light, are polyfluorenes derivatives [34–37] and Si-based polymers [38–40]. The latter, however, were mainly used as active layers for PLEDs, not as down-converter.

Unlike organic dyes used for down conversion, conjugated polymers are relatively free from concentration quenching effects, so that high luminophore densities can be used. Among conjugated polymers, the polyfluorenes show the most interesting properties, in particular, high fluorescence, high chemical purity and reproducibility, large quantum efficiency, and possibility to be tuned in the entire visible spectrum [41].

An HWLED using the polyfluorene copolymer containing 3% green-emitting {4-(2-[2,5-dibromo-4-{2-(4-diphenylamino-phenyl)-vinyl}-phenyl]-vinyl)-phenyl}-diphenylamine (DTPA) and 2% red-emitting 2-{2-(2-[4-{bis(4-bromo-phenyl)amino}-phenyl]-vinyl)-6-tert-butyl-pyran-4-ylidene}-malononitrile (TPDCM) was found to exhibit a maximum brightness of 820 cd m^{-2} with CIE coordinates of (0.33, 0.35) [34]. In this case, white emission was obtained using a single-layer polymer blending system as a down converter. Today, this is the most common practice, but the first attempts to obtain broadband white light LEDs consisted in using multilayer systems of two or more active layers, each layer emitting a different color [42].

Another blend used to produce HWLEDs consisted of: poly(9,9-dioctylfluorene-co9,9-di(4-methoxy)phenylfluorene) (F8DP) (blue emitter), poly(9,9-dioctylfluorene-co-benzothiadiazole) (F8BT) (green emitter), and the Sumitomo-proprietary copolymer Red F (red emitter). The pumping LED emitted at 370 nm where only F8DP fully absorbs the radiation of the primary source. However, thanks to the overlap between the absorption and emission spectra of these polymers, the blend could support incomplete non-radiative resonant energy transfer thus allowing all three colors to be emitted [35].

Using polyfluorene functionalized with azide groups (PFA), other researchers obtained high-quality white emission, by pumping with a 378-nm LED [36]. Azide groups provide intentionally cross-linking of polymer chains when cast into solid-state form by heat treatment or exposition to UV light. The down-conversion process generated a very broad photoluminescence spectrum across the entirety of the visible spectrum, yielding high CRI up to 91 and with a quantum efficiency of 0.86.

It is worth to mention a particular type of energy-transfer occurring between quantum wells of a semiconductor and an overhead layer that is the Förster resonant energy transfer (FRET), successfully used to fabricate HWLEDs. To exhibit efficient FRET, it is required that both the overhead layer absorbs the energy emitted by the quantum wells and the two materials be placed in close proximity (a few nm). In FRET-based HWLEDs,

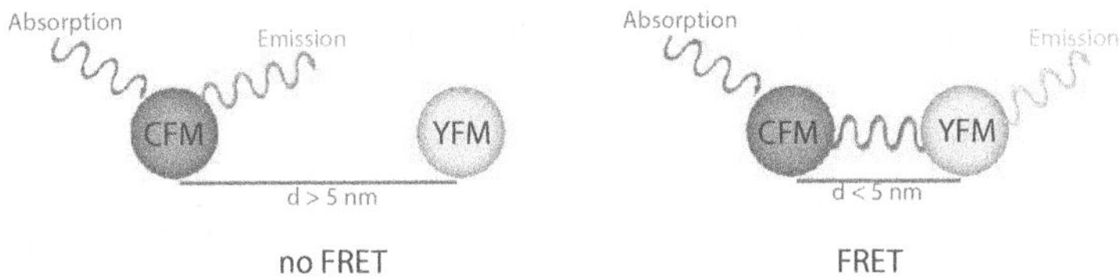

Figure 8.4 Sketch of FRET between a Cyan Fluorescent Molecule (CFM) and a Yellow Fluorescent Molecule (YFM).

a non-radiative electronic energy transfer occurs between InGaN/GaN multiple quantum wells of an exciting blue/UV LED and an absorbing organic layer as a color-conversion medium. More generally, for III-V semiconductors, several hundreds of picoseconds are necessary to transfer resonant electronic energy from an excited semiconductor quantum well to a close absorbing organic medium. In absence of such a transfer, this time is meaningfully higher than the semiconductor excitation lifetime, so that a large fraction of the excitation energy can be transferred to the organic layer [43]. The mechanism of FRET is illustrated in Figure 8.4. In recent times, FRET technique was primarily used with quantum dots as absorbing mediums both for white light source fabrication [44] and micro-display applications [45]. FRET-based HWLEDs were fabricated by Belton *et al.* [35] as follows: a ~2-μm thickness layer of high-temperature GaN was grown, followed by InGaN/GaN quantum wells of nominal thickness 2.5 nm, then a thin cap GaN layer, the latter used as an interspace between quantum wells and organic layer. Structures for FRET studies were prepared by spin coating a polymer blend (98.4% F8DP, 1% F8BT, and 0.6% RedF) onto InGaN/GaNQWstructures with nominal cap layer thicknesses of 15, 4, and 2.5 nm. In all these cases, and for excitation values ranging from 360 to 920 μJ cm^{-2}, the hybrid emission lies close to the (0.33, 0.33) standard illuminant E point.

8.4.2 Molecular Luminescent Dyes

From the first works on organic wavelength converters to date, many and different molecular dyes were used to fabricate HWLEDs. Perylene, fluorescein, porphyrins, and metal complexes are just some examples of molecules that were used as down-converters. In contrast with polymers that can be easily blended and deposited by a liquid phase, molecular dyes are in general available as powders, crystals, and in general, in a solid form. As discussed in the previous section, these dyes are needed to be dispersed in an opportune host matrix to be fluorescent. The characteristic of this matrix—in general, a polymer—may influence luminescence features of the dyes, so the choice of the appropriate host material should be carefully done.

Perylene: Perylene derivatives was the first organic dyes used in an HWLED as down converters. After the first studies led by Schlotter [29, 46], the enthusiasm for perylene—and in particular for organic converters—goes rapidly down, due to the better photostability of Ce:YAG. Nevertheless, the huge interest created by OLEDs (together with the limited quantum yield of Ce:YAG and the high-costs and sustainability of rare-earths) acted as a driver for a renewed interest in perylene as a down-converter in HWLEDs. Moreover, perylene was successfully used as luminescence down-shifting layer to improve Si solar cell efficiency [47].

A group of University of Palermo, Italy, picked up the perylene again to prepare HWLEDs and reported on tunability and stability of Lumogen® F Yellow 083 (derivative of perylene, commercialized by BASF) when pumped by a wafer-level 450-nm LED [7, 48, 49]. The dye was dissolved in a polymethyl methacrylate (PMMA) solution (0.5 mg/ml), dissolved in turn in ethyl acetate. A suitable viscosity was obtained with a concentration of PMMA of molecular weight 350,000 from 8% to 16% (w/w percent). The authors preferred to use ethyl acetate as a solvent, even if the best performances are reported using chloroform or trichloroethylene, given its low toxicity to avoid carcinogenic risks. The dye solution was either spinned or dipped on the rear side of the sapphire substrate where the 450-nm LED had been grown. High values of 9.37 lm of luminous flux and 118.23 lm W^{-1} of luminous efficiency were achieved (with 20 mA of driving current) [48]. These high values of high luminous efficiency are produced by the higher fluorescence quantum yield of Lumogen® Yellow (91%) compared with the one of Ce:YAG (75%, typical). Blending Lumogen® F Yellow 083 with another perylene-derivative, Lumogen® F Red 305, in different weight proportions (and dissolving the blend in solutions with equal amounts of PMMA in ethyl acetate), the authors tested the color tunability of the HWLEDs by dipping the wafer LEDs in each solution. High values of 8.03 lm of luminous flux and 116.11 lm W^{-1} of optical efficiency were achieved, but in this case with higher values of CRI (90.1). Also, a 395-nm LED was tested as a primary source but the results were poorer [7]. Figure 8.5 shows pictures of (a) the wafer-level 450 nm LED, (b) the HWLED obtained after the dipping process of the 450-nm LED in a solution of PMMA and Lumogen F Yellow 083 (0.5 mg/ml) (HWLED 1), and (c) in a solution of PMMA, Lumogen F Yellow 083 (0.5 mg/ml), and Lumogen F Red 305 (0.05 mg/ml) (HWLED 2); (d) PMMA solutions with different

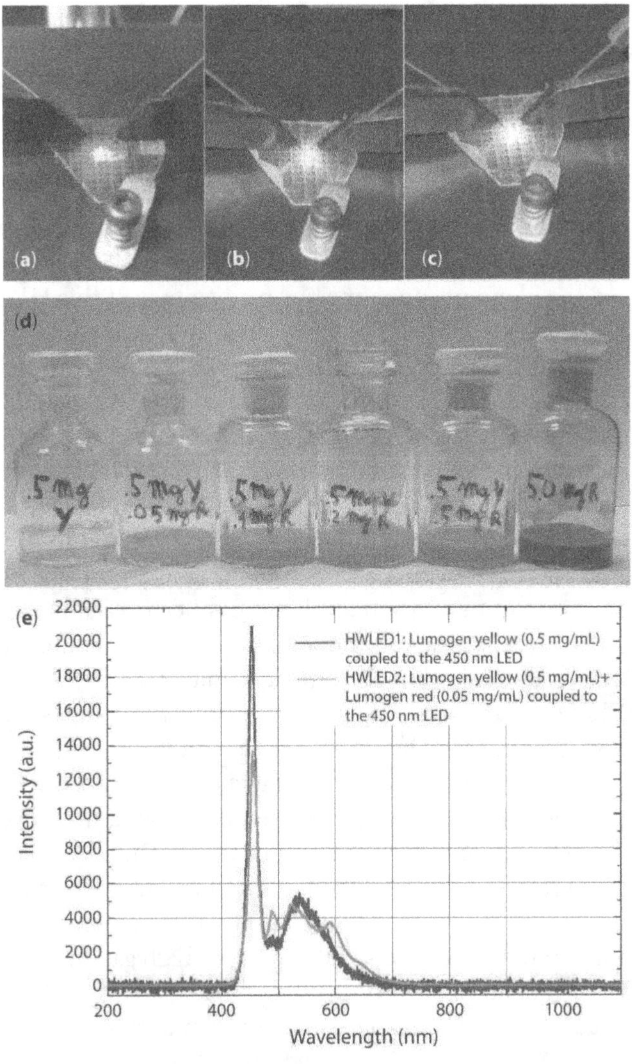

Figure 8.5 Pictures of the blue LED emitting at 450 nm (a) before and after the dipping process (b) in a solution of PMMA and Lumogen F Yellow 083 (0.5 mg/ml) (HWLED 1), and (c) in a solution of PMMA, Lumogen F Yellow 083 (0.5 mg/ml), and Lumogen F Red 305 (0.05 mg/ml) (HWLED 2); (d) color palette obtained mixing different concentrations of yellow and red dyes in the PMMA solution. From left to right, concentrations of yellow and red dyes in PMMA have the following pairs of values (mg/ml of Lumogen Yellow; mg/ml of Lumogen Red): (0.5; 0), (0.5; 0.05), (0.5; 0.1), (0.5; 0.2), (0.5; 0.5), (0; 0.5); (e) emission spectra of the cold-white HWLED 1 (blue plot) and the warm-white HWLED 2 (red plot). Adapted from M. Mosca, F. Caruso, L. Zambito, R. Macaluso, C. Calì, and E. Feltin, Hybrid LEDs_Pave_Way_to New Lighting Applications, *Photonics Spectra* (April 2013); with permission of Photonics Media, Pittsfield MA.

concentrations of yellow and red dyes; (e) electroluminescence spectra of the HWLED 1 and HWLED 2.

Kozma *et al.* [50] prepared high-efficient HWLED using luminescent linear polystyrene decorated with randomly distributed pendants of perylene diimide. This exhibited a nearly four-fold increase in quantum efficiency compared to a blend with the same polystyrene/dye ratio. Since this copolymer is red-emitting, another green emitter [fluorene-thieno-pyrrolo-dione (TPD-FLU)] was used together with the copolymer; a-450 nm commercial LED pumped the blend of copolymers. An advantage of this HWLED was that the gap between the blue pumping peak and the yellow emission is filled thanks to the TPD-FLU luminescence, so that the CRI is high. Stable white light was obtained with CIE chromatic coordinates (0.31, 0.34) and efficiency of 28 lm W^{-1}. Perylene diimide, dispersed into a polymer matrix at high temperature, were also used by He *et al.* [51], to fabricate HWLEDs with improved features: CRI up to 92 and efficiency up to 82 lm W^{-1} were obtained, with the possibility to tune white light from cold to warm.

An original approach, which successfully faces the problem of molecular aggregation in luminescence quenching, has been recently proposed by Galeotti *et al.* [52] by encapsulating fluorescent perylene nanofibers (as down-converting material) in flexible elastomeric semitransparent discs. Versatile tunable HWLEDs are fabricated by simply adjusting the composition and the number of discs placed onto a blue GaN LED, providing a luminous efficiency of 70 lm W^{-1}. A similar approach, shown in Figure 8.6, was followed by Guner *et al.* [53] who used PTE as a green emitter, perylene diimides as a yellow emitter, and a new red-emitting perylene derivative (DiPhAPDI); these dyes where dispersed respectively in solution, in drop-casted films, in polystyrene (PS) fibers, and in PS fibers embedded in poly (dimethyl siloxane) (PDMS). The researchers showed that PS:dye fibers prevent aggregation to some extent and improve the quantum yield of the dyes. The best HWLEDs were fabricated by stacked PS:dye fibers and compressed in a mold; then, they were

Figure 8.6 Sketch of an HWLED prepared as a stack of individual fibers. Reprinted from: T. Guner, E. Aksoy, M. M. Demir, and C. Varlikli, Perylene-embedded electrospun PS fibers for white light generation, *Dyes and Pigments,* 160 (2019) 501–508; with permission of Elsevier.

embedded in PDMS cured at 100°C for 15 min. Primary source was a 455-nm LED. HWLEDs were found to show high CRI (≥90), adjustable CCTs (7,500 K–5,000 K), and a maximum luminous efficiency as high as 220 lm W^{-1}.

Fluorescein: Fluorescein is a dye commonly used in ophthalmology and optometry but also in cosmetics. It is highly fluorescent in solution but, like other organic molecules, not in solid state since the aggregation induces quenching. Fluorescein was initially taken into consideration as a promising material for HWLEDs by a group of Chonbuk National University, South Korea. Synthesizing low molecular weight fluorescent polyesters containing fluorescein units, Uthirakumar *et al.* [54] observed that these polymers were thermally more stable than fluorescein dye, due to the presence of rigid aromatic units in their main chains. They proposed that the aggregation of fluorescein units in the polymers, could be prohibited by protecting the OH end groups with benzoyl groups. The same group prepared an HWLED using yellow-light emitting polymer particles prepared from a solution of polyester bearing fluorescein dye units in tetrahydrofuran. Since intermolecular interactions in organic

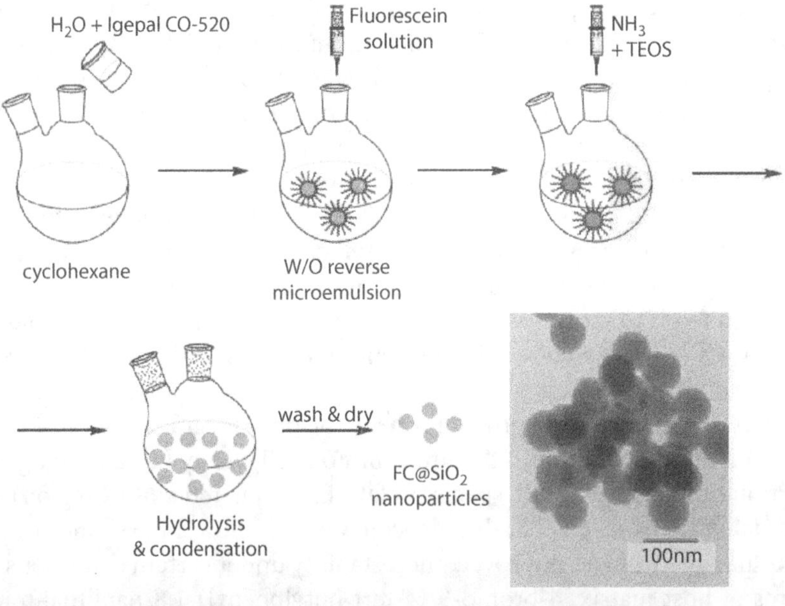

Figure 8.7 Schematic diagram for the formation of fluorescein-silica nanoparticles from a reverse microemulsion *via* hydrolysis and condensation. TEM image and the corresponding particle size distribution of the nanoparticles. Reprinted from R. Boonsin, G. Chadeyron, J.-P. Roblin, D. Boyer, and R. Mahiou, Silica encapsulated fluorescein as a hybrid dye for blue-LED based lighting devices. *J. Mater. Chem. C*, 4 (2016) 6562–6569; with permission of The Royal Society of Chemistry.

materials causes the formation of excimer which should be avoided in light-emission application, the aim of the researchers was to prepare copolymers containing fluorescein units, where the fluorescent dye was placed between non-fluorescent moieties in the polymer chain, in order to avoid such intermolecular interactions. Using a 460-nm LED as a primary source, they obtained white light with a luminous flux of 1.81 lm at 20 mA, and color coordinates (0.281, 0.357) [55].

An improvement of the luminous features was obtained by using fluorescein-silica nanoparticles with a core-shell structure, synthesized by reverse microemulsion method, sketched in Figure 8.7. These nanoparticles, embedded in silicone, under 450-nm irradiation exhibited a white emission with warm color temperature (>6,000 K) and a luminous efficiency of 26 lm W^{-1} [25]. However, luminous features in fluorescein dyes result to be poorer than in perylene.

Porphyrins: Porphyrins and phthalocyanines were reviewed by Armstrong [56] and proposed as materials for emerging technologies. Consequently, they have attracted a moderate interest as possible luminescent converter for HWLEDs. Xiang *et al.* [57] used a near-UV 393-nm LED to obtain white light using platinum(II) meso-tetrakis(pentafluorophenyl) porphyrin complex ($PtF_{20}TPP$) and aluminum tris(8hydroxyquinolinate) (Alq_3), the first dye emitting in the red region with a peak at 650 nm and the second one in the green region at 520 nm. The absorption peak of the dyes fell in the near-UV region, close to the pumping wavelength. The chromatic coordinates at 40 mA of the HWLED were (0.32, 0.31), close to those of the illuminant E, and it exhibited a moderate sensitivity to driving current. Furthermore, the luminous efficiency augmented linearly with increasing input current while for standard inorganic white LEDs, the blue emission tends to saturate at high driving current. Under a forward bias of 20 mA, CCT was 6,800 K, CRI 90.6, and luminous efficiency 10 lm W^{-1}. In spite of the promising optical properties, porphyrin complexes were put aside as down converters, probably due to the photodegradation (50% degradation after 1,000 h) and the low luminous efficiency (a non-negligible portion of the $PtF_{20}TPP$ PL spectrum falls in the far-red and IR region).

Naphthalimides: Naphthalimide compounds show high fluorescence quantum yield and an improved photostability under certain conditions in terms of host matrix. 4-bromo-9-(4-tert-butylphenyl)-1,8-naphthalimide (DBN) was dispersed in cured epoxy resin and in PMMA and deposited on a 460-nm LED chip. DBN has a broadband PL spectrum centered at 560 nm. At 20 mA a bright white light of 3.4 lm with CIE chromaticity coordinates of (0.32, 0.33) and conversion efficiency of 82% was obtained. The luminescence decreased rapidly for the LED with DBN dispersed in epoxy resin, but when was dispersed in PMMA, the decrease was much slower [58]. To improve the stability, DBN was immobilized into PMMA chains by copolymerization; a polymerizable DBN (P-DBN) was successfully synthesized and then polymerized with methyl methacrylate to obtain the PMMA-co-DBN copolymer. While at 160°C, a phase separation occurs in simple DBN/PMMA blend, the copolymer does not undergo the same detriment. Unfortunately, however, the fluorescence quantum yield of the copolymer film results to be much lower than that of P-DBN/PMMA film (0.36, instead of 0.65) [59].

Chalcones: Chalcone derivatives have attracted significant attention for their high optical non-linearities resulting from the significant delocalization of the electron clouds. The relatively small HOMO-LUMO energy gaps promote these crystals for optoelectronic applications. The fluorescence quantum yields of chalcones strongly depend on the solvent (or the matrix), the highest values being obtained for a class of solvents including anisole, ethyl acetate, acetone, acetophenone, *N*,*N*-dimethylformamide, and acetonitrile [60]. First experiments on chalcone molecules in HWLED applications were led in 2003 by Ermakov *et al.* [61]. They prepared several samples of 4-dimethylaminochalcone (DMAC) absorbing at 450 nm and emitting at 510 nm and Nile Red (NR) absorbing at about 500 nm and emitting at about 600 nm. The HWLEDs were made by dissolving the organic molecules in epoxy and placing it in a reflector together with the LED chip. Using a single layer of DMAC dissolved in epoxy at a concentration of 0.5 g/L, the emission appeared as blue-green [chromatic coordinates: (0.197, 0.420)]; nevertheless, increasing the concentration up to 5 g/L, the emitted light looked yellow-green [chromatic coordinates: (0.400, 0.590)]. Inserting a second layer containing NR on the top of the DMAC one, the resulting light was characterized by chromatic coordinates of (0.268, 0.357) and finally appeared as yellowish-white.

Another promising chalcone molecule for HWLED applications is the yellow-emitting 3-(4-(diphenylamino)phenyl)-1-phenylprop-2-en-1-one (DPPO). A color-conversion film was prepared dissolving 1 g of polyethylene glycol and 10 mg of DPPO into 10 ml of dichloromethane and depositing it on a quartz substrate. Exposing the film to a 465-nm LED radiation under the bias voltage of 3 V, a white light of chromatic coordinates (0.30, 0.33) was obtained with a CCT of 5,700 K and a luminous efficacy as high as 275 lm W^{-1} [62]. DPPO, as a *p*-conjugated molecular material with fused rings, has a high stability structure [63]; therefore, no detailed data about HWLED stability were published so far.

BODIPY: A new class of luminophores, characterized by a high-luminescence quantum yield, is constituted by the boron-dipyrromethene derivatives, known under the BODIPY tradename. This dye consists of a dipyrromethene complex with a disubstituted boron center, typically BF_2. BODIPY was synthesized for the first time in 1968 by Treibs and Kreuzer [64]. Nucleophilic and electrophilic reaction sites occurring at different positions of the molecule, enable substitutions directly on the frame of the BODIPY luminophores. This is considered to cause the low stability of the core compound. Due to their intrinsic instability, unsubstituted BODIPY dyes were not synthesized until 2009 [65–67]. A group of University of Strathclyde, UK, studied four linear oligofluorene-BODIPY systems as possible down-converters for light applications: two *meso*-substituted oligofluorene-BODIPYs, *meso*-TFBOD and *meso*-QFBOD, and two beta-substituted oligofluorene-BODIPYs, beta-TFBOD and beta-QFBOD. These systems differ in both the length of oligofluorene chain (either three or four fluorenes) and the substitution position on the BODIPY (*meso*- vs. beta-). The best candidate revealed to be the *meso*-QFBOD which was dissolved in toluene and deposited on top of a 365-nm LED to test its suitability as an organic down converter. The large emission peak at 585 nm, obtained pumping the dye at 365 nm, came directly from the organic material. These systems were unable to absorb the blue radiation, so they could be used only as down-conversion layers for UV light, resulting in yellow-emitting hybrid LEDs [68]. Given the high absorptivity in the blue region and the high photoluminescence of the 4,4-difluoro-4-borata-3*a*-azonia-4*a*-aza-*s*-indacene BODIPY-containing unit $((BODFluTh)_2FB$, the group of Strathclyde designed and synthesized $(BODFluTh)_2FB$ specifically to be used as a down-converter using blue LEDs as primary sources. The dye was embedded in a 1,4-cyclohexanedimethanol divinyl ether (CHDV) matrix to prevent aggregation and quenching of the emission and providing it protection from the environment and deposited (1% w/v) on a 445-nm LED. A strong emission in the yellow spectrum occurred and a white light was perceived due to the partial absorption of the blue radiation by the organic dye leading to an emission at around 560 nm. The chromatic coordinates were close to those of the standard illuminant E. The efficacy of the HWLED LED increased by a factor 4.25 after adding the BODIPY dye and resulted to be 13.6 lm W^{-1}. The CCT of the LED encapsulated with the 1% organic converter was 5,137 K and remained stable after about 200 h of continuous operation at 25 mA (while efficacy decreased less than 10%) [69]. The same group of researchers carried out a more detailed study both on dependence of the HWLED performances on $(BODFluTh)_2FB$ concentration (cf. Figure 8.8a) and the effects the volume of the down-converting layer, using a 444-nm LED as a pumping source [27]. They found that by increasing the concentration of $(BODFluTh)_2FB$

in the transparent CHDV matrix (0.25%, 0.5%, 1%, 2%, and 4% w/v), the blue peak in the resulting emission spectrum decreased (because more blue light was absorbed); therefore, also the yellow peak decreased (probably due to the aggregation of the organic molecules at high concentration, which caused the blue radiation to be getting adsorbed but not converted

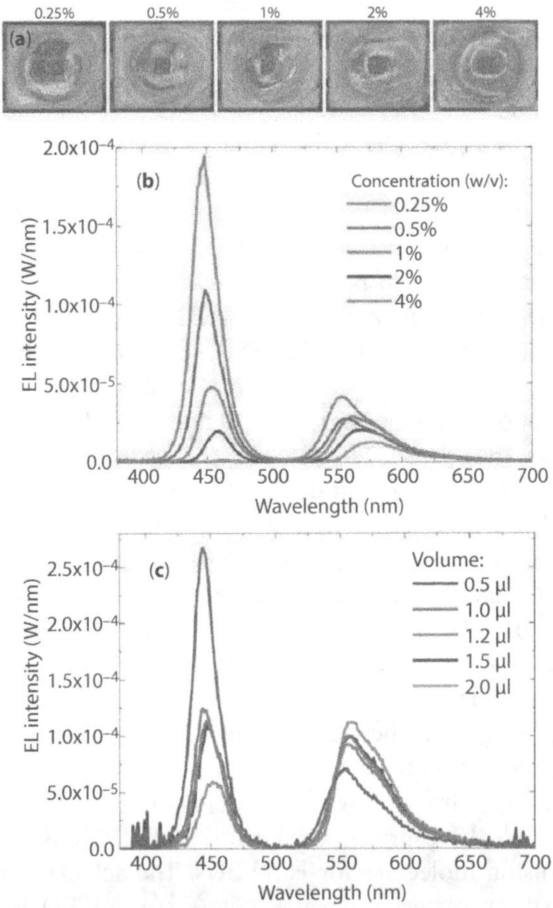

Figure 8.8 (a) Optical microscope images of LEDs coated with the color converter at concentrations 0.25%–4% (w/v). (b) EL spectra of the hybrid LEDs with concentrations ranging from 0.25% to 4% (w/v) of the fluorophores. (c) EL spectra of the LEDs coated with a down-converter layer having volumes ranging from 0.5 to 2.0 μl at a concentration of 1% (w/v). All the LEDs works at a continuous forward current of 25 mA. Reprinted from: J. Bruckbauer, C. Brasser, N. J. Findlay, P. R. Edwards, D. J. Wallis, P. J. Skabara, and R. W. Martin, Color tuning in white hybrid inorganic/organic light-emitting diodes. *J. Phys. D: Appl. Phys.*, 49 (2016) 405103; with permission of IOP Publishing Ltd.

in yellow light) (cf. Figure 8.8b). The aggregation was as well responsible of a redshift in the emission spectrum (due to a convolution of emission peaks from molecules in different aggregation states). By increasing the volume of the down converter layer, the blue peak decreased (because of the larger blue light absorbed) but this time, the peak intensity related to the organic layer remained roughly constant (probably either for re-absorption of the emitted yellow light, or by photo-induced absorption of blue light by the accumulated triplet states) (cf. Figure 8.8c). The CCT of the light generated ranged from 2,770 K (warm light) to 7,680 K (cool light); however, CRI was poor due to missing green light caused by self-absorption of the organic layer [27].

In order to improve the CRI, a larger spectral emission including the green region is desirable. To this purpose, a new generation of BODIPY-containing molecules was developed [70]. They contain two fluorene-triphenylamine arms, connected to either a benzothiadiazole or bisbenzothiadiazole core; in particular, these molecule utilizes either an electron-deficient 2,1,3-benzothiadiazole (BT) core ($(TPA\text{-}Flu)_2BT$), or the corresponding dimer ($(TPA\text{-}Flu)_2BTBT$), together with fluorene and triphenylamine donor arms. Two different kinds of HWLEDs based on this new generation of molecules were tested: one used $(TPA\text{-}Flu)_2BT$ as a down-converter, the other used $(TPA\text{-}Flu)_2BTBT$; the dyes were embedded in a CHDV matrix solution and deposited on a 444 LED wafer by drop-cast method. These HWLEDs offered a significant improvement compared to the previous BODIPY-containing analogs, since their absorption and emission spectra had less overlap and, consequently, less self-absorption in the green region. CCT could be easily tuned from cool to warm white regions (10,000 K to 2,400 K) of the chromaticity diagram by simple variation of the concentration of the dyes (cf. Figure 8.9), and the CRI was reasonably higher, ranging between 60 and 70. $(TPA\text{-}Flu)_2BT$ was the most performing molecule, both in terms of quality and efficiency: the luminous efficacy was 41 lm W^{-1}, while for $(TPA\text{-}Flu)_2BTBT$ was only 10 lm W^{-1} [70]. Due to these promising results, BODIPY-based dyes appear to be as promising molecules for HWLEDs. The achievements obtained as down-converter materials created interest as OLED emitters. It is worth to mention that a BODIPY molecule (boron dipyrromethene: 4,4-difluoro-4-bora-3a,4a-diaza-s-indacene) has been recently exploited for the fabrication of OLEDs [71].

Cyanates: Cyanates like dicyanomethylene [(4-(dicyanomethylene)-2-methyl-6-(4dimethylaminostyryl)-4H-pyran) (DCM)] have initially achieved a certain interest as red down converters when incorporated in a silica matrix [72]. Another cyanate (4-(dicyanomethylene)-2-(t-butyl)-

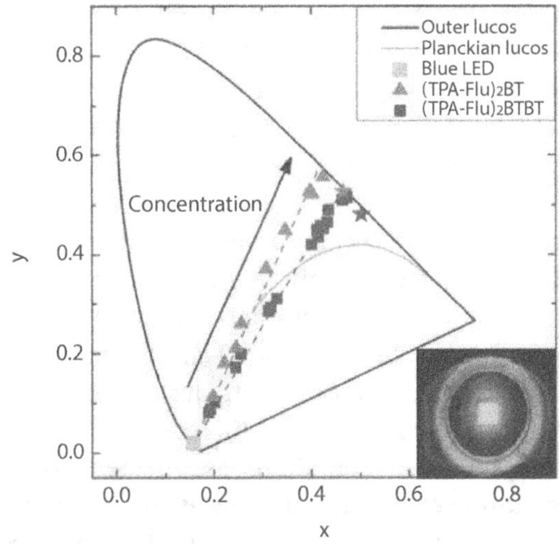

Figure 8.9 CIE chromaticity diagram illustrating the development of coordinates with concentration for both (TPA-Flu)2BT and (TPA-Flu)2BTBT on a blue LED. The arrow indicates the direction of increasing concentration for both converters. Stars represent the chromaticity coordinates of each of the converter materials without the LED. The inset shows a photograph of the HWLED with the converter deposited on the blue LED operating at 25 mA. The size of the LED chip is 1 mm^2. Reprinted from E. Taylor-Shaw, E. Angioni, N. J. Findlay, B. Breig, A. R. Inigo, J. Bruckbauer, D. J. Wallis, P. J. Skabara, and R. W. Martin, Cool to warm white light emission from hybrid inorganic/organic light-emitting diodes. *J. Mater. Chem. C*, 4 (2016) 11499–11507; with permission of The Royal Society of Chemistry.

6(1, 1, 7, 7-tetramethyljulolidyl-9-enyl)-4H-pyran) (DCJTB) was successfully used in a white OLED, where DCJTB worked as a red emissive layer, being inserted between two blue phosphorescence emissive layers. Efficiencies as high as 9.6 cd A^{-1} were reported [73]. These results encouraged other researchers to design new cyanate molecules, such as DCDCC2 and DCDDC2, with fluorescent properties for OLED applications [74]. Recently, Nyalosaso *et al.* [75] used DCM as down-converter layer to fabricate HWLEDs. Since DCM emits in the red portion of the spectrum (peak at λ = 624 nm), pyranine (HPTS) (peak at 533) was coupled with the former one to cover a larger portion of the visible spectrum. The two dyes were dispersed in two zinc hydroxyacetate $Zn_5(OH)_8(CH_3COO)_2{\cdot}2H_2O$ single-layered hydroxide (Zn-SLH) inorganic matrices. In order to ensure good dispersion and avoid aggregation of the molecules, the composite, kept in its wet form, was successively embedded in a second silicon matrix. Pumping the layer with the blue radiation of a 454 nm LED, a white-light emission with excellent photometric parameters was obtained: CCT of

5,409 K and CRI of 81. The best results were obtained with the superposition of the HPTS film over that of DCM arranged in a remote phosphor configuration [75].

AIEgens: As well known, in the aggregate state, luminophores may show reduced, or quenched emission, in comparison to their dilute solutions. The phenomenon of aggregation-caused quenching (ACQ) has been discovered by Forster in 1954 [76] and is referred to as concentration quenching because the emission from luminophore solution is quenched with an increase in concentration. Conventional luminophores molecules often emit strongly when isolated but show ACQ effect when they are aggregated or clustered. The aggregation causes the molecules to be located one close to the other. The aromatic rings of the molecules are subjected to strong intermolecular π-π stacking interactions, as reported in previous sections. Such aggregates show excited states that decay to the ground state through non-radiative paths, resulting in quenching of the luminophore emission. Also, the formation of excimers, resulting in observed ACQ effect, is often involved in these π-π stacking interactions.

As odd as it may sound, there are a class of luminophores where the luminescence, not only is not quenched, but is also significantly enhanced only after aggregation. This phenomenon is called aggregation-induced emission (AIE). This term was coined by Tang's group (Hong Kong University) in 2001 [77] who discovered and observed this effect for the first time. While the noun "luminophores" is referred to the emissive chromophores which emits as molecular species but quenches in aggregate forms, "luminogens" are instead named those non-emissive as molecules but emissive as aggregates. The luminogens, exhibiting AIE properties, are named *AIEgens*.

It is now worth to consider one typical ACQ luminophore and another AIE luminogens. An example of a typical ACQ luminophore is fluorescein whose fluorescence becomes weak and weak when a poor solvent, as well as acetone is gradually mixed with water. If the fraction of acetone is too high, the solvating power of the water/acetone mixture is so scarce that fluorescein molecules begins to form aggregates. As a result, the light emission of fluorescein gets completely quenched. For this reason, fluorescein in state of powder cannot emit light at all, showing a marked ACQ effect in the solid state. On the contrary, hexaphenylsilole (HPS) is a typical AIEgen, being non-emissive when dissolved in a good solvent—tetrahydrofuran (THF), for example—and emissive in a poor solvent, such as water. Dissolving HPS in THF and adding water in the solution, one can observe that if the fraction of water becomes too high, the fluorescence turns on, thanks to the aggregation of the HPS molecules in the aqueous solution with scarce solvating power. Figure 8.10 shows several photographs of the

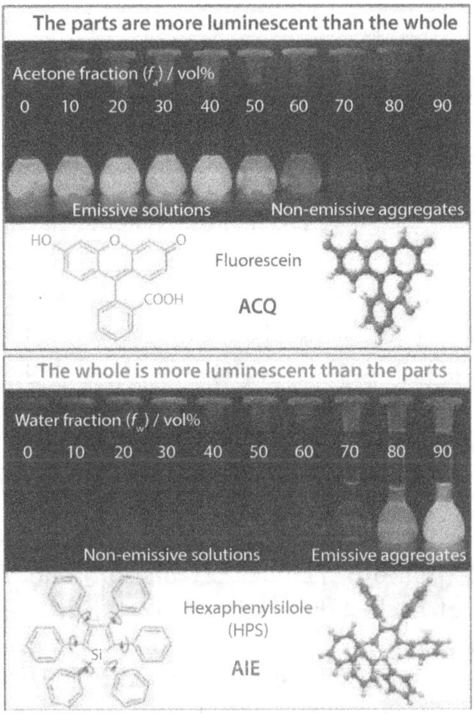

Figure 8.10 Solutions and suspensions fluorescence images of (upper panel) fluorescein (15 μM) in water/acetone mixtures with different fractions of acetone (f_w) and (lower panel) hexaphenylsilole (HPS; 20 μM) in THF/water mixtures with different fractions of water (f_w). Reprinted from J. Mei , Y. Hong, J. W. Y. Lam, A. Qin, Y. Tang, and B. Z. Tang, Aggregation-Induced Emission: The Whole Is More Brilliant than the Parts, *Adv. Mater.*, 26 (2014) 5429–5479; with permission of John Wiley and Sons.

two solutions with different content of the poor solvent and evidences the different behavior of the two molecules.

The comprehension of AIE mechanisms is not easy because the causes can be different and depending on the type of molecule considered, and also, more mechanisms may be involved at the same time. Restriction of Intramolecular Rotations (RIR) was proposed as the main reason of the AIE behavior in HPS [78]. In a single HPS molecule, the silole core is single-bonded with six phenyl rings; due to the steric repulsion between the neighboring phenyl rings, the molecule results to be conformationally flexible and highly twisting. Consequently, a dense face-to-face packing structure is avoided and no π-π stacking interaction results in the solid state. Thanks to the long distances between the silole cores, no strong chromophore interactions are present.

While the behavior of many AIEgens can be interpreted by the RIR mechanism, other AIE systems, such as the 10,10′,11,11′-tetrahydro-5,5′-bidibenzo[*a*,*d*][7]annulenylidene (THBA) cannot be fully explained by the RIR process, because it has no rotatable units, as its phenyl rings are locked by a pair of ethylene tethers; however, it exhibits AIE activity [79]. The origin of this activity should lie in the flexibility of the flexure connecting the phenyl rings, which allows them to dynamically bend or vibrate in the solution state. Therefore, the flexure acts as a relaxation pathway for the excited states to non-radiatively decay. In the aggregate state, due to the physical constraint in the solid state, the intramolecular vibrations are restricted. In the light of this, the Restriction of Intramolecular Vibrations (RIV) is considered to be involved for the AIE activity of THBA. As a consequence, the non-emissive pathway is blocked and RIV enables THBA to decay radiatively in the aggregate state.

If both RIR and RIV are together involved in an AIEgen system, the working mechanism is called Restriction of Intramolecular Motions (RIM), where M (motion) includes R (rotation) and V (vibration). A typical example of such AIEgens is the 11,11,12,12-tetracyano-9,10-anthraquinodimethane [80]. More extensive discussions on AIEgen mechanisms have been outstandingly reviewed in [79, 81–85]. The three mechanisms are illustrated in Figure 8.11.

Initially, AIEgen systems were considered to be appropriate as emitting layers in an OLED. N^4, N^4, $N^{4'}$,$N^{4'}$-tetraphenylbiphenyl-4,4'-diamine,

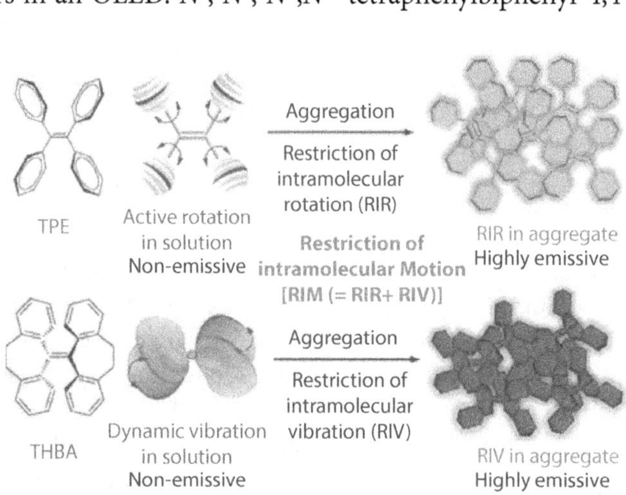

Figure 8.11 Illustration of fluorescence turn-on process of AIEgens by RIR (TPE) and RIV (THBA). Reprinted from J. Mei, Y. Hong, J. W. Y. Lam, A. Qin, Y. Tang, and B. Z. Tang, Aggregation-Induced Emission: The Whole Is More Brilliant than the Parts, *Adv. Mater.*, 26 (2014) 5429–5479; with permission of John Wiley and Sons.

a triphenylamine (TPA) dimer (DTPA) was used to prepare an AIEgen for OLED applications. DTPA is an ACQ luminophore, showing a high fluorescence quantum yield in terahydrofuran (THF) solution (75.6%) but a 5.5-fold lower efficiency in film state (13.7%). DTPA and tetraphenylethene (TPE) are the components of the 3TPETPA and 4TPEDTPA luminogens prepared according to the synthetic routes reported in [86]. Fluorescence quantum yield of 3TPETPA and 4TPEDTPA in THF are 0.42% and 0.55%, respectively, while those of their solid films are as high as 91.6% (3TPETPA) and 100% (4TPEDTPA), proving the AIEgen effect of these systems. 3TPETPA and 4TPEDTPA exhibited absorption peaks at 370 and 375 nm, respectively, and, as emitting layer sandwiched in an OLED standard structure, showed a peak in the EL spectrum at 492 nm. Moreover, the results proved not only that the AIE luminogens are highly emissive in the solid state, but also that present excellent hole-transport properties, making them suitable as both emitting and hole-transport layers in OLEDs [86].

The first attempt to fabricate an HWLED by an AIEgen-based down-converter layer was reported in 2013 by the group of the Hong Kong University led by Prof. Tang [87, 88]. The HWLED was fabricated by simply capping a commercial blue LED emitting at 460 nm with a 4,7-Bis[4-(1,2,2-triphenylvinyl)phenyl]benzo-2,1,3-thiadiazole (BTPETD) yellow-emitting AIEgen compound. In order to improve the stability, the device was successively encapsulated with an epoxy coating. The HWLED exhibited CIE coordinates of (0.32, 0.33), very close to those of standard illuminant E, and 45.4% of optical extraction efficiency from BTPETD layer. The luminous efficacy of the device was as high as 123.8 lm W^{-1} under a driving current of 30 mA and the luminance was measured to be 2.69×10^5 cd m^{-2} (at 80 mA). However, as usual in this kind of devices, at high current, the white light tended to be bluish due to the increase of the blue peak intensity while the yellow fluorescence tended to be saturated [87, 88].

Coumarins: Coumarins (1-benzopyran-2-one) are chemical compounds in the benzopyrone class of organic compounds. They occur in nature in some essential oils and essences, such as tonka and lavender. Fluorescence activity in certain coumarin derivatives was already reported in 1940 [89] and coumarin-6 was firstly used in an OLED as an emitting layer in 1995 [90]. As a green-converter for HWLEDs application, coumarin-6 was used for the first time by Kim *et al.* in 2010 [91]. Coumarin-6 was mixed in a PMMA matrix dissolved in THF and α-terpineol was added to control the viscosity of the solution. The HWLED was fabricated by using a simple roll-laminating technique, between the blue LED chip (λ = 460 nm) and the wavelength converter film a diffusion sheet was inserted. Due to lack of red emission, the chromatic coordinates of the HWLED were (0.229, 0.308), corresponding to a cool white. The luminous efficiency was measured to be 34.6 lm W^{-1}. The diffusion sheet, together with a light-guide panel (which limited the intensity of the blue LED on the converter film), were used to protect the organic film by the heat generated by the LED chip; however, the relative wavelength-converting intensity decreased by about 10% after 1 h [91].

Another coumarin-based HWLED was designed by means of coumarin-30 and N-alkylated pyrrolopyrrole-2,5-dione with thiophene substituents (DPPth) using a violet LED (405 nm) [92]. Coumarin-30 in PMMA shows a quantum yield of 0.92, emits at 470 nm, and absorbs at 400 nm, while DPPth in PMMA exhibits a quantum yield of about 1 and emits at 580 nm. The absorption spectrum of DPPth falls in the emission band of coumarin-30. At 405 nm, absorption is very low for DPPth and very high for coumarin-30, thus DPPth can be excited *via* absorption of light emitted by coumarin-30 and by FRET mechanism. The two dyes were embedded in mass polymerized PMMA. Incorporation in PMMA

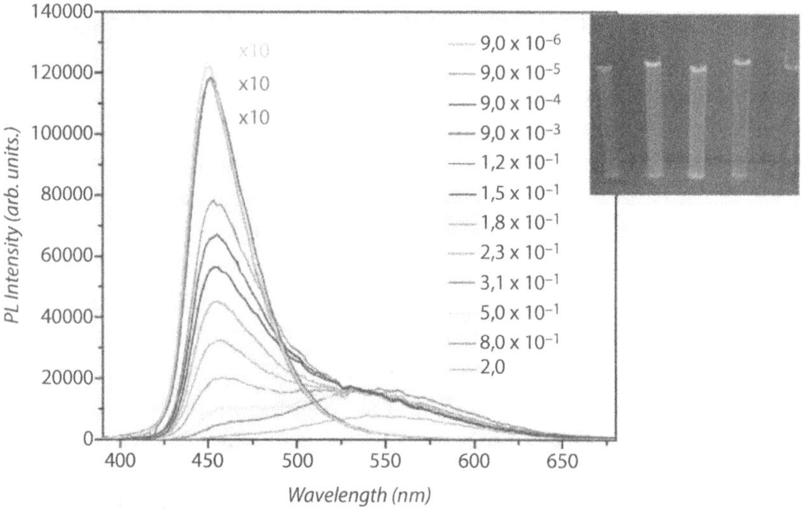

Figure 8.12 Emission spectra of coumarin in $CHCl_3$ (excitation wavelength at 405 nm); the main bands (9×10^{-4} M, 9×10^{-5} M, and 9×10^{-6} M) are divided for a factor 10. The quantum yields for these solutions were estimated in 1. The picture reports the naked eye emissions from representative solutions, from right to left: 1M, 10^{-1}M, 10^{-2}M, 10^{-3}M, 10^{-4}M. Reprinted from A. Luridiana, G. Pretta, D. Chiriu, C. M. Carbonaro, R. Corpino, F. Secci, A. Frongia, L. Stagi, and P. C. Ricci, A facile strategy for new organic white LED hybrid devices: design, features and engineering. *RSC Adv.*, 6 (2016) 22111–22120; with permission of The Royal Society of Chemistry.

increased both the photostability and the efficiency of the parental molecules. By varying of both the thickness of the PMMA shell and the concentration of the dopant dyes, it was possible to obtain various colors, the chromatic coordinates spanning from the green/blue to the yellow/red region of the chromaticity space. White light with chromatic coordinates of (0.32, 0.33) was attained, with a CRI of 80. Short-term stability was tested and it was found that the emitted light intensity decreased by 10% after 72 h of continuous operation [92].

Better performances were achieved by Luridiana *et al.* [93] in terms of both color tuning and photostability. In this case, the converter was based on both a push-pull–based coumarin-1 and the ethyl 7-(diethylamino) coumarin-3-carboxylate (20) and [2-[2-[4(dimethylamino)phenyl]ethenyl]-6-methyl-4H-pyran-4-ylidene]-propanedinitrile (DCM). The dyes were dispersed at different concentrations in a polycarbonate/$CHCl_3$ solution; then, a commercial 405-nm LED was immersed for several hours in the polycarbonate dispersion. By changing the concentration of the luminophores, it was possible to achieve a fine tuning of the white light emitted, from cold to white; CCT ranged between 2,779 K and 5,980 K and CRI from 65 to 81. The variation of the photoluminescence intensity with the concentration is shown in Figure 8.12.

Concerning the photostability, after 10 days of continuous operation, the intensity of the photoluminescence in continuous operation at 15 mA of driving current and at room temperature did not appreciably change. At 50°C, thermally-induced degradation was less than 5%, while at 80°C was considerably higher (53%) [93]. This suggests these molecules to be more thermal-sensitive, than photo-sensitive.

8.4.3 Biomaterials and Biomolecules

A new frontier for down converter materials in HWLEDs is that of the biomaterials and the biomolecules which opens exciting perspective in terms of luminous efficiency and photostability. Biomaterials are substances or compounds of natural or human-made origin, engineered to interact with biological systems for a medical purpose and suitable for interfacing with living tissue and artificial devices; biomolecules, instead, involve molecules, ions and compounds naturally occurring in living organisms that are essential to one or more typically biological processes, including proteins, metabolites, nucleic acids, or carbohydrates. In the last five years, many progress have been made in solid-state lighting technologies based on biomaterials and biomolecules. Many biomaterials, like silk, cellulose, starch, chitin, were explored for their potential usage in LED technologies, as well as biomolecules like mucin, bovine serum albumin, deoxyribonucleic acid (DNA), and fluorescent proteins (FPs). So far, biomaterials and biomolecules have been mostly used to develop matrices to encapsulate chromophores in down-converting-based LEDs or simply as substrates for organic electronics [94–97]. Instead, biomolecules such as DNA and FPs were used to produce biophosphors suitable for HWLEDs.

Curcumin: Biological-based HWLEDs (Bio-HWLEDs) based on natural deoxyribonucleic acid (DNA)-curcumin complexes with cetyltrimethylammonium (CTMA) in bio-crystalline form were proposed and fabricated by the Yeungnam University in South Korea (2016) [98]. The main idea was to bound curcumin to the DNA double helix structure. Curcumin is a diarylheptanoid, belonging to the group of curcuminoids, a bright yellow chemical produced by *Curcuma longa* plants (a member of the ginger family); it is non-toxic neither in large doses (cf. Figure 8.13). Moreover, curcumin absorbs in a wide wavelength region, from UV to visible, producing fluorescence. In the first work of the Yeungnam University group [98], curcumin was extracted from turmeric as a chromophore, generating efficient luminescence due to the tightly bound curcumin chromophore to DNA duplex. An aqueous DNA-curcumin solution activates quenching phenomena, making it not suitable for the commercial production of HWLEDs; therefore, the best strategy for solving this problem is to convert the solution to the crystalline form. Bio-HWLED were fabricated by coating a crystalline DNA-curcumin on a lens put on a 365-nm UV LED. The optima values of both concentration and film thickness were 1 g of curcumin to DNA and PMMA at a constant film thickness of 100 μm. By increasing the current of the UV LED from 10 to 250 mA, the color coordinates moved from greenish (0.39, 0.56) to near yellow emission (0.37, 0.49). The solid bio-crystals confined the activating bright luminescence with a quantum yield of 62%. The luminous drop rate was as low

Figure 8.13 Pictures of raw curcuma, curcuma powder, and curcuminoids. Reprinted from M. Al Shafouri, N. M. Ahmed, Z. Hassan, M. A. Almessiere, and M. Jumaah, Optical and structural properties of curcuminoids extracted from *Curcuma longa* L. for hybrid white light diode, *Eur. Phys. J. Appl. Phys.*, 84 (2018) 10501; with permission of EDP Sciences.

as 0.0551 s^{-1}, meaning a moderate stability of luminous efficiency during working time. Unfortunately, the quantum yields decreases with increasing temperature owing to the loss of binding interaction between the DNA and curcumin [98].

Another group of researchers demonstrated that the way in which curcumin is extracted from turmeric can have some effects on the optical properties of the material and on the performances of Bio-HWLEDs using curcumin as down-converter [99]. A combination of two different methods of extraction, such as the normal extraction method (based on a successively heat-and-decantation steps of curcuma powder in methanol), and the extraction by a Soxhlet extractor (shown in Figure 8.14) for 5 h, turned out to be the most effective approach.

Increasing the concentration of curcumin increased the value of CCT, because the luminophores attained the saturation and hence could no longer produce yellow light. The best results exhibited CIE values of (0.333, 0.3151) with CCT of 5,405 K, and a CRI of 61.2 [99].

A considerable improvement is reported in [100] where the authors fabricated a Bio-HWLED using coumarin, curcumin, and sulforhodamine as the chromophores, and a cassava crystalline thick film as the host material.

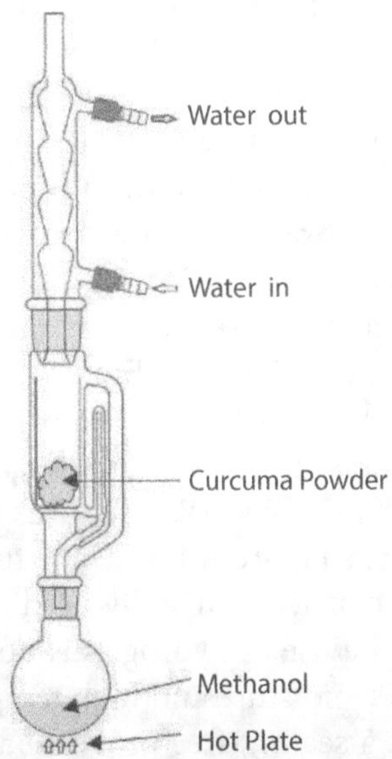

Figure 8.14 Soxhlet apparatus setup. Reprinted from M. Al Shafouri, N. M. Ahmed, Z. Hassan, M. A. Almessiere, and M. Jumaah, Optical and structural properties of curcuminoids extracted from *Curcuma longa L.* for hybrid white light diode, *Eur. Phys. J. Appl. Phys.*, 84 (2018) 10501; with permission of EDP Sciences.

Cassava (extracted from tapioca root) is composed predominantly of cellulose fibers and residual starch and was selected as a chromophore host material because of its low-cost, abundance, and eco-friendliness, as well as for its thermal stability and solubility in organic solvents. A bright white light was emitted by the Bio-HWLED through a dual FRET process. The dual-FRET takes place from coumarin to sulforhodamine *via* the curcumin chromophore. The researchers observed that an aqueous solution of the cassava-based trichromophores exhibited an aggregation-induced quenching phenomenon; consequently, it could not be appropriate for LED fabrication, while in crystalline form and as a thick film (500 mm), the aggregation-induced quenching effect successfully decreased. The luminous drop rate was as low as 0.00069 s^{-1}, that is two order of magnitude less than using only curcumin, as in [98]. The longer lifetime of the Bio-HWLED was attributed to the crystalline structure of the cassava-based trichromophores although its stability was still insufficient due to easy oxidization of cassava [101]. The Bio-HWLED showed chromatic coordinates which moved from (0.33, 0.32) at 10 mA to (0.29, 0.27) at 100 mA. The photoluminescence quantum yields in the thick film forms was as high as 81%, confirming the better performances of the cassava-based trichromophores Bio HWLEDs [100].

Fluorescence proteins (FPs): Since 2015, a group of young and dynamic researchers of the IMDEA Materials Institute (Spain) has been developing the use of FPs as a chromophore for Bio-HWLED applications and their integration in optoelectronics. FPs are one of the most efficient compounds used by nature to produce light and their use as down-converter materials in HWLEDs open new and exciting perspectives in white LEDs market.

In 2015, Costa and his group of the IMDEA Institute first proposed how FPs can be used as novel color-converting materials [102]. They combined both violet (390 nm) and blue (450 nm) LEDs with blue, green, and red fluorescent protein-based rubber materials. The proteins used for this work were: mTagBFP (blue-fluorescent), eGFP (green), and mCherry (red). The architecture of these FP-based Bio-HWLEDs was based on a cascade-like encapsulation, as shown in Figure 8.15a. Unfortunately, FPs are incompatible with most of the matrices currently used in both inorganic LEDs and hybrid LEDs, while the use of hydrogels is not desired due to its low mechanical stability under moisture and high temperatures. To avoid this problem, they developed a sealing-free protein-based gel that transforms into a rubber-like material under moderate vacuum conditions, mixing a branched and a linear polymer, i.e., trimethylolpropane ethoxylate and polyethylene oxide, respectively (shown in Figure 8.15b). The terminal hydroxyl groups provide a high compatibility with the protein solution,

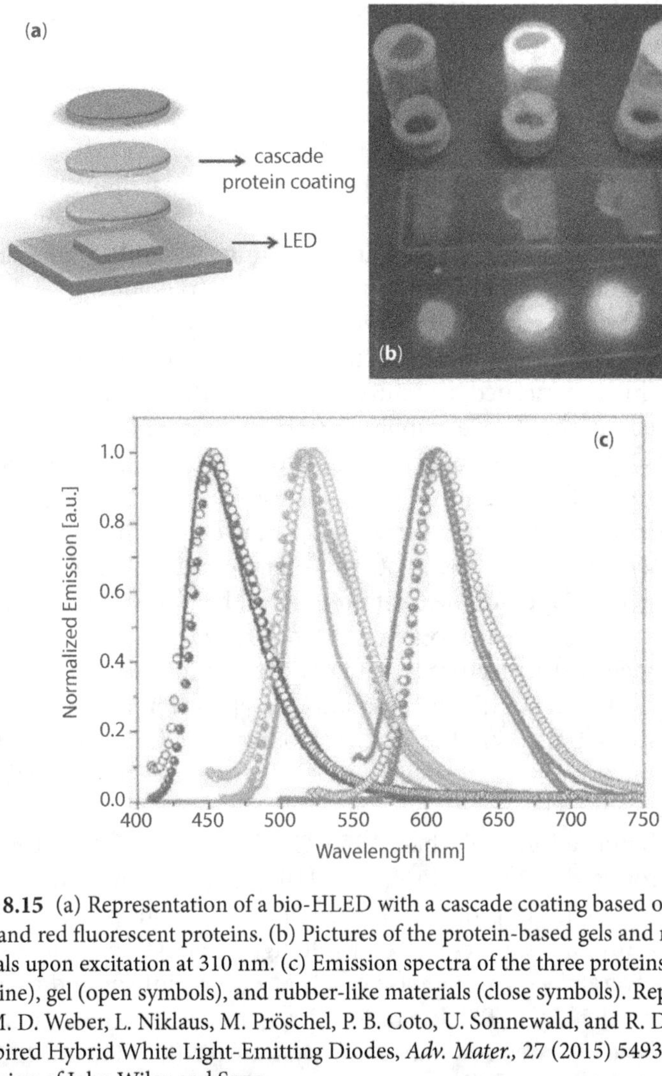

Figure 8.15 (a) Representation of a bio-HLED with a cascade coating based on blue, green, and red fluorescent proteins. (b) Pictures of the protein-based gels and rubber-like materials upon excitation at 310 nm. (c) Emission spectra of the three proteins in solution (solid line), gel (open symbols), and rubber-like materials (close symbols). Reprinted from M. D. Weber, L. Niklaus, M. Pröschel, P. B. Coto, U. Sonnewald, and R. D. Costa, Bioinspired Hybrid White Light-Emitting Diodes, *Adv. Mater.*, 27 (2015) 5493–5498; with permission of John Wiley and Sons.

retaining the necessary moisture into the gel network. This matrix stabilizes FPs over years when it is further transformed into an elastomeric compound and featured a transmittance larger than 90% over the whole visible spectrum. Moreover, the combined system was stable up to temperatures close to 100°C. Optimized devices made of 390 nm-LED/mTagBFP(300 µm)/eGFP(400 µm)/mCherry(500 µm) and 450 nm-LED/eGFP(400 µm)/mCherry(800 µm) showed stable white light with chromathic coordinates of (0.32, 0.33) and CRI as high as 80. However, the stability of the

Bio-HWLEDs needed to be improved since the luminous efficiency starts to break down after 70 h of operation, probably due to the oxidative stress caused by the formation of OH^- or peroxide radicals that oxidize the FPs [102]. The emission spectra of the three proteins in solution, gel, and rubber-like materials are shown in Figure 8.15c.

The drawback of this multilayered cascade-like architecture is that the thickness of the whole structure overcomes 1 mm, limiting the efficiency due to the low optical transmittance. Thus, submillimeter coatings, where the location of the proteins is well-defined, would be highly desired. Another work of the same authors reported on the fabrication of FP-based Bio-HWLEDs using a single FP thin films (<300 μm) as a down-converter, featuring a micropatterned structure to improve the luminous efficiency without affecting the device stability [103]. The patterning was obtained using the 3D printing technique at the hydrogel form. Different micro-patterns were studied, such as lines, open-grids, and closed-grids. The micropatterned coatings were successively transformed into a rubber-like material after a vacuum process at room temperature. The authors used the same green and red luminescent proteins of their previous work (eGFP and mCherry); as a primary source a 480-nm LED was used. The line-like micropatterned coating featured 6 lm W^{-1}, chromatic coordinates of (0.33, 0.33), and CCT of 5,500 K, showing a 5-fold enhancement of the device efficiency without affecting the device stability. As additional advantage, the size of the micropatterned coatings could be scaled to be suitable for display applications [103].

The first application of FP-based Bio-HWLEDs as a backlight for displays was proposed in 2016 by Press *et al.* [104]. They synthesized and purified eGFP and mCherry proteins, then loaded an aqueous protein solution on two 465-nm blue LED chips for white light generation. Finally, these white LEDs were placed at the base of a box with a diffusor and used as a backlight for a 3.5" LCD TV. The ratio of proteins was selected to optimize the efficiency of Förster resonance energy transfer (η = 0.273). In addition to the FRET process, the red color conversion was further amplified thanks to the increased confinement of light due to the increased refractive index mismatch. This enhanced the absorption path of mCherry for eGFP emission; therefore, color-conversion of red was improved. The backlight exhibited a high luminous efficacy of 248 lm W^{-1}. The light emitted by the LEDs appeared to be as a cool white with a CCT of 8,440 K, the EL spectrum being shown in Figure 8.16a. The RGB color coordinates of the backlight were (0.10, 0.17), (0.22, 0.64), and (0.69, 0.31), respectively, and the area of the gamut covered 80% of the NTSC color space, which turns out to be larger than a yellow phosphor-based color conversion LED. Using

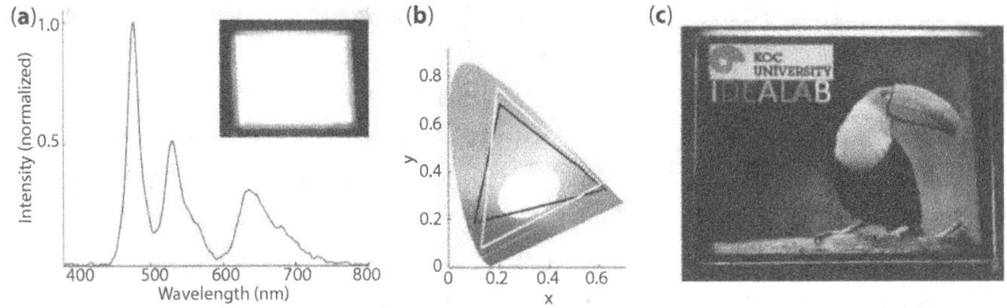

Figure 8.16 (a) The spectrum and photograph (in the inset) of fluorescent protein integrated backlight at 20 mA after the diffuser part of the LCD TV. (b) The color gamut triangle of the backlight (black), the theoretical backlight (grey) using a 440-nm pump LED, eGFP, and mPlum proteins, and the NTSC color space (white). (c) A photograph of a 3.5″ LCD display image illuminated *via* fluorescent protein integrated backlight. Reprinted from D. A. Press, R. Melikov, D. Conkar, E. N. Firat-Karalar, and S. Nizamoglu, Fluorescent protein integrated white LEDs for displays, *Nanotechnology*, 27 (2016) 45LT01; with permission of IOP Publishing, Ltd.

mPlum instead of mCherry on a 440-nm blue LED, the color gamut could be further expanded, covering the NTSC color space of 114% (Figure 8.16b). The resultant colors and objects in the image on the display could be well observed and distinguished, as shown in Figure 8.16c [104].

Concerning FP-based Bio-HWLED stability, no many data are reported in previous papers. Some progress have been announced by the IMDEA group in 2019 [105]. They identified two mechanisms responsible of the degradation: a thermal quenching effect, which can be avoided by changing the device architecture, and a photoinduced deactivation, at the beginning slow and reversible, then quick and irreversible. More details about FP-based Bio-HWLED stability will be given in the next section.

Several in-depth analysis about FP-based Bio-HWLED and their perspectives in LED market are well reviewed in [5, 106]; these reports are recommended to whom wishes to examine in depth the topic.

8.4.4 Metal-Organic Frameworks

Metal-organic frameworks (MOFs) are crystalline materials made of transition metal cations and polydentate organic ligands. They are rapidly emerging as a unique type of porous and organic/inorganic hybrid materials which can be simply self-assembled from their corresponding inorganic metal with organic linkers. MOFs feature microscopic pores or channels, which can accommodate guest molecules. In terms of porosity, they are even superior to other well-known porous materials such as zeolites and carbon materials. For this reason, they are very attractive as color-converters in HWLED applications since chromophore molecules can be embedded into MOF as guests, being accommodated in the pores; therefore, one can obtain a fine tuning of the emission spectra [107]. Furthermore, considering metal ions and organic linkers as simple building blocks to be arranged in an appropriate framework topology, there is a very huge number of assortment of available building blocks, which can lead to an enormity of designed MOFs, as illustrated in Figure 8.17. Given, the ample possibility of designing, the chemical features of MOFs were proven to be stochastically foreseeable by machine learning methods [108].

A particular class of chromophores which can be exploited as guest molecules in MOFs is that of AIEgens; they did arise a huge interest in scientific community, as reported in [109].

As organic/inorganic self-assembled hybrid materials, one can take advantages of the physical and chemical properties of both organic and inorganic components. For example, the pores within MOFs can be engineered so that the pore spaces can be efficiently utilized for gas storage. Moreover, they can encapsulate a large number of different species of diverse functions, so a variety of functional MOF/composite materials can be readily synthesized. The intrinsic optical properties of metal ions,

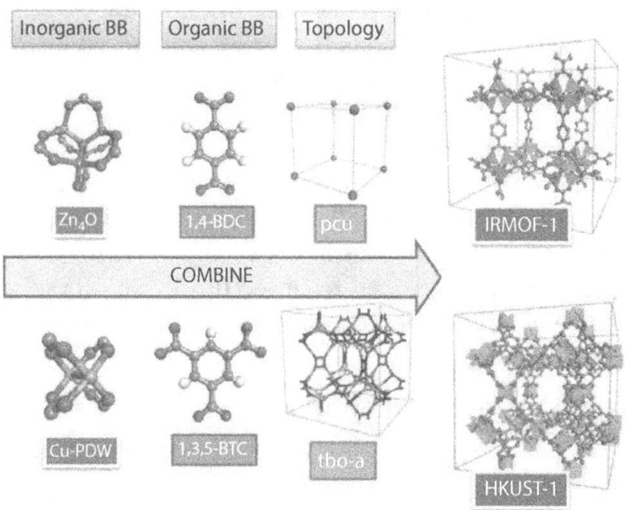

Figure 8.17 Structure of MOFs: examples of arrangements of inorganic and organic building blocks (BB) in a selected framework topology. Reprinted from G. Borboudakis, T. Stergiannakos, M. Frysali, E. Klontzas, I. Tsamardinos, and G. E. Froudakis, Chemically intuited, large-scale screening of MOFs by machine learning techniques. *Npj Comput. Mater.*, 3 (2017) 40; with permission of Springer Nature.

organic ligands, and possible guest molecules (or ions) can be effectively exploited to fabricate a series of novel MOF-based applications, such as radiometric luminescent thermometers, O_2 sensors, nonlinear optical devices, and, evidently, white-light-emitting LEDs [107].

First attempts to obtain white light by color-conversion using a MOF structure date back to 2007 when Wang *et al.* [110] from Chinese Academy of Science used a borate-based hybrid open framework. By illuminating the structure with near-UV radiation, they obtained white light, although they did not fabricate any HWLED prototype. Since the first demonstration of MOF-based HWLED, shown by the same research group in 2009 [111], MOF structures captured rapidly the interests of the researchers working in this field, basically for the following reasons: i) they prevent dye molecules from aggregation-induced quenching, ii) they efficiently transfer energy to the dye, and iii) they allow achieving huge possibility of color tuning and tailoring the luminescence properties. Due to these properties, in the last years MOF structures were massively proposed as down converter materials in HWLEDs. To date, dye-encapsulated MOFs are the most investigated organic structure to fabricate HWLEDs. The researchers used many diverse organic ligands, metals, and dyes; nevertheless, it is not easy to compare the different structures and extract a course of action (or a rigorous protocol) to follow for a progress prediction.

An overview of the most important results attained by MOF-based HWLEDs and reported in literature are summarized in Table 8.1.

MOF structures as down-converting materials show emission lifetimes significantly shorter than the lifetime of standard phosphors in commercial WLEDs. As a consequence, switching speed of the MOF-based HWLEDs is higher than that of commercial WLEDs. This property makes the MOF-based HWLEDs suitable for visible-light communication (VLC), which uses room lighting to carry transmitted signals as an alternative wireless communication technology [112, 113].

As results from Table 8.1, most of works does not report on luminous properties of the HWLEDs, as well as their thermal and photo-stability, proving the low level of ripeness of the technology. The highest luminous efficacy value reported is 156 lm W^{-1}, obtained using a YAG+Rh@bio-MOF-1 structure, which is an excellent result if compared with commercial inorganic LEDs [114]. From the point of view of chromaticity, the confinement of the guest dyes within the pores of MOFs not only provides a fine tuning of the white light, but also prevents the ACG process improving the fluorescence quantum efficiency [115].

Using a vast assortment of ligands synthesis of MOFs allows constructing various structures to down-convert a blue or near-UV LED radiation in a

Table 8.1 Overview of the MOF-based HWLEDs reported in literature.

Year	Organic ligands	Guest dyes, dopant	Source wavelength (nm)	Luminous flux, efficacy, quantum yield	CIE coordinates, CCT, CRI	Stability	Notes	Ref.
2019	$\left[Pb_2X_3^+\right]\left[-O_2C(C_6H_4)CO_2^-\right]_2\left[(CH_3)_2NH_2^+\right]_3$ (X=Cl, Br, or I)	Eu3+	340, 380	not reported	CCT: 4,258–7,993 K CRI: 89	not reported	Transition of Eu3+ at 593, 616, and 700 nm along with the broadband photoemission from the MOF	[117]
2019	$Gd(pic)_3$, $La(pic)_3$	Eu^{3+}/Tb^{3+}	365	QY: 85%	(0.305, 0.346), (0.301, 0.345) CCT: 6755 K, 7001 K CRI: 71.8, 72.9	not reported	Gd3+ complex serves as a better host than La3+	[118]
2019	LIFM-WZ-6, $[(Zn_{2-}(TATZTPE)Bpy{\cdot}8DMA)]_n$	Rhodamine B (RhB^+), Basic red 2 ($BR\text{-}2^+$), Basic red 46 ($BR46^+$), 4-(p-dimethylaminostyryl)-1-methylpyridinium) (DSM^+), Astrazon pink FG ($APFG^+$)	365	not reported	RhB^+: (0.33, 0,35) CCT: 4745 K CRI: 77; $BR\text{-}2^+$: (0.33, 0,36) CCT: 5524 K CRI: 69; $APFG^+$: (0.34, 0,36) CCT: 4436 K CRI: 83	not reported	Dual-way photon energy conversion	[119]

(Continued)

Table 8.1 Overview of the MOF-based HWLEDs reported in literature. (*Continued*)

Year	Organic ligands	Guest dyes, dopant	Source wavelength (nm)	Luminous flux, efficacy, quantum yield	CIE coordinates, CCT, CRI	Stability	Notes	Ref.
2018	Rh@bio-MOF-1	green: $(Ba,Sr)_2SiO_4$:Eu2$^+$, yellow: $Y_3Al_5O_{12}$:Ce^{3+}, red: $CaAlSiN_3$:Eu^{2+}	450	At 10 mA : YAG+Rh@ bio-MOF-1: LE: 156 lm W^{-1}; BSSO+Rh@ bio-MOF-1: LE: 125 lm W^{-1}; BSSO+CASN+Rh@ bio-MOF-1: LE: 96 lm W^{-1};	YAG+Rh@ bio-MOF-1: (0.339, 0.339) CCT: 5225 K CRI: 80 BSSO+Rh@ bio-MOF-1: (0.340, 0.356) CCT: 5203 K CRI: 89 BSSO+CASN+Rh@ bio-MOF-1: (0.390, 0.373) CCT: 3713 K CRI: 96	not reported	Shorter lifetimes (3–7 ns) than the conventional inorganic phosphors	[114]
2018	ZJU-28	Coumarin 6 (Cou-6), Rhodamine 6G (R6G), Rhodamine 101 (R101)	460	QY: 82.9% LE: 126 lm W^{-1}	(0.36, 0.34) CCT: 4446 K CRI: 88	Brightness decreases of 5% during first 10 h of work; then is constant during next 60 h	Good thermostability	[120]

(*Continued*)

Table 8.1 Overview of the MOF-based HWLEDs reported in literature. (*Continued*)

Year	Organic ligands	Guest dyes, dopant	Source wavelength (nm)	Luminous flux, efficacy, quantum yield	CIE coordinates, CCT, CRI	Stability	Notes	Ref.
2018	bio-MOF-101	1,2-bis(4-(4-carboxyphenyl))-1,2-diphenylethene (H_2BCPPE)	375	not reported	(0.35,0.36) at 3 V	not reported	SALE (solvent-assisted ligand exchange) protocol to incorporate H_2BCPPE into MOF	[116]
2017	Zr-TCBPE-MOL (metal-organic layer: 2D-MOF)	4',4",4"',4""(ethene-1,1,2,2-tetrayl)tetrakis([1,1'-biphenyl]-4-carboxylic acid) (H4TCBPE)	450	not reported	(0.37, 0.41)	30% drop of the output power after 168 h, possibly due to decomposition of the TCBPE ligand under intense light in the presence of oxygen	Fast switching speed, suitable for visible-light communication (VLC)	[112]

(*Continued*)

Table 8.1 Overview of the MOF-based HWLEDs reported in literature. (*Continued*)

Year	Organic ligands	Guest dyes, dopant	Source wavelength (nm)	Luminous flux, efficacy, quantum yield	CIE coordinates, CCT, CRI	Stability	Notes	Ref.
2017	RhB@Al-DBA (DBA: 9,10-dibenzoate anthracene)	Rhodamine B (RhB)	395	0.89 mW/cm^2 at 450 mA	(0.32, 0.30)	not reported	Fast switching speed. Modulating frequency of 3.6 MHz for VLC, six times that of commercial WLEDs	[113]
2017	MCM-41	blue: Triethoxysilylpropyl-substituted derivatives of perylene; green: Benzofurazane red: Nile red	UV	QY: 4.6%	(0.33, 0.33) CCT: 5,500 K, 10,700 K, 41,100 K	not reported	Large color tuning	[121]

(*Continued*)

Table 8.1 Overview of the MOF-based HWLEDs reported in literature. (*Continued*)

Year	Organic ligands	Guest dyes, dopant	Source wavelength (nm)	Luminous flux, efficacy, quantum yield	CIE coordinates, CCT, CRI	Stability	Notes	Ref.
2017	HSB-W1 (HSB = hydrogenated schiff base)	fluorescent brighteners (CBS-127, CBS-X, and KSN), coumarin derivatives (Coumarin 6, C6; 7-(diethylamino)3-(2-thienyl)coumarin, C6a; etc.), 4-(dicyanomethylene)-2-methyl-6-(p-dimethylaminostyryl)-4H-pyran (DCM), 4-(p-dimethylaminostyryl)-1-methylpyridinium (DSM)	365	QY: 26%	(0.33, 0.33) CRI: 92	not reported	Light tunable by changing the contents or types of the dye guests	[122]
2016	H_3CPOMBA (4,4′-(((5carboxy-1,3-phenylene) bis(oxy))bis(methylene)) dibenzoic acid)	Eu^{3+}, Tb^{3+}	365	not reported	(0.326, 0.343) CCT: 5733 K CRI: 73.4	color emission after heating for 48 h at 150°C does not change under the excitation of 365 nm	LaEuTb-CPOMBA was also tested as MOF material	[123]

(*Continued*)

Table 8.1 Overview of the MOF-based HWLEDs reported in literature. (*Continued*)

Year	Organic ligands	Guest dyes, dopant	Source wavelength (nm)	Luminous flux, efficacy, quantum yield	CIE coordinates, CCT, CRI	Stability	Notes	Ref.
2015	ZJU-28	red dye: 4-(p -dimethylaminostyryl)1-methylpyridinium (DSM), green dye: acriflavine (AF)	365	QY: 17.4%	(0.34, 0.32) CCT: 5,327 K CRI: 91	not reported	10-methyl-acridine (AcrM) and NOTT-210 were also tested as guest dyes	[115]
2015	[Zn_2(tcbpe)·*x*DMA, DMA = N, N-dimethylacetamide] or LMOF-231	no guest molecules	455–460	LE: 58.9±1.5 lm W^{-1}	not reported	Heating for 12 h from 100 to 160°C causes a decrease in emission intensity within 15%	Very efficient light absorber (~100%)	[124]
2015	UiO type MOF bearing the 2,2'-bipyridyl moiety (UiO-67-bpydc, 1).	Eu^{3+}@1	375	LE: 32 lm W^{-1}	(0.348, 0.329) CCT: 4,914 K CRI: 75	not reported	Eu^{3+}/Tb^{3+}@1 nanocrystals are used here to develop a colorimetric thermometer	[125]

(*Continued*)

Table 8.1 Overview of the MOF-based HWLEDs reported in literature. (*Continued*)

Year	Organic ligands	Guest dyes, dopant	Source wavelength (nm)	Luminous flux, efficacy, quantum yield	CIE coordinates, CCT, CRI	Stability	Notes	Ref.
2014	$[Zn_6(btc)_4(tppe)_2(DMA)_2]$ (btc = benzene-1,3,5-tricarboxylate, tppe = 1,1,2,2-tetrakis(4-(pyridin-4-yl)phenyl)ethene, DMA = dimethylacetamide)	DMA, H_2O	455–460	LE: 47.4 lm W^{-1}	not reported	Iinternal QY is 58.1% originally, and 57.6% after being exposed to air for 3 months (<1% decrease). Material is both thermally and optically stable up to 180°C	Highest internal quantum yield reported, among all rare earth-free phosphors: 90.7% (excitation at 400 nm)	[126]
2014	MOF-PEMA: MOF-253 with an uncoordinated bipyridyl group (2,2'-bipyridyl), modified with TTA (TTA= 2-thenoyltrifluoro-acetone), functionalized with ethyl methacrylate (EMA) and polymerized (PEMA)	Eu^{3+} (1% and 8 % molar ratio)	395	not reported	(0.39, 0.38) CCT: 3,742 K CRI: 87.34	not reported	White light could be adjusted by altering the content of MOF-EMA in the polymer.	[127]
2013	$[(CH_3)_2NH_2]_{15}[(Cd_2Cl)_3(TATPT)_4] \cdot 12DMF \cdot 18H_2O$	$[Ir(ppy)_2(bpy)]^+$	370	QY: 20.4%	(0.30, 0.35) CCT: 5,409K CRI: 84.5	Relatively stable only below 115°C.	Eu^{3+} and Tb^{3+} co-doping was also tested	[128]

(*Continued*)

Table 8.1 Overview of the MOF-based HWLEDs reported in literature. (*Continued*)

Year	Organic ligands	Guest dyes, dopant	Source wavelength (nm)	Luminous flux, efficacy, quantum yield	CIE coordinates, CCT, CRI	Stability	Notes	Ref.
2012	$CH_3SCH_2CH_2S$- and (*S*)-$CH_3(OH)CHCH_2S$- (2-hydroxylpropyl) are each attached as secondary groups to the 2,5- positions of 1,4 benzene-dicarboxylic acid (bdc), and the resultant molecules (L1 and L2, respectively) are crystallized with Pb(II) into the topologically similar 3D nets of PbL1 and PbL2, both consisting of interlinked Pb-carboxyl chains.	no guest molecules	360	not reported	Cool white	stable for months in air	PbL1 shows photoluminescence in yellowish green, PbL2 features bright white emission	[129]
2009	$[AgL]_n \cdot nH_2O$ (L = 4-cyanobenzoate)	no guest molecules	325–350	QY: 10.86%	(0.31, 0.33)	not reported	First HWLED reported, based on MOF	[111]

white light. Unfortunately, the metal ions and ligands are often self-assembled in an unpredictable way, and it is not possible to create the desired structures and functionalities directly [116]. This is still a considerable weakness for the commercialization of MOF-based HWLEDs. However, as mentioned before, machine learning methods can be helpful for chemical design [108].

8.4.5 Carbon Dots

Recently, quantum dots (QDs), being new luminescent materials, have been used as color-converters in white LEDs. In 2004, a paper reporting the first attempt to fabricate a white LED using QDs was published by Park *et al.* [130]. The authors generated the white light by combining the blue luminescence of an organic polymer with the green and red luminescence of two-sized (3 and 7 nm) CdSe quantum dots [130]. From then on, outstanding performances have been reported, using CdSe [131] ZnS [132], CdSe/ZnS [133], CdSe/CdS [134], ZnCdTe/CdSe [135], InP/ZnS [136], InP/CdZnTe [137], ZnO/ZnS [138], Cu–In–Zn–S/ZnS [139]. Unfortunately, QDs are prepared by a complex process and often involve environmental- and human-unfriendly (and expensive) chemicals. Among zero-dimensional semiconductor nanostructures, fluorescent carbon dots (CDs) could go beyond the above-mentioned disadvantages: in fact, they exhibit intriguing properties, such as high fluorescence, elemental abundance, photostability, biocompatibility, chemical inertness, low cost, and low toxicity, which make them promising color-converters for HWLEDs.

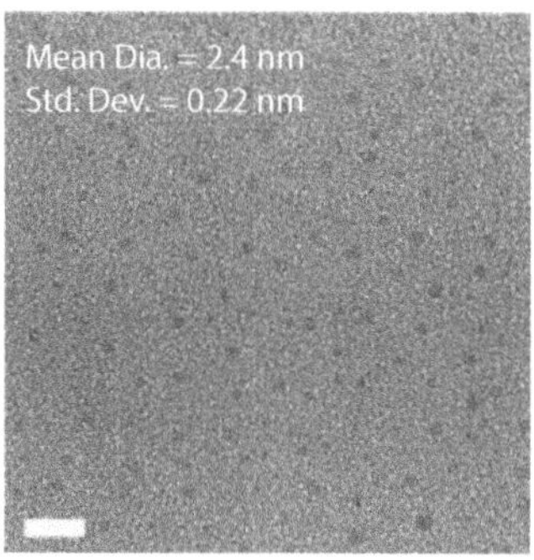

Figure 8.18 TEM images of CDs dispersed in a PMMA matrix of molecular weight 10,000 (scale bars = 10 nm). Reprinted from W. Kwon, S. Do, J. Lee, S. Hwang, J. K. Kim, and S.-W. Rhee, Freestanding Luminescent Films of Nitrogen-Rich Carbon Nanodots toward Large-Scale Phosphor-Based White-Light-Emitting Devices, *Chem. Mater.*, 25 (2013) 1893–1899; with permission of The American Chemical Society.

CDs can be produced inexpensively on a large scale by many ways, like arc-discharge methods, laser-ablation methods, electrochemical synthesis, combustion (from a burning candle), use of supports to localize the growth by blocking nanoparticle agglomeration, synthesis by microwaves and in aqueous solution [140]. A TEM image of 2.4 nm-sized CDs is displayed in Figure 8.18.

The first CD-based HWLED was reported in 2012 [141], where CDs were prepared *via* a one-step pyrolysis of poly(styrene-co-glycidylmethacrylate) photonic crystals. Silicone was mixed with the CDs and put in a vacuum chamber to remove the bubbles; then, the mixture was placed into the cup-shaped void on a 370-nm LED chip and thermally cured at 150°C for 1 h. Almost at the same time, Tang *et al.* [142] prepared graphene quantum dots (GQDs) by microwave-assisted hydrothermal method. GODs with size of 1.65 nm exhibited deep ultraviolet emission at 302 nm. The properties of the GQDs were dependent on the excitation wavelength as expected in carbon-based quantum dots. However, the emission wavelength was independent of the size of the GQDs differently from the CDs prepared by conventional methods. The unique optical properties of the GQDs were attributed to the self-passivated layer on the surface of the GQDs which was responsible of the formation of surface states between the π^* band and π energy level. If deposited onto a commercial 410-nm blue light emitting diode, these GQDs were able of converting blue light into white light. The primary source excited a broad band peaked at ~510 nm. The emission was perceived as a cool white light with chromatic coordinates (0.282, 0.373) [142].

Despite unique optoelectronic properties of GQDs, water-soluble GQDs were not suitable for HWLEDs applications. Luk *et al.* [143] went a step beyond by fabricating HWLEDs based on a GQDs/agar composite which demonstrated an excellent optical stability. White light was emitted by applying the GQDs/agar composite on a blue LED. The luminous efficiency was as high as 42.2 lm W^{-1}, corresponding to a light conversion efficiency of 61.1%. No luminescence quenching was observed and the light conversion efficiency appeared to be stable for over 100 h of continuous operation [143].

Kwon *et al.* [144] fabricated luminescent PMMA films of nitrogen-rich CDs obtained from the emulsion-templated carbonization of polyacrylamide, which were further applied as color-converting phosphors to construct HWLEDs. The PMMA matrix dispersed the CDs to prevent solid-state quenching. A 400-nm blue LED was coated by a film with 2.4 nm-sized nitrogen-rich CDs and at 50 mA produced a cool white light with chromatic coordinates of (0.366, 0.448) and CCT of 5,080.4 K. The luminous efficacy was as high as 108.19 lm W^{-1}, comparable to that of inorganic phosphors, and did not change during over 12 h operating, exhibiting an excellent stability [144].

Other CD-based HWLEDs were fabricated with silica-based CDs/CdS nanocrystals by co-hydrolyzing the blue fluorescent CDs with yellow-emitting CdS nanocrystals. CDs surface was passivated by (3-aminopropyl) triethoxysilane (APTES) *via* a hydrothermal method from citric acid and APTES. These CDs exhibit high blue fluorescence under UV illumination. Coating a 380-nm LED chip with the CDs/CdS nanocrystals blended with silicone, a cool white light with chromatic coordinates of (0.27, 0.32), a CCT of 9,051 K and a CRI of 71 (at 350 mA) was attained [145].

Solvents used in the formation of CDs can strongly influence the emitting properties of the material itself: while under UV illumination CDs pyrolyzed from poly(acrylic acid) (PAA) in water were blue emitting, surface-passivated CDs pyrolyzed from PAA and glycerol exhibited a white fluorescence. Moreover, these latter showed a higher carbonization degree and an improved quantum yield due to the participation of glycerol in the reaction. In fact, glycerol not only accelerated the carbonization of PAA but also passivated the surface of the CDs. Coating a 380-nm LED chip by a CD/silicone mixture, the as-fabricated HWLED produced cool white light with chromatic coordinates located at (0.26, 0.30) with a CRI of 78.2 at 350 mA, comparable to commercial white LEDs. The emission spectrum of this HWLED was broad and covered the entire visible range [146] as shown in Figure 8.19.

In order to reinforce the intensity of luminescence produced by CD-based HWLEDs, it was proposed to exploit surface plasmon resonance of Ag nanoparticles (NPs) [147]. By adjusting the volume ratio between Ag NPs and long wavelength CDs (LWCDs) solution, it is possible to maximize the enhancement of the luminescence. Spin-coating technology was utilized to

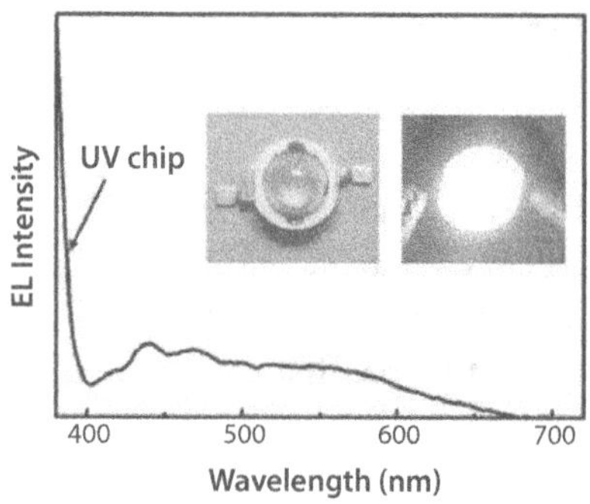

Figure 8.19 EL spectrum of the white LED based on CDs operated at 350 mA, inset: photographs of the white LED under daylight and operation, respectively. Reprinted from L.-H. Mao, W.-Q. Tang, Z.-Y. Deng, S.-S. Liu, C.-F. Wang, and S. Chen, Facile Access to White Fluorescent Carbon Dots toward Light-Emitting Devices. *Ind. Eng. Chem. Res.*, 53 (2014) 6417–6425; with permission of The American Chemical Society.

deposit Ag-LWCDs hybrid films onto a 460-nm blue LED chip. A broad yellow-red emission band ranging from 500 nm to 700 nm was produced by the Ag-LWCDs hybrid films. With an optimal thickness of 0.6 mm for the hybrid film, the HWLED emitted a white light of chromatic coordinates (0.3340, 0.3453), CCT of 5,435 K, and CRI of 74.6. CRI could be increased till 81.6 by reducing the thickness of the film as low as 0.4 mm: the downside was that CCT dramatically increased to 11,663 K. The effectiveness of the Ag NPs on the LWCDs luminescent properties was proved by comparing the luminous efficacy for two different HWLEDs: one with a 0.6 mm Ag-LCDs hybrid film, the other one with a 0.6 mm pure LWCDs film. With the hybrid film the luminous efficacy increased from 32.63 lm W^{-1} (pure LWCDs) to 41.26 lm W^{-1} (hybrid Ag-LWCDs) [147].

Inspired by mussel adhesive, a group of the Northwestern University, USA, found a method to form multifunctional polymer coatings through simple dip-coating of objects in an aqueous solution of dopamine [148]. Li *et al.* exploited this method to prepare a polymer functionalized N-doping CDs for HWLED applications [149]. The advantage of mussel-inspired chemistry is that a simple mixing at room temperature is enough to activate the reaction and preserve the properties of the substrates. A hydrophilic poly(poly(ethylene glycol) methyl ether methacrylate (PEGMA) and hydrophobic polystyrene (PS) terminated with catechol groups were grafted onto the surface of N-doping blue CDs. The obtained CD@PEGMA and CD@PS under solid state showed improved optical properties as compared with pure CDs. This suggested that grafting polymer could passivate the surface state and also increase the distance between the adjacent CDs preventing self-quenching. An HWLED was fabricated by *in situ* polymerization with CD@PS as blue chromophore, CdSe/ZnS QDs as red component, and 5-(2-methacryloyl-ethyloxymethyl)-8-quinolinol (MQ) coordinated QDs surface state as green component, using a 380-nm LED as the excitation source. The HWLED showed CIE coordinates located at (0.32, 0.31) and a CCT of 6,250 K [149].

8.5 Color Tuning and Rendering

Huge color tuning and potential ability to render a large palette of colors are two of the key strengths of the hybrid inorganic-organic technology to manufacture white LEDs. In fact, the list of luminescent dyes which can be synthesized is potentially infinite, and consequently, the color palette in terms of different hue, saturation, and brightness is hugely vaster than the choice available with inorganic phosphors. Therefore, no surprise is expected about the potential larger area of the color gamut achievable in the CIE chromaticity

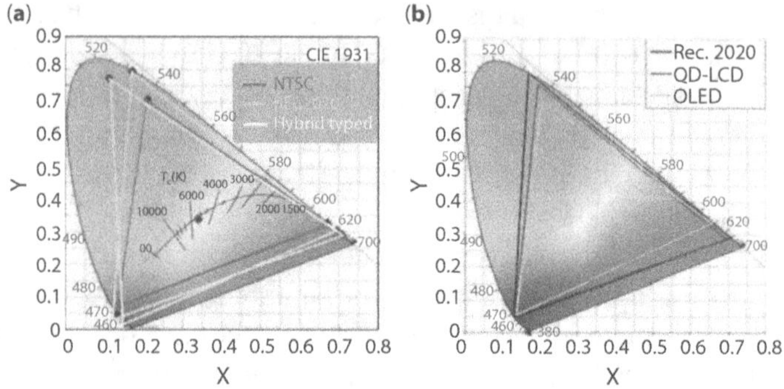

Figure 8.20 (a) CIE color gamut for the NTSC system, the Rec. 2020 standard, and a hybrid solid-liquid-type based on CdSe/ZnS QDs and perovskite QDs. Adapted from C.-H. Lin *et al.*, Ultrawide Color Gamut Perovskite and CdSe/ZnS Quantum-Dots-Based White Light-Emitting Diode with High Luminous Efficiency, *Nanomaterials*, 9 (2019) 1314; with permission of MDPI. (b) CIE color gamut for OLED and QD-LCD displays compared to the Rec. 2020 standard. Reprinted from H. Chen, R. Zhu, K. Käläntär, and S.-T. Wu, Quantum Dot-Enhanced LCDs with Wide Color Gamut and Broad Angular Luminance Distribution, *SID Symp. Dig. Tech.*, 47 (2016) 1413–1416; with permission of John Wiley and Sons, Society for Information Display.

diagram by mixing organic luminophores. Just as an example, in Figure 8.20a, different CIE color gamuts are compared: NTSC, the standard for analog television, is a color gamut developed by the National Television Standards Committee of US (close to that of Adobe RGB) and corresponds to the color gamut of recent liquid-crystal display (LCD) monitors backlighted by inorganic-based white LEDs; Rec. 2020 standard is an ITU-R Recommendation which defines the so-called wide color gamut, for 4k and 8k TV resolutions, apt to be exploited by the new QD and OLED technologies; Hybrid typed is related to a LED consisting of solid CdSe/ZnS red QDs and combined with liquid green perovskite QDs (notice that here the term "hybrid" concerns the antinomy solid-liquid, not organic-inorganic) [150]. In Figure 8.20b, it is shown the gamut for both an LCD-backlighted display with a QD enhancement film and an OLED display compared to the Rec. 2020 wide color gamut [151]. As shown, the palette of organic dyes can compete the challenge against the so-far most performing technology of QDs and, in any case, the color gamut covers an area largely broader of NTSC system.

In terms of color rendering, all the organic materials used as down converter layers show high values of the CRI, between 70 and 90 for most of luminescent polymers and molecules, with highest peaks >90 for polyfluorene, perylene, and porphyrins. The highest value of CRI reported for Bio-HWLEDs is as high as 80 and is attained using FPs [102]. The CRI of

MOF-based HWLEDs generally ranged between 70 and 90 with the highest value (=96) reported in [114] for a Rh@bio-MOF-1. Finally, CDs can exhibit values of CRI as high as 80. Obviously, the values of CRI depend on the reference used; most of times the reference consists in the standard illuminant E but this information is not always reported.

Concerning luminous flux, luminous efficacy, and luminous efficiency, it should be said that most of the time they are not reported. When calculated, in general, they are related to high-CCT cool white LEDs. The highest value of luminous flux reported is ~9 lm for perylene-based white LEDs for which a significant value of ~118 lm W^{-1} of luminous efficiency is also reported [48]. Values of 220 lm W^{-1} are also obtained using perylene but embedded in in PS fibers [53]. The highest values of luminous efficacy reported are: 275 lm W^{-1} for chalcones [62], 248 lm W^{-1} for FP-based Bio-HWLEDs [104], 156 lm W^{-1}, for a YAG+Rh@bio-MOF-1 structure [114], ~124 lm W^{-1} for AIEgen compound [87, 88], ~108 lm W^{-1} for CDs [144].

8.6 Stability

Stability of the down-converting dye is the Achilles' heel of the HWLEDs. If organic dyes benefited from a higher stability—both thermal and photo—a scenario where HWLEDs took the place of inorganic phosphor-converted LED would be plausible. Unfortunately, all the organic materials suffer from a bad stability, which reduces the lifetime of the devices. From this point of view, inorganic phosphors have no equal, having lifetimes many order of magnitude more than organic dyes. Nevertheless, given the ever-growing interest of the optoelectronics market segment for the organic luminescent materials (mainly driven by the OLED TV market), both chemical manufacturing and academic research on organic fluorophores put their foot on the accelerator with a major interest on long-term stability of the devices.

In general, polymers are more stable than monomers and small molecules (because, from a thermodynamic point of view, the totality of the energies of the polymer is less than the sum of the energies of the individual monomers [152]); however, they are not free from weathering problems. This means that not only thermal and photo-instability but also moisture, oxygen, solar radiation, and atmospheric pollutants play a role on the long-term performances of the polymer-based HWLEDs.

High-power HWLEDs could be mostly affected by thermal stability, since their junction temperature reaches higher values than in low-power devices. However, polymers are usually degraded at temperatures higher

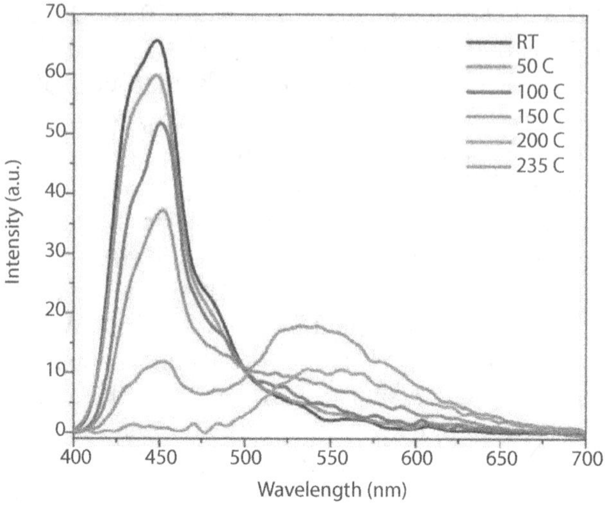

Figure 8.21 Fluorescence spectra of PFA after progressive heat treatment of PFA thin film in air, from room temperature (RT) to 235°C. Reprinted from I. O. Huyal, T. Ozel, U. Koldemir, S. Nizamoglu, D. Tuncel, and H. V. Demir, White emitting polyfluorene functionalized with azide hybridized on near-UV light emitting diode for high color rendering index. *Opt. Express*, 16 (2008) 1115; with permission of The Optical Society (OSA).

than in normal working operations. For example, Figure 8.21 shows how the photoluminescence spectrum of the azide-functionalized polyfluorene PFA (excited at 378 nm) is modified by an annealing treatment in air (50-235°C), After a harsh treatment (≥220°C), the blue components are strongly reduced while a green band rises and the same behavior is also reproduced under argon environment. Obviously, the chromatic coordinates of an HWLED coupled to PFA as a wavelength converter would move towards the top of the chromatic diagram. However, a PFA-based HWLED exhibits for 15 h the same emitted power value when driven at a current level of 30 mA (and an electrical power as low as 135 mW), meaning that for low-power operation, junction temperature is so low that does not compromise the polymer stability [36].

Thermal instability also involves down converters based on small molecules or monomers, indeed more seriously since they are intrinsically less stable than polymers. Thermal degradation of the organic dyes strongly depends on the required host matrix used to encapsulate the fluorescent molecules that would otherwise exhibit aggregation-induced quenching phenomena. For example, Kim *et al.* [58] working with naphthalimides such as DBN, observed that when DBN was dispersed in epoxy as wavelength converter, the resulting HWLED, operating at current as high as

200 mA, was less thermal stable than embedded in PMMA (in epoxy the luminous flux decreased to 17% of the initial value after only 3 h of working, while in PMMA the reduction was only 67% of the initial value after 19 days of operation). Jin *et al.* [59] improved this result by copolymerization with methylmethacrylate (MMA) to produce the corresponding copolymer (PMMA-co-DBN). Compared to a normal DBN/PMMA blend film, the copolymer PMMA-co-DBN did not undergo phase separation upon heating and the luminous flux of the corresponding HWLED was maintained stable during 12 days of continuous operation at 20 mA. The efficacy of PMMA as a host matrix on photochemical stability was also successfully tested to embed dyes, such as coumarin-30 and a DPPth derivative [92].

Heat transfer can be minimized by inserting a spacer between the emitting surface of the inorganic LED and the color converter. Kim *et al.* [91] inserted a transparent diffusion sheet before depositing a coumarin-6-based green-emitting converter embedded in a PMMA matrix; Caruso *et al.* [48, 49] used perylene-based Lumogen® F Yellow 083 dispersed in PMMA as a down

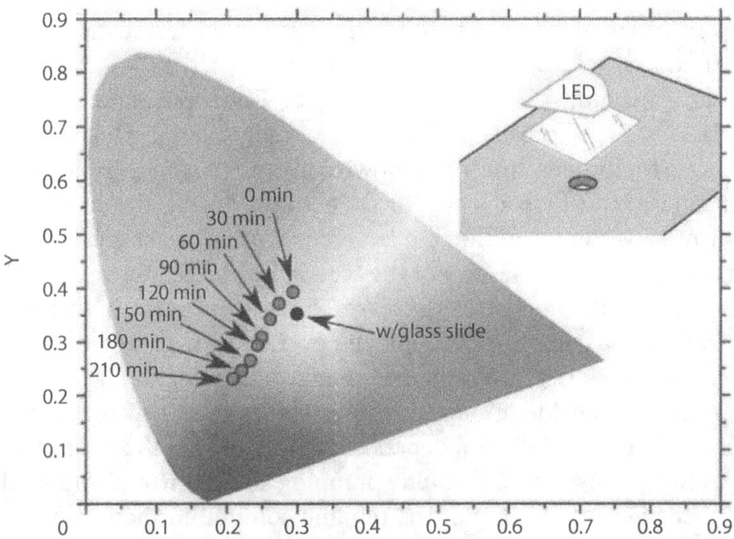

Figure 8.22 Chromatic coordinates in CIE diagram of a Lumogen® F Yellow 083-based HWLED, without (red dots) and with (black dot) the insertion of the glass slide with the luminescent film. Inset: Coupling of the slide and the wafer-level LED: light is collected from the hole below. Reprinted from M. Mosca, F. Caruso, L. Zambito, R. Macaluso, C. Calì, and E. Feltin, Hybrid LEDs_Pave_Way_to New Lighting Applications, *Photonics Spectra* (April 2013); with permission of Photonics Media, Pittsfield MA.

converter, coupled with a 453-nm wafer-level LED: while during 3 h, the chromatic coordinates of the HWLED rapidly migrated towards the blue region of the CIE diagram, the coordinates remained unchanged if the film was deposited on a microscope glass slide and the latter was placed onto the blue LED from the uncoated glass side (cf. Figure 8.22). Air—instead of a solid material—can be better used as a spacer to reduce the thermal effect on the organic layer: in fact, it should be considered that air has a thermal conductivity as low as 0.2 W m^{-1} K^{-1} (in glass, used in microscope slides, thermal conductivity is 0.9 W m^{-1} K^{-1}). Bio-HWLEDs using curcuminoids as wavelength converters exhibit a reduction in the spreading of chromatic coordinates during continuous working time if the LED source and the bio-layer are equally spaced by air, instead glass [99].

An intriguing way to measure junction temperature of the pumping LED is to acquire the *I-V* characteristic of the LEDs and compare it with the *I-V* characteristics of the same LED plotted after heating the LED wafer at different temperatures. Caruso *et al.* verified that the temperature of a blue wafer-level LED used to irradiate the above reported PMMA-embedded Lumogen®-coated glass slide was around 25°C (and well low 45°C) after 1 h of continuous operation at 20 mA [153]. Consequently, this experiment confirms that low-power HWLEDs do not seem to be affected by thermal degradation due to the Joule effect.

A more insidious "enemy", common for both low- and high-power HWLEDs, is the photo-instability of the organic dyes, which in turn is responsible for the bleaching of the down converter coating. Photobleaching is particularly sneaky for white LEDs because during the irradiation of the down converter layer, the film becomes more transparent, then for an external observer the intensity of the blue light source increases while the fluorescence of the converter decreases. This means that both the chromatic coordinates and the CCT change over time, making the HWLEDs unusable, well before that thermal or chemical degradation intervenes.

As a general consideration, the photostability of an organic compound exposed to a generic radiation depends on the energy of the radiation itself. For HWLEDs, this means that blue pumping sources would allow achieving higher photo-durability of the organic compound than using an UV LED (which, in general, leads to slightly higher values of CRI). However, the decay of luminous flux is slower using blue light but it is not zero [57].

Practically, all the HWLEDs show photobleaching. Probably, from this point of view, the best performances are reported by CDs. Kwon *et al.* reported no changes in the emission spectrum of a CD-based HWLED over 12 h of continuous operation [144]; Mao *et al.* measured CD photoluminescence for 30 h under high power UV radiation (~1 W at 365 nm) and under an oxidic or reductive environment, concluding that no significantly variations in the PL intensity could be appreciated [146]. However, stability should be measured in a longer lapse of time to be significative.

FP-based Bio-HWLEDs deserve a separate discussion. Unlike other HWLEDs, FP-based Bio-HWLEDs are more responsive to thermal stability due to the FP deactivation process at high driving currents occurring in such LEDs. Thermal deactivation already arises when the FP layer reaches temperatures of around 60°C, corresponding to the melting temperature of the host matrix, due to the continuous excitation of the FPs [105]. Unfortunately, it is necessary to use a special polymeric matrix, merging a branched and a linear polymer (like trimethylolpropane ethoxylate and polyethylene oxide) since hydrogels to embed FPs are not appropriate due to their poor properties in terms of mechanical stability and strength to high temperatures and moisture [102]. The melting process of the polymer fades the interactions between the FPs and the host matrix and produces several irreversible modifications of the charge and the structure of the luminophores. At high currents (200 mA), the stability of the Bio-HWLED is less than 1 min. In order to increase the thermal stability, one can space as more as possible the pumping LED by the FP converter, as previously reported for other organic compounds. For instance, if the FP layer is directly placed on the LED, a temperature of ~60°C is reached in 10 min, while with a remote configuration the maximum temperature achieved depends on the distance: 40°C for 0.5 cm, 33°C for 1 cm, and 27°C for 2 cm. In the last case, the stability of the Bio-HWLED becomes more than 300 h at the same working conditions. Adopting a remote chip-coating arrangement, only photoinduced deactivation can degrade the FP-based Bio-HWLEDs. Unfortunately, as in many biological molecules, photostability here is difficult to control. The deactivation mechanism is grounded on a pair of slow and reversible partial dehydration processes, followed by a rapid and irreversible deactivation of the highly emissive ionic form, which remains in the non-emissive neutral form. It is worthwhile to point out that in order to stabilize the ionic form of the FP luminophores, the use of buffer solutions, such as a phosphate buffer (which was reported to have a beneficial impact on the thermal stability on green FPs [154]), did not prove to have a real effect on the stability in FP-based Bio-HWLEDs [105].

With a view to improving the photobleaching of the organic converters, several methods have been proposed, although none of them turned out to be resolutive. Most of these methods start from the consideration that photoinduced degradation involves oxygen, consequently to reduce (or delay) the photodegradation, one should remove oxygen atoms from the host matrix and prevent atmospheric oxygen to diffuse inside. While in the dark oxidation phenomena progress slowly, they are activated by light and accelerated by many orders of magnitude in the reactions with singlet oxygen. Thus, one way to reduce photobleaching is to quench singlet oxygen. This can be carried out by using specific singlet oxygen scavengers, like tetraphenylcyclopentadienone (TPC), or quenchers, like 1,4-diazobicyclo[2.2.2]-octane (DABCO) [155, 156]. Furthermore, since the singlet oxygen is similarly quenched by the hydroxyl groups, a good practice is to choose a host matrix with high concentrations of hydroxyl groups (e.g., cyclodextrin): for instance, it was demonstrated that adding β-cyclodextrin to an aqueous solution of 7-amino-4-methylcoumarin (C120) significantly improves the photochemical stability of C120 [157].

In order to avoid the migration of the chromatic coordinates in the CIE diagram due to the photobleaching, it was also proposed to use "thick" converters that completely absorb the light coming from the primary source and generate white lights only by additive color mixing of the light produced by the organic chromophores. Clearly, it is not possible to use a blue light as a primary source since it is not transmitted by the converter; therefore, it is necessary to use a LED emitting higher energy [33].

Other methods to reduce the photobleaching and to enhance the overall stability are not focused on the organic chromophore features, but on the characteristics on the polymer matrix. To comprehend whether the LED source radiation interacts with the matrix, a simple experiment can be a case in point: exposing a pure PMMA film (preparing without any dye) to the 450-nm light of a blue LED for 100 h (at 20 mA), a decreasing in the absorbance spectrum (~3%) occurs at around 450 nm. Fourier-transform infrared (FT-IR) spectroscopic analyses show that the exposition to blue light causes the PMMA to suffer from stretching of COO– of ester group, C–O and C–H, respectively [153]. The modification of the absorbance and the chemical structure of the polymer suggests that the photobleaching phenomena are strictly linked to the features of the host matrix. Hence, the stability of the converter can be improved by trying to reduce the influence of the pumping light on the optical properties of the polymer matrix. Three methods were proved to enhance the photostability of an HWLED based on PMMA-embedded Lumogen®: i) UV curing treatment of the matrix (in solution form); ii) using a polymer for the matrix with a higher average molecular weight; iii) using a higher polymer concentration in the prepared solution. Concerning the first method, after curing a PMMA solution (molecular weight: 350,000) in ethyl acetate with a UV lamp for 1 h and fabricating an HWLED with it, a stabilization effect is evident and the chromatic coordinates exhibit a shift $\Delta(x;y)$ equal to $-(0.03;0.08)$ after 210 min of continuous operation at 20 mA. The second method consists in using PMMA having a higher molecular weight (M_w = 996,000) at 11% (w/w) of concentration; here, the chromatic coordinates are more stable and the shift $\Delta(x;y)$ is $-(0.01;0.04)$ (after 210 min at 20 mA). Finally, in the third method, the concentration of the PMMA solution (M_w = 350,000) is increased as high as to 18% (w/w); after 210 min at 20 mA, the emission spectrum does not exhibit significant changes and the chromatic shift $\Delta(x;y)$ is even smaller as $-(0.01;0.02)$ [153]. While working with high concentrations and high molecular weights of polymer involves some difficulties (dyes incorporation, solution viscosity, etc.), these procedures open interesting perspectives for the improvement of the HWLED photostability.

8.7 Conclusions

Materials, technologies, characterizations, issues, and challenges concerning HWLEDs have been reviewed in this chapter. State of the art, details on process, open scenarios, and all the information reported in this chapter have been intended to be as exhaustive as possible. After an introduction aimed at defining the subject and contextualizing it in the academic and industrial research scene (Introduction), some vagueness in the terminology and other possible ambiguities on the topic are disclosed to the reader (Hybrid Devices and Other Ambiguities). The importance to use a host matrix to embed the organic luminophores is successively revealed (Necessity of a Host Matrix). Then, a long overview reports on all the different materials used in the last 20 years to fabricate HWLEDs; reported details concerns not only the chemistry of the materials but also a set of complementary information such as optical properties, color rendering, device lifetime, etc. (Materials for Hybrid LEDs). Further considerations on the color availability, mixing, and rendering (Color Tuning and Rendering), as well as thermal and photoinduced stability of the fabricated devices (Stability) complete the chapter and conclude it ("Conclusions").

References

1. Zissis, G. and Bertoldi, P., Status of LED-Lighting world market in 2017, Joint Research Centre Technical Report, European Commission, Ispra, Italy, 2018.
2. Pattison, M. *et al.*, *2018 Solid-State Lighting R&D Opportunities*, DOE/EE-1907, U.S. Department of Energy (DoE), Office of Energy Efficiency & Renewable Energy, Southwest Washington, D.C. Jan. 2019.
3. Schubert, E.F., *Light Emitting Diodes*, 2nd Edition, Cambridge University Press, New York, US, 2006.

4. Cao, X.A., Li, S., Li, X.M., Liu, L.Y., Hybrid White Light-Emitting Diodes Utilizing Radiative or Nonradiative Energy Transfer for Wavelength Conversion. *IEEE Trans. Electron Devices*, 65, 11, 4891–4896, Nov. 2018.
5. Fresta, E., Fernández-Luna, V., Coto, P.B., Costa, R.D., Merging Biology and Solid-State Lighting: Recent Advances in Light-Emitting Diodes Based on Biological Materials. *Adv. Funct. Mater.*, 28, 24, 1707011, Jun. 2018.
6. Massari, S. and Ruberti, M., Rare earth elements as critical raw materials: Focus on international markets and future strategies. *Resour. Policy*, 38, 1, 36–43, Mar. 2013.
7. Mosca, M. *et al.*, Warm white LED light by frequency down-conversion of mixed yellow and red Lumogen, in: *Presented at the SPIE Microtechnologies*, Grenoble, France, p. 87670L, 2013.
8. Krotkus, S. *et al.*, Towards Deep-Blue Materials with Efficient Triplet Harvesting. *SID Symp. Dig. Tech.*, 49, 1, 239–242, May 2018.
9. Deepak Kumar, N., Joshi, M.P., Friend, C.S., Prasad, P.N., Burzynski, R., Organic–inorganic heterojunction light emitting diodes based on poly(p-phenylene vinylene)/cadmium sulfide thin films. *Appl. Phys. Lett.*, 71, 10, 1388–1390, Sep. 1997.
10. L.-S. Hung and J.K. Madathil, Organic Electroluminescent Device with Inorganic Electron Transporting Layer, 6,096,442, 30-May-2000.
11. Kayunkid, N., Keawprajak, A., Jaruwanawat, A., Nukeaw, J., Blue Emission Mechanism of NPB/ZnSe Hybrid Structure, in: *2007 2nd IEEE International Conference on Nano/Micro Engineered and Molecular Systems*, pp. 643–646, 2007.
12. Vasilopoulou, M. *et al.*, Flexible organic light emitting diodes (OLEDs) based on a blue emitting polyfluorene. *Phys. Status Solidi C*, 5, 12, 3658–3662, 2008.
13. Jiang, D.F., Yin, Y., Ran, G.Z., Zhang, B., Hybrid organic light-emitting device with inorganic mobility-tunable hole transport layer of silicon-rich silicon oxide. *J. Opt. Soc. Am. B*, 28, 1, 89–92, Jan. 2011.
14. Sessolo, M. and Bolink, H.J., Hybrid Organic–Inorganic Light-Emitting Diodes. *Adv. Mater.*, 23, 16, 1829–1845, 2011.
15. Chou, P.-T., Chi, Y., Chung, M.-W., Lin, C.-C., Harvesting luminescence via harnessing the photophysical properties of transition metal complexes. *Coord. Chem. Rev.*, 255, 21, 2653–2665, Nov. 2011.
16. Fu, H., Cheng, Y.-M., Chou, P.-T., Chi, Y., Feeling blue? Blue phosphors for OLEDs. *Mater. Today*, 14, 10, 472–479, Oct. 2011.
17. Giebink, N.C. *et al.*, Intrinsic luminance loss in phosphorescent small-molecule organic light emitting devices due to bimolecular annihilation reactions. *J. Appl. Phys.*, 103, 4, 044509, Feb. 2008.
18. Khanna, V.K., *Fundamentals of Solid-State Lighting. LEDs, OLEDs, and Their Applications in Illumination and Displays*, CRC Press, Taylor & Francis Group, Boca Raton, FL, US, 2014.
19. Khanh, T.Q., Bodrogi, P., Vinh, Q.T., Winkler, H., *LED Lighting. Technology and Perception*, Wiley-VCH, Weinheim, Germany, 2015.

20. Rea, M.S. and Freyssinier, J.P., Color rendering: Beyond pride and prejudice. *Color Res. Appl.*, 35, 6, 401–409, Dec. 2010.
21. Gennusa, M.L., Macaluso, R., Mosca, M., Scaccianoce, G., Massaro, F., Cirrincione, L., An experimental study on relationship between LED lamp characteristics and non image-forming, in: *2017 IEEE International Conference on Environment and Electrical Engineering and 2017 IEEE Industrial and Commercial Power Systems Europe (EEEIC/I&CPS Europe)*, Milan, Italy, pp. 1–6, 2017.
22. Cirrincione, L., Macaluso, R., Mosca, M., Scaccianoce, G., Costanzo, S., Study of Influence of the LED Technologies on Visual and Subjective/Individual Aspects, in: *2018 IEEE International Conference on Environment and Electrical Engineering and 2018 IEEE Industrial and Commercial Power Systems Europe (EEEIC/I&CPS Europe)*, Palermo, Italy, pp. 1–5, 2018.
23. Aksoy, E., Demir, N., Varlikli, C., White LED light production using dibromoperylene derivatives in down conversion of energy. *Can. J. Phys.*, 96, 7, 734–739, Feb. 2018.
24. Karapire, C., Timur, C., İçli, S., A comparative study of the photophysical properties of perylenediimides in liquid phase, PVC and sol-gel host matrices. *Dyes Pigm.*, 56, 2, 135–143, Feb. 2003.
25. Boonsin, R., Chadeyron, G., Roblin, J.-P., Boyer, D., Mahiou, R., Silica encapsulated fluorescein as a hybrid dye for blue-LED based lighting devices. *J. Mater. Chem. C*, 4, 27, 6562–6569, 2016.
26. El-Daly, S.A. and Fayed, T.A., Photochemistry of N, Ń-ditridecyl-3,4:9,10-perylenetetracarboxylic diimide in chloromethane solvents. *J. Photochem. Photobiol. A Chem.*, 137, 1, 15–19, Oct. 2000.
27. Bruckbauer, J. *et al.*, Colour tuning in white hybrid inorganic/organic light-emitting diodes. *J. Phys. D: Appl. Phys.*, 49, 40, 405103, Oct. 2016.
28. Biju, V., Yamauchi, M., Ishikawa, M., Distribution of single molecules in polymer thin films. *J. Photochem. Photobiol. A Chem.*, 140, 3, 237–241, May 2001.
29. Schlotter, P., Schmidt, R., Schneider, J., Luminescence conversion of blue light emitting diodes. *Appl. Phys. A: Mater. Sci. Process.*, 64, 4, 417–418, Apr. 1997.
30. Hide, F., Kozodoy, P., DenBaars, S.P., Heeger, A.J., White light from InGaN/conjugated polymer hybrid light-emitting diodes. *Appl. Phys. Lett.*, 70, 20, 2664–2666, May 1997.
31. Zhang, C. and Heeger, A.J., Gallium nitride/conjugated polymer hybrid light emitting diodes: Performance and lifetime. *J. Appl. Phys.*, 84, 3, 1579–1582, Jul. 1998.
32. Nakamura, S., Present performance of InGaN-based blue/green/yellow LEDs, in: *Proc. SPIE*, San Jose, CA, US, vol. 3002, p. 26, 1997.
33. Zak, P.P., Lapina, V.A., Pavich, T.A., Trofimov, A.V., Trofimova, N.N., Tsaplev, Y.B., Luminescent materials for modern light sources. *Russ. Chem. Rev.*, 86, 9, 831–844, Sep. 2017.

34. Lee, S.K. *et al.*, The Fabrication and Characterization of Single-Component Polymeric White-Light-Emitting Diodes. *Adv. Funct. Mater.*, 15, 10, 1647–1655, Oct. 2005.
35. Belton, C.R. *et al.*, New light from hybrid inorganic–organic emitters. *J. Phys. D: Appl. Phys.*, 41, 9, 094006, May 2008.
36. Huyal, I.O., Ozel, T., Koldemir, U., Nizamoglu, S., Tuncel, D., Demir, H.V., White emitting polyfluorene functionalized with azide hybridized on near-UV light emitting diode for high color rendering index. *Opt. Express*, 16, 2, 1115, 2008.
37. Chen, K.-J. *et al.*, Efficient hybrid white light-emitting diodes by organic-inorganic materials at different CCT from 3000K to 9000K. *Opt. Express*, 23, 7, A204, Apr. 2015.
38. Paik, K.L., Baek, N.S., Kim, H.K., Lee, J.-H., Lee, Y., White Light-Emitting Diodes from Novel Silicon-Based Copolymers Containing Both Electron-Transport Oxadiazole and Hole-Transport Carbazole Moieties in the Main Chain. *Macromolecules*, 35, 18, 6782–6791, Aug. 2002.
39. Carlos, L.D., Sá Ferreira, R.A., Pereira, R.N., Assunção, M., de Zea Bermudez, V., White-Light Emission of Amine-Functionalized Organic/Inorganic Hybrids: Emitting Centers and Recombination Mechanisms. *J. Phys. Chem. B*, 108, 39, 14924–14932, Sep. 2004.
40. Luponosov, Y.N. *et al.*, Nanostructured Organosilicon Luminophores for Effective Light Conversion in Organic Light Emitting Diodes. *Org. Photonics Photovolt.*, 3, 1, 148–155 Jan. 2015.
41. Bernius, M.T., Inbasekaran, M., O'Brien, J., Wu, W., Progress with Light-Emitting Polymers. *Adv. Mater.*, 12, 23, 1737–1750, Dec. 2000.
42. Wang, Y.Z., Sun, R.G., Meghdadi, F., Leising, G., Epstein, A.J., Multicolor multilayer light-emitting devices based on pyridine-containing conjugated polymers and para-sexiphenyl oligomer. *Appl. Phys. Lett.*, 74, 24, 3613–3615, Jun. 1999.
43. Agranovich, V.M., Basko, D.M., La Rocca, G.C., Bassani, F., New concept for organic LEDs: Non-radiative electronic energy transfer from semiconductor quantum well to organic overlayer. *Synth. Met.*, 116, 1–3, 349–351, Jan. 2001.
44. Zhuang, Z. *et al.*, High Color Rendering Index Hybrid III-Nitride/Nanocrystals White Light-Emitting Diodes. *Adv. Funct. Mater.*, 26, 1, 36–43, Jan. 2016.
45. Huang Chen, S.-W. *et al.*, Full-color monolithic hybrid quantum dot nanoring micro light-emitting diodes with improved efficiency using atomic layer deposition and nonradiative resonant energy transfer. *Photonics Res.*, 7, 4, 416, Apr. 2019.
46. Schlotter, P. *et al.*, Fabrication and characterization of GaN/InGaN/AlGaN double heterostructure LEDs and their application in luminescence conversion LEDs. *Mater. Sci. Eng.: B*, 59, 1–3, 390–394, May 1999.

47. Ahmed, H.A., Walshe, J., Kennedy, M., Confrey, T., Doran, J., Sarah., J., McCormack, Enhancement in solar cell efficiency by luminescent down-shifting layers. *Adv. Energy Res.*, 1, 2, 117–126, Jun. 2013.
48. Caruso, F., Mosca, M., Macaluso, R., Feltin, E., Calì, C., Generation of white LED light by frequency downconversion using perylene-based dye. *Electron. Lett.*, 48, 22, 1417, 2012.
49. Mosca, M., Caruso, F., Zambito, L., Macaluso, R., Calì, C., Hybrid LEDs Pave Way to New Lighting Applications. *Photonics Spectra*, 47, 4, 60–64, 2013.
50. Kozma, E., Mróz, W., Galeotti, F., A polystyrene bearing perylene diimide pendants with enhanced solid state emission for white hybrid light-emitting diodes. *Dyes Pigm.*, 114, 138–143, Mar. 2015.
51. He, J., Yang, S., Zheng, K., Zhang, Y., Song, J., Qu, J., One-pot synthesis of dispersible thermally stable organic downconversion materials under DBU catalyzation for high performance hybrid-LED lamps. *Green Chem.*, 20, 15, 3557–3565, Jul. 2018.
52. Galeotti, F., Mróz, W., Catellani, M., Kutrzeba-Kotowska, B., Kozma, E., Tailorable perylene-loaded fluorescent nanostructures: A multifaceted approach enabling their application in white hybrid LEDs. *J. Mater. Chem. C*, 4, 23, 5407–5415, 2016.
53. Guner, T., Aksoy, E., Demir, M.M., Varlikli, C., Perylene-embedded electrospun PS fibers for white light generation. *Dyes Pigm.*, 160, 501–508, Jan. 2019.
54. Uthirakumar, P., Suh, E.-K., Hong, C.-H., Lee, Y.-S., Synthesis and characterization of polyesters containing fluorescein dye units. *Polymer*, 46, 13, 4640–4646, Jun. 2005.
55. Uthirakumar, P., Hong, C.-H., Suh, E.-K., Lee, Y.-S., Yellow light-emitting polymer bearing fluorescein dye units: Photophysical property and application as luminescence converter of a hybrid LED. *React. Funct. Polym.*, 67, 4, 341–347, Apr. 2007.
56. Armstrong, N.R., Phthalocyanines and porphyrins as materials. *J. Porphyr. Phthalocya.*, 4, 4, 414–417, 2000.
57. Xiang, H.-F., Yu, S.-C., Che, C.-M., Lai, P.T., Efficient white and red light emission from GaN/*tris*-(8-hydroxyquinolato) aluminum/platinum(II) *meso*-tetrakis(pentafluorophenyl) porphyrin hybrid light-emitting diodes. *Appl. Phys. Lett.*, 83, 8, 1518–1520, Aug. 2003.
58. Kim, H.-J., Jin, J.-Y., Lee, Y.-S., Lee, S.-H., Hong, C.-H., An efficient luminescence conversion LED for white light emission, fabricated using a commercial InGaN LED and a 1,8-naphthalimide derivative. *Chem. Phys. Lett.*, 431, 4–6, 341–345, Nov. 2006.
59. Jin, J.-Y., Kim, Y.-M., Lee, S.-H., Lee, Y.-S., Synthesis of an acrylic copolymer bearing fluorescent dye pendants and characterization as a luminescence conversion material in fabrication of a luminescence conversion light-emitting diode. *Synth. Met.*, 159, 17–18, 1804–1808, Sep. 2009.
60. Bakhshiev, N.G., Gularyan, S.K., Dobretsov, G.E., Yu. Kirillova, A., Sarkisov, O.M., Yu. Svetlichny, V., Nature of the influence of a solvent on the

fluorescence quantum yield of solutions of the fluorescent probe 4-dimethylaminochalcone. *Opt. Spectrosc.*, 109, 6, 913–916, Dec. 2010.
61. Ermakov, O.N., Kaplunov, M.G., Efimov, O.N., Yakushchenko, I.K., Belov, M.Y., Budyka, M.F., Hybrid organic–inorganic light-emitting diodes. *Microelectron. Eng.*, 69, 208–212, 2003.
62. Zhang, L., Li, B., Lei, B., Hong, Z., Li, W., A triphenylamine derivative as an efficient organic light color-conversion material for white LEDs. *J. Lumin.*, 128, 1, 67–73, Jan. 2008.
63. Zainuri, D.A., Razak, I.A., Arshad, S., Molecular structure, DFT studies and Hirshfeld analysis of anthracenyl chalcone derivatives. *Acta Crystallogr. E Cryst. Commun.*, 74, 6, 780–785, Jun. 2018.
64. Treibs, A. and Kreuzer, F.-H., Difluorboryl-Komplexe von Di- und Tripyrrylmethenen. *Justus Liebigs Ann. Chem.*, 718, 1, 208–223, 1968.
65. Schmitt, A., Hinkeldey, B., Wild, M., Jung, G., Synthesis of the Core Compound of the BODIPY Dye Class: 4,4′-Difluoro-4-bora-(3a,4a)-diaza-s-indacene. *J. Fluoresc.*, 19, 4, 755–758, Jul. 2009.
66. Arroyo, I.J., Hu, R., Merino, G., Tang, B.Z., Peña-Cabrera, E., The Smallest and One of the Brightest. Efficient Preparation and Optical Description of the Parent Borondipyrromethene System. *J. Org. Chem.*, 74, 15, 5719–5722, Aug. 2009.
67. Tram, K., Yan, H., Jenkins, H.A., Vassiliev, S., Bruce, D., The synthesis and crystal structure of unsubstituted 4,4-difluoro-4-bora-3a,4a-diaza-s-indacene (BODIPY). *Dyes Pigm.*, 82, 3, 392–395, Sep. 2009.
68. Findlay, N.J. *et al.*, Linear oligofluorene-BODIPY structures for fluorescence applications. *J. Mater. Chem. C*, 1, 11, 2249, 2013.
69. Findlay, N.J. *et al.*, An Organic Down-Converting Material for White-Light Emission from Hybrid LEDs. *Adv. Mater.*, 26, 43, 7290–7294, Nov. 2014.
70. Taylor-Shaw, E. *et al.*, Cool to warm white light emission from hybrid inorganic/organic light-emitting diodes. *J. Mater. Chem. C*, 4, 48, 11499–11507, 2016.
71. Merkushev, D.A. *et al.*, BODIPY associates in organic matrices: Spectral properties, photostability and evaluation as OLED emitters. *Mater. Chem. Phys.*, 187, 104–111, Feb. 2017.
72. Cui, Y., Yu, J., Gao, J., Wang, Z., Qian, G., Synthesis and luminescence behavior of inorganic–organic hybrid materials covalently bound with pyran-containing dyes. *J. Sol-Gel Sci. Technol.*, 52, 3, 362–369, Dec. 2009.
73. Lei, G. *et al.*, White organic light emitting devices with hybrid emissive layers combining phosphorescence and fluorescence. *J. Phys. D: Appl. Phys.*, 41, 10, 105114, May 2008.
74. Xiao-Jun, L. *et al.*, TDDFT Investigation and Design for Fluorescent Molecules with Push-Pull Structures. *Acta Phys. –Chim. Sin.*, 28, 6, 1337–1346, 2012.
75. Nyalosaso, J.L. *et al.*, Towards rare-earth-free white light-emitting diode devices based on the combination of dicyanomethylene and pyranine

as organic dyes supported on zinc single-layered hydroxide. *Beilstein J. Nanotechnol.*, 10, 760–770, Mar. 2019.

76. Förster, T. and Kasper, K., Ein Konzentrationsumschlag der Fluoreszenz. *Z. Phys. Chem. (Muenchen, Ger.)*, 1, 275–277, 1954.
77. Luo, J. *et al.*, Aggregation-induced emission of 1-methyl-1,2,3,4,5-pentaphenylsilole. *Chem. Commun.*, 18, 1740–1741, 2001.
78. Chen, J. *et al.*, Synthesis, Light Emission, Nanoaggregation, and Restricted Intramolecular Rotation of 1,1-Substituted 2,3,4,5-Tetraphenylsiloles. *Chem. Mater.*, 15, 7, 1535–1546, Apr. 2003.
79. Mei, J., Hong, Y., Lam, J.W.Y., Qin, A., Tang, Y., Tang, B.Z., Aggregation-Induced Emission: The Whole Is More Brilliant than the Parts. *Adv. Mater.*, 26, 31, 5429–5479, Aug. 2014.
80. Liu, J. *et al.*, Aggregation-induced emission enhancement based on 11,11,12,12,-tetracyano-9,10-anthraquinodimethane. *Chem. Commun.*, 49, 12, 1199–1201, 2013.
81. Mei, J., Leung, N.L.C., Kwok, R.T.K., Lam, J.W.Y., Tang, B.Z., Aggregation-Induced Emission: Together We Shine, United We Soar! *Chem. Rev.*, 115, 21, 11718–11940, Nov. 2015.
82. Hong, Y., Lam, J.W.Y., Tang, B.Z., Aggregation-induced emission: Phenomenon, mechanism and applications. *Chem. Commun.*, 29, 4332–4353, 2009.
83. Li, Q. and Li, Z., The Strong Light-Emission Materials in the Aggregated State: What Happens from a Single Molecule to the Collective Group. *Adv. Sci.*, 4, 7, 1600484, Jul. 2017.
84. Mao, L.-C., Zhang, X.-Y., Wei, Y., Recent Advances and Progress for the Fabrication and Surface Modification of AIE-active Organic-inorganic Luminescent Composites. *Chin. J. Polym. Sci.*, 37, 4, 340–351, Apr. 2019.
85. Zhao, E. and Tang, B.Z., AIE Luminogens: A Family of New Materials with Multifaceted Functionalities, *Sigma-Aldrich.* [Online]. Available: https://www.sigmaaldrich.com/technical-documents/articles/biology/aie-luminogens.html.
86. Yuan, W.Z. *et al.*, Changing the Behavior of Chromophores from Aggregation-Caused Quenching to Aggregation-Induced Emission: Development of Highly Efficient Light Emitters in the Solid State. *Adv. Mater.*, 22, 19, 2159–2163, Mar. 2010.
87. Yue, Z., Cheung, Y.F., Choi, H.W., Zhao, Z., Tang, B.Z., Wong, K.S., White light emission from InGaN/organic molecule light-emitting diode, in: *2013 Conference on Lasers and Electro-Optics Pacific Rim (CLEOPR)*, Kyoto, pp. TuPH–6, 2013.
88. Yue, Z., Cheung, Y.F., Choi, H.W., Zhao, Z., Tang, B.Z., Wong, K.S., Hybrid GaN/Organic white light emitters with aggregation induced emission organic molecule. *Opt. Mater. Express*, 3, 11, 1906–1911, Nov. 2013.
89. Rangaswami, S. and Seshadri, T.R., A note on certain constitutional factors controlling visible fluorescence in compounds of the benzo-pyrone group. *Proc. Indian Acad. Sci. (Math. Sci.)*, 12, 4, 375–380, Oct. 1940.

90. Suzuki, H., Meyer, H., Hoshino, S., Haarer, D., Electroluminescence from multilayer organic light-emitting diodes using poly(methylphenylsilane) as hole transporting material. *J. Appl. Phys.*, 78, 4, 2684–2690, Aug. 1995.
91. Kim, S.W., Yoo, H.S., Kwon, B.-H., Kang, D.S., Suh, M., Jeon, D.Y., Organic wavelength-converting-film-based hybrid planar white light-emitting diodes. *J. Soc. Inf. Disp.*, 18, 12, 1104–1110, 2010.
92. Di Martino, D., Beverina, L., Sassi, M., Brovelli, S., Tubino, R., Meinardi, F., Straightforward fabrication of stable white LEDs by embedding of inorganic UV-LEDs into bulk polymerized polymethyl-methacrylate doped with organic dyes. *Sci. Rep.*, 4, 1, 4400, May 2015.
93. Luridiana, A. *et al.*, A facile strategy for new organic white LED hybrid devices: Design, features and engineering. *RSC Adv.*, 6, 26, 22111–22120, 2016.
94. Meredith, P., Bettinger, C.J., Irimia-Vladu, M., Mostert, A.B., Schwenn, P.E., Electronic and optoelectronic materials and devices inspired by nature. *Rep. Prog. Phys.*, 76, 3, 034501, Mar. 2013.
95. Sun, M. *et al.*, Towards efficient solid-state photoluminescence based on carbon-nanodots and starch composites. *Nanoscale*, 6, 21, 13076–13081, Sep. 2014.
96. Hendler, N., Belgorodsky, B., Mentovich, E.D., Gozin, M., Richter, S., Efficient Separation of Dyes by Mucin: Toward Bioinspired White-Luminescent Devices. *Adv. Mater.*, 23, 37, 4261–4264, Oct. 2011.
97. Benson, K., Ghimire, A., Pattammattel, A., Kumar, C.V., Protein Biophosphors: Biodegradable, Multifunctional, Protein-Based Hydrogel for White Emission, Sensing, and pH Detection. *Adv. Funct. Mater.*, 27, 43, 1702955, Nov. 2017.
98. Reddy, M.S.P. and Park, C., Bright luminescence from pure DNA-curcumin-based phosphors for bio hybrid light-emitting diodes. *Sci. Rep.*, 6, 1, 32306, Aug. 2016.
99. Al Shafouri, M., Ahmed, N.M., Hassan, Z., Almessiere, M.A., Jumaah, M., Optical and structural properties of curcuminoids extracted from Curcuma longa L. for hybrid white light diode. *Eur. Phys. J. Appl. Phys.*, 84, 1, 10501, Oct. 2018.
100. Reddy Mallem, S.P., Im, K.-S., Lee, J.-H., Park, C., Bathalavaram, P., Trichromophore-doped cassava-based biopolymer as low-cost and eco-friendly luminous material for bio hybrid white-light-emitting diodes by dual-FRET process. *Opt. Mater.*, 95, 109270, Sep. 2019.
101. de Oliveira, N.R., Fornaciari, B., Mali, S., Carvalho, G.M., Acetylated Starch-Based Nanoparticles: Synthesis, Characterization, and Studies of Interaction With Antioxidants. *Starch - Stärke*, 70, 3–4, 1700170, Mar. 2018.
102. Weber, M.D., Niklaus, L., Pröschel, M., Coto, P.B., Sonnewald, U., Costa, R.D., Bioinspired Hybrid White Light-Emitting Diodes. *Adv. Mater.*, 27, 37, 5493–5498, Oct. 2015.
103. Niklaus, L. *et al.*, Micropatterned Down-Converting Coating for White Bio-Hybrid Light-Emitting Diodes. *Adv. Funct. Mater.*, 27, 1, 1601792, Jan. 2017.

104. Press, D.A., Melikov, R., Conkar, D., Firat-Karalar, E.N., Nizamoglu, S., Fluorescent protein integrated white LEDs for displays. *Nanotechnology*, 27, 45, 45LT01, Nov. 2016.
105. Fernández-Luna, V., Sánchez-de Alcázar, D., Fernández-Blázquez, J.P., Cortajarena, A.L., Coto, P.B., Costa, R.D., Deciphering Limitations to Meet Highly Stable Bio-Hybrid Light-Emitting Diodes. *Adv. Funct. Mater.*, 29, 42, 1904356, Oct. 2019.
106. Fernández-Luna, V., Coto, P.B., Costa, R.D., When Fluorescent Proteins Meet White Light-Emitting Diodes. *Angew. Chem. Int. Ed.*, 57, 29, 8826–8836, Jul. 2018.
107. Cui, Y., Li, B., He, H., Zhou, W., Chen, B., Qian, G., Metal–Organic Frameworks as Platforms for Functional Materials. *Acc. Chem. Res.*, 49, 483–493, 2016.
108. Borboudakis, G., Stergiannakos, T., Frysali, M., Klontzas, E., Tsamardinos, I., Froudakis, G.E., Chemically intuited, large-scale screening of MOFs by machine learning techniques. *Npj Comput. Mater.*, 3, 1, 40, Dec. 2017.
109. Gui, B., Yu, N., Meng, Y., Hu, F., Wang, C., Immobilization of AIEgens into metal-organic frameworks: Ligand design, emission behavior, and applications. *J. Polym. Sci. Part A: Polym. Chem.*, 55, 11, 1809–1817, Jun. 2017.
110. Wang, M.-S. *et al.*, A White-Light-Emitting Borate-Based Inorganic–Organic Hybrid Open Framework. *Angew. Chem. Int. Ed.*, 46, 21, 3909–3911, May 2007.
111. Wang, M.-S. *et al.*, A Direct White-Light-Emitting Metal–Organic Framework with Tunable Yellow-to-White Photoluminescence by Variation of Excitation Light. *J. Am. Chem. Soc.*, 131, 38, 13572–13573, Sep. 2009.
112. Hu, X. *et al.*, Two-Dimensional Metal-Organic Layers as a Bright and Processable Phosphor for Fast White-Light Communication. *Chem. Eur. J.*, 23, 35, 8390–8394, Jun. 2017.
113. Wang, Z. *et al.*, Warm-White-Light-Emitting Diode Based on a Dye-Loaded Metal–Organic Framework for Fast White-Light Communication. *ACS Appl. Mater. Interfaces*, 9, 40, 35253–35259, Oct. 2017.
114. Chen, W. *et al.*, Color-Tunable and High-Efficiency Dye-Encapsulated Metal–Organic Framework Composites Used for Smart White-Light-Emitting Diodes. *ACS Appl. Mater. Interfaces*, 10, 22, 18910–18917, Jun. 2018.
115. Cui, Y., Song, T., Yu, J., Yang, Y., Wang, Z., Qian, G., Dye Encapsulated Metal-Organic Framework for Warm-White LED with High Color-Rendering Index. *Adv. Funct. Mater.*, 25, 30, 4796–4802, Aug. 2015.
116. Zhao, S.-S., Zhang, H., Wang, L., Chen, L., Xie, Z., Facile preparation of a tetraphenylethylene-doped metal–organic framework for white light-emitting diodes. *J. Mater. Chem. C*, 6, 43, 11701–11706, 2018.
117. Peng, C., Song, X., Yin, J., Zhang, G., Fei, H., Intrinsic White-Light-Emitting Metal–Organic Frameworks with Structurally Deformable Secondary Building Units. *Angew. Chem. Int. Ed.*, 58, 23, 7818–7822, 2019.

118. Qu, M., Li, H., Zhao, Y., Zhang, X.-M., Single-Component Color-Tunable Gd(pic)$_3$: Eu^{3+} Phosphor Based on a Metal–Organic Framework for Near-UV White-Light-Emitting Diodes. *ACS Omega*, 4, 2, 3593–3600, Feb. 2019.
119. Wang, Z. *et al.*, White-Light Emission from Dual-Way Photon Energy Conversion in a Dye-Encapsulated Metal-Organic Framework. *Angew. Chem. Int. Ed.*, 58, 29, 9752–9757, Jul. 2019.
120. Tang, Y., Xia, T., Song, T., Cui, Y., Yang, Y., Qian, G., Efficient Energy Transfer within Dyes Encapsulated Metal-Organic Frameworks to Achieve High Performance White Light-Emitting Diodes. *Adv. Opt. Mater.*, 6, 24, 1800968, Dec. 2018.
121. Börgardts, M. and Müller, T.J.J., Energy down converting organic fluorophore functionalized mesoporous silica hybrids for monolith-coated light emitting diodes. *Beilstein J. Org. Chem.*, 13, 768–778, Apr. 2017.
122. Wen, Y. *et al.*, Introduction of Red-Green-Blue Fluorescent Dyes into a Metal–Organic Framework for Tunable White Light Emission. *Adv. Mater.*, 29, 37, 1700778, 2017.
123. Zhao, Y.-W., Zhang, F.-Q., Zhang, X.-M., Single Component Lanthanide Hybrids Based on Metal–Organic Framework for Near-Ultraviolet White Light LED. *ACS Appl. Mater. Interfaces*, 8, 36, 24123–24130, Sep. 2016.
124. Hu, Z. *et al.*, Achieving exceptionally high luminescence quantum efficiency by immobilizing an AIE molecular chromophore into a metal-organic framework. *Chem. Commun.*, 51, 15, 3045–3048, 2015.
125. Zhou, Y. and Yan, B., Lanthanides post-functionalized nanocrystalline metal–organic frameworks for tunable white-light emission and orthogonal multi-Readout thermometry. *Nanoscale*, 7, 9, 4063–4069, 2015.
126. Gong, Q. *et al.*, Solution Processable MOF Yellow Phosphor with Exceptionally High Quantum Efficiency. *J. Am. Chem. Soc.*, 136, 48, 16724–16727, Dec. 2014.
127. Lu, Y. and Yan, B., Lanthanide organic–inorganic hybrids based on functionalized metal–organic frameworks (MOFs) for a near-UV white LED. *Chem. Commun.*, 50, 97, 15443–15446, Nov. 2014.
128. Sun, C.-Y. *et al.*, Efficient and tunable white-light emission of metal–organic frameworks by iridium-complex encapsulation. *Nat. Commun.*, 4, 1, 2717, Dec. 2013.
129. He, J., Zeller, M., Hunter, A.D., Xu, Z., White Light Emission and Second Harmonic Generation from Secondary Group Participation (SGP) in a Coordination Network. *J. Am. Chem. Soc.*, 134, 3, 1553–1559, Jan. 2012.
130. Park, J.H., Kim, J.Y., Chin, B.D., Kim, Y.C., Kim, J.K., Park, O.O., White emission from polymer/quantum dot ternary nanocomposites by incomplete energy transfer. *Nanotechnology*, 15, 9, 1217–1220, Jul. 2004.
131. Bowers, M.J., McBride, J.R., Rosenthal, S.J., White-Light Emission from Magic-Sized Cadmium Selenide Nanocrystals. *J. Am. Chem. Soc.*, 127, 44, 15378–15379, Nov. 2005.

132. Jang, E.-P., Song, W.-S., Lee, K.-H., Yang, H., Preparation of a photodegradation-resistant quantum dot–polymer composite plate for use in the fabrication of a high-stability white-light-emitting diode. *Nanotechnology*, 24, 4, 045607, Jan. 2013.
133. Shen, C., Li, K., Hou, Q., Feng, H., Dong, X., White LED Based on YAG: Ce,Gd Phosphor and CdSe–ZnS Core/Shell Quantum Dots. *IEEE Photon. Technol. Lett.*, 22, 12, 884–886, Jun. 2010.
134. Xuan, Y., Pan, D., Zhao, N., Ji, X., Ma, D., White electroluminescence from a poly(N-vinylcarbazole) layer doped with CdSe/CdS core–shell quantum dots. *Nanotechnology*, 17, 19, 4966–4969, Sep. 2006.
135. Wang, R., Zhang, J., Xu, X., Wang, Y., Zhou, L., Li, B., White LED with high color rendering index based on Ca8Mg(SiO4)4Cl2:Eu2+ and ZnCdTe/CdSe quantum dot hybrid phosphor. *Mater. Lett.*, 84, 24–26, Oct. 2012.
136. Kim, K., Jeong, S., Woo, J.Y., Han, C.-S., Successive and large-scale synthesis of InP/ZnS quantum dots in a hybrid reactor and their application to white LEDs. *Nanotechnology*, 23, 6, 065602, Jan. 2012.
137. Ryckaert, J. *et al.*, Selecting the optimal synthesis parameters of InP/$Cd_xZn_{1-x}Se$ quantum dots for a hybrid remote phosphor white LED for general lighting applications. *Opt. Express*, 25, 24, A1009–A1022, Nov. 2017.
138. Tiwari, N., Nishad, K.K., Pandey, R.K., Hybrid Heterostructured LEDs Based on Superstrate Architecture of ZnO and ZnS Quantum Dots. *IEEE J. Quantum Electron.*, 55, 2, 1–7, Apr. 2019.
139. Ilaiyaraja, P., Mocherla, P.S.V., Srinivasan, T.K., Sudakar, C., Synthesis of Cu-Deficient and Zn-Graded Cu–In–Zn–S Quantum Dots and Hybrid Inorganic–Organic Nanophosphor Composite for White Light Emission. *ACS Appl. Mater. Interfaces*, 8, 19, 12456–12465, May 2016.
140. Baker, S.N. and Baker, G.A., Luminescent Carbon Nanodots: Emergent Nanolights. *Angew. Chem. Int. Ed.*, 49, 38, 6726–6744, 2010.
141. Guo, X., Wang, C.-F., Yu, Z.-Y., Chen, L., Chen, S., Facile access to versatile fluorescent carbon dots toward light-emitting diodes. *Chem. Commun.*, 48, 21, 2692–2694, Feb. 2012.
142. Tang, L. *et al.*, Deep Ultraviolet Photoluminescence of Water-Soluble Self-Passivated Graphene Quantum Dots. *ACS Nano*, 6, 6, 5102–5110, Jun. 2012.
143. Luk, C.M., Tang, L.B., Zhang, W.F., Yu, S.F., Teng, K.S., Lau, S.P., An efficient and stable fluorescent graphene quantum dot–agar composite as a converting material in white light emitting diodes. *J. Mater. Chem.*, 22, 42, 22378–22381, Oct. 2012.
144. Kwon, W., Do, S., Lee, J., Hwang, S., Kim, J.K., Rhee, S.-W., Freestanding Luminescent Films of Nitrogen-Rich Carbon Nanodots toward Large-Scale Phosphor-Based White-Light-Emitting Devices. *Chem. Mater.*, 25, 9, 1893–1899, May 2013.

145. Chen, J., Liu, W., Mao, L.-H., Yin, Y.-J., Wang, C.-F., Chen, S., Synthesis of silica-based carbon dot/nanocrystal hybrids toward white LEDs. *J. Mater. Sci.*, 49, 21, 7391–7398, Nov. 2014.
146. Mao, L.-H., Tang, W.-Q., Deng, Z.-Y., Liu, S.-S., Wang, C.-F., Chen, S., Facile Access to White Fluorescent Carbon Dots toward Light-Emitting Devices. *Ind. Eng. Chem. Res.*, 53, 15, 6417–6425, Apr. 2014.
147. Lin, S. *et al.*, Easy synthesis of silver nanoparticles-orange emissive carbon dots hybrids exhibiting enhanced fluorescence for white light emitting diodes. *J. Alloys Compd.*, 700, 75–82, Apr. 2017.
148. Lee, H., Dellatore, S.M., Miller, W.M., Messersmith, P.B., Mussel-Inspired Surface Chemistry for Multifunctional Coatings. *Science*, 318, 5849, 426–430, Oct. 2007.
149. Li, L., Zhao, F., Zhang, T., Lü, C., A facile method to prepare polymer functionalized carbon dots inspired by the mussel chemistry for LED application. *Dyes Pigm.*, 162, 845–854, Mar. 2019.
150. Lin, C.-H. *et al.*, Ultrawide Color Gamut Perovskite and CdSe/ZnS Quantum-Dots-Based White Light-Emitting Diode with High Luminous Efficiency. *Nanomaterials*, 9, 9, 1314, Sep. 2019.
151. Chen, H., Zhu, R., Käläntär, K., Wu, S.-T., Quantum Dot-Enhanced LCDs with Wide Color Gamut and Broad Angular Luminance Distribution. *SID Symp. Dig. Tech.*, 47, 1, 1413–1416, 2016.
152. Scientific Principles: Polymers - Department of Materials Science and Engineering - University of Illinois Urbana-Champaign, *MAST, Materials Science and Technology.* [Online]. Available: http://matse1.matse.illinois.edu/polymers/prin.html.
153. Caruso, F. *et al.*, Frequency-Downconversion Stability of PMMA Coatings in Hybrid White Light-Emitting Diodes. *J. Electron. Mater.*, 45, 1, 682–687, Jan. 2016.
154. Penna, T.C.V., Ishii, M., Junior, A.P., Cholewa, O., Thermal stability of recombinant green fluorescent protein (GFPuv) at various pH values. *Appl. Biochem. Biotechnol.*, 113–116, 469–483, 2004.
155. Sokołowska, J., Czajkowski, W., Podsiadły, R., The photostability of some fluorescent disperse dyes derivatives of coumarin. *Dyes Pigm.*, 49, 3, 187–191, Jun. 2001.
156. Enko, B., Borisov, S.M., Regensburger, J., Bäumler, W., Gescheidt, G., Klimant, I., Singlet Oxygen-Induced Photodegradation of the Polymers and Dyes in Optical Sensing Materials and the Effect of Stabilizers on These Processes. *J. Phys. Chem. A*, 117, 36, 8873–8882, Sep. 2013.
157. Nowakowska, M., Smoluch, M., Sendor, D., The Effect of Cyclodextrins on the Photochemical Stability of 7-Amino-4-methylcoumarin in Aqueous Solution. *J. Incl. Phenom. Macro.*, 40, 213–219, 2001.

Index

Also of Interest

Forthcoming and published related titles from Scrivener Publishing

Photoelectrochemical Solar Cells
Edited by Nurdan Demirci Sankir and Mehmet Sankir
Published 2019. ISBN 978-1-119- 45993-4

Flexible Glass
Enabling Thin, Lightweight, and Flexible Electronics
Edited by Sean Garner
Published 2017. ISBN 978-1-118-94636-7

Printable Solar Cells
Edited by Nurdan Demirci Sankir and Mehmet Sankir
Published 2016. ISBN 978-1-119-28371-3

Advanced Magnetic and Optical Materials
Edited by Ashutosh Tiwari, Parameswar K. Iyer, Vijay Kumar and Hendrik Swart
Published 2016. ISBN 978-1-119-24191-1

Advanced Electrode Materials
Edited by Ashutosh Tiwari, Filiz Kuralay and Lokman Uzun
Published 2016. ISBN 978-1-119-24252-9

Advanced Molecularly Imprinting Materials
Edited by Ashutosh Tiwari and Lokman Uzun
Published 2016. ISBN 978-1-119-33629-7

Advanced Materials Interfaces
Edited by Ashutosh Tiwari, Hirak K. Patra and Xumei Wang
Published 2016. ISBN 978-1-119-24245-1

Patent Engineering
A Guide to Building a Valuable Patent Portfolio and Controlling the Marketplace
By Donald S. Rimai
Published 2016. ISBN 978-1-118-94609-1

Roll-to-Roll-Vacuum Deposition of Barrier Coatings, 2nd edition
By **Charles A. Bishop,**
Published 2015. ISBN 978-1-118-94614-5

Atomic Layer Deposition: Operation and Practical Applications
By **Tommi Kääriäinen, David Cameron, Marja-Leena Kääriäinen** and **Arthur Sherman**
Published 2013. ISBN 978-1-118-06277-7

www.scrivenerpublishing.com

www.ingramcontent.com/pod-product-compliance
Lightning Source LLC
LaVergne TN
LVHW011523200926
848484LV00001B/135
9781119654605